August Koehler

Practical botany

Structural and systematic

August Koehler

Practical botany
Structural and systematic

ISBN/EAN: 9783337224394

Printed in Europe, USA, Canada, Australia, Japan

Cover: Foto ©berggeist007 / pixelio.de

More available books at **www.hansebooks.com**

PRACTICAL BOTANY

STRUCTURAL AND SYSTEMATIC

THE LATTER PORTION BEING

AN ANALYTICAL KEY TO THE WILD FLOWERING PLANTS
TREES, SHRUBS, ORDINARY HERBS, SEDGES
AND GRASSES OF THE

NORTHERN AND MIDDLE UNITED STATES
EAST OF THE MISSISSIPPI

BY

AUGUST KOEHLER M.D.

Professor of Botany in the College of Pharmacy of the City of New York.

COPIOUSLY ILLUSTRATED.

NEW YORK
HENRY HOLT AND COMPANY
1876.

S. W. GREEN,
PRINTER AND ELECTROTYPER,
NEW YORK.

PREFACE.

It is hoped that this brief volume may supply some new facilities to the student who will meet them with reasonable intelligence and faithfulness, for the rapid acquisition of a practical knowledge of Botany. It is also hoped that to the person already somewhat versed in the science, it will prove a more ready instrument for a large class of his identifications than any heretofore offered. The first part merely summarizes topics that are the common material of the science, in the common way. A glance at the table of contents, and through the headings of the pages, will sufficiently indicate the nature of all the divisions of the work except the "Key." In that, the author has good authority for believing that he has introduced a method hitherto not applied in American treatises on the science, and also for hoping that the method has (as is not always the case with innovations) enough usefulness to justify the novelty. For some explanation of the uses of this Key, both as an aid and stimulus to the unlearned, and a labor-saver to those already somewhat acquainted with the science, the reader's patient attention is requested.

The study of Botany can not become truly profitable until a number of plants have been identified by the student, and their images received into his memory.

To this end, an analytical key, pointing out the orders of the plants apt to claim the student's attention, is prefixed to the American standard manuals of Chapman, Gray, and Wood. These contrivances may be successfully employed, though with a loss of much time, by more advanced students; but, notwithstanding their indisputable value, they will usually prove far from answering the wants of beginners. For them I have elaborated the analytical key, which makes the second part of this book. By a limited number of antitheses, it points out the very genus to which any phænogamous species whose name we want to know, belongs.

This contrivance is based on what is called the dichotomal method, and thus far agrees with the keys indicated above; but by pointing out directly the genera, instead of the orders, it is, of course, fitted for a comparatively rapid identification.

The dichotomal method was introduced by De la Marck, a contemporary and friend of De Candolle. He suggested and proved that he who wants to elaborate a most useful botanical key, must not take hold of some one part or organ of the vegetable exclusively, regarding its peculiar state as a sufficient differential character of the plant, when brought contrasted to another; that he must not care whether he always associates plants, which naturally seem to be allied, or whether he contrasts them, provided this contrast helps to characterize them beyond all doubt; and, lastly, that opposite things become more evident by their juxtaposition (*contraria juxta se posita magis elucescunt*). This method has been adopted by many for the purpose of making intelligent young people fond of botany.

Suppose we have a dozen flowering plants before us.

We at once notice that they all have perfect flowers, and, therefore, consult § 4 of this key. Directed from there to § 5, we separate our dozen into two heaps, on account of the number of their stamens. Six of them have more than 20 stamens, and six less.

Let us attack first the half dozen with over 20 stamens. In § 6 we ascertain that four of them have their stamens distinct, while the other two have them united. To learn the names of the four plants with distinct stamens, we consult § 7, and there are led to observe that three of them have their stamens inserted on the receptacle, while one has his attached to the calyx. The plant with the stamens inserted into the calyx must be looked for in 48; and from there, as it has 3-12 ovaries, we proceed to 57. The number of the perianth-pieces being definite, namely, only five, we are directed to 58, where, ascertaining that the fruit is a follicle, etc., we establish that the plant under consideration is a *Spiræa*.

As to the three other plants with their distinct stamens inserted on the receptacle, we are in § 8 advised to ascertain the number of pistils. We find that two of the plants have one pistil only, while one of them has five pistils. As regards the latter, we are directed to 25, and verifying there that the pistils are distinct, we proceed to 26; and as the pistils are not sunk in hollows of the receptacle, but free on its surface, to 27. The leaves being not peltate (or shield-shaped), we are shown to 28, and from there, as the plant is an herb, to 29. The ovary being not one-ovuled, etc., we pass on to 38, and from there, the petals being present, to 41. The petals being spurred, we consult § 42, where, the petals proving to be all alike, it becomes evident that our plant is *Aquilegia Canadensis*, L.

To identify the plants with one pistil (§ 8) and regular flowers (§ 9), we must proceed to § 10. Here we are made aware that one of them, containing a yellow, milky juice, has a two-sepalled calyx, while the other has more than two sepals. In § 11, being directed to 12 and 13, we find that our plant, having prickly leaves, is *Argemone Mexicana*, L. As to the other 1-pistillate, 5-sepalled plant with regular flowers, we first apply to 16, and as it is not a tree, to 17, from where, the plant having simple, pitcher-form leaves, we proceed to 19, to learn that the plant is called *Sarracenia*.

To identify the last two plants of our first half dozen, we return to § 6; and, whereas the stamens are united, we are shown to § 66. The filaments being united in several bundles, we proceed to 74. One of the plants has the leaves opposite—it is *Hypericum*; the other one has them alternate, and must be looked for in 75. The plant under consideration has the peduncles of its flower-clusters adnate to the midrib of a leaf-like bract; and is therefore *Tilia*, the *Linden*.

And now for the second half dozen!

In § 5 we are required to consult § 77. Here we ascertain that one of the six plants has a double perianth, while three others, one of them grass-like, have a simple set, and the two last plants, one of them also grass-like, no perianth at all, or only a poor substitute of it in form of bristles.

Our plant with a double perianth obliges us to proceed to § 78, from where, as the corollas of our specimen are gamopetalous, we are directed to 320. We ascertain there that, as each flower has the ovary inferior, we must proceed to 507. As our plant has no tendrils, we consult § 508. We notice that there are three stamens in each flower with a regular corolla, whereupon we look for further information in 534, where, as the leaves are in whorls of four, etc., we make certain that it is called *Galium trifidum*, L.

Returning to § 77, to learn how to ascertain the names of the three plants with one set of perianth, we are directed to 537. Noticing that one of the three plants under consideration has three to five flowers in a calyx-like, funnel-form involucre, and that these flowers have a bell-shaped perianth each, we find in 538 that it is *Oxybaphus nyctagineus*, Nees. Of the two remaining plants with a simple perianth, we may attack the grass-like one first. Its flowers consist of an ovary, three stamens, and two pales (the perianth of grasses). We must start from § 537, from where, as there is no involucre to several flowers, we proceed to 539. From here we are directed to 564, where we are informed that we have a true grass before us, and are required to consult § 565. Here, as each spikelet contains a solitary, perfect flower, we are advised to ask for further information in 566. Two glumes being present, we proceed to 568, and, the flower having two pales, to 570. The spikelets being not imbedded in hollows of the stem, we consult 571. As our grass, with the spikelets of one sort, has four spikes placed digitately on the summit of the culm, we proceed to 573, where we ascertain that it is *Cynodon Dactylon*, Pers. Resuming § 539 (from where we must start in behalf of the rest of our plants, as they have all incomplete flowers), we inquire regarding the last plant with perianth-bearing flowers. As it has a very conspicuous perianth, we proceed to 564. As it is no grass, we must consult 639. The number of stamens not exceeding 12, we proceed to 640, and from there, the perianth being irregular, to 755. As the ovary is under the perianth, we pass on to 757. The perianth being tubular, and bent in form of an S, etc., we decide that it is *Aristolochia*.

As the last two plants of our half dozen have no perianth at all, or only a few bristles supplying its place, we are from 539 directed to 540, where we learn that one of them is a sedge. Being directed to 551, and ascertaining there that this sedge is not monœcious nor diœcious, we pass on to 552. The glumes being two-rowed, we apply to 553, where, as our sedge has its spikes on solitary, axillary peduncles, etc., we verify that it is *Dulichium spathaceum*, Nees.

To identify the very last plant, we were directed from 539 to 540. As it is not grass-like, we proceed to 541. It being not a stemless or scapeless plant, we pass on to 542. It being further no maritime plant, we apply to 544, where, as it is a leafless, succulent, fleshy plant, with a jointed stem, and closely appressed (or, as it were, imbedded) flowers, always three together, forming club-shaped spikes, etc., we ascertain that it is *Salicornia*.

The young botanist, after having a few times used the key, will be aware that it is not always necessary to begin an investigation on the first page. First of all, he must, whenever a phænogamous plant is to be identified, begin with the questions: Are there several flowers in a common involucre, and, if so, are the anthers united? Then: Are the flowers diclinous? These questions answered in the negative, he will continue asking: Are there more or less than 20 stamens in each flower? In most cases he will see fit to start at once from § 77.

After having some practice in using the key, the student will be capable of at once attacking one or the other of the chief divisions, namely:

A. Corolla polypetalous...................(79–319).
B. Corolla gamopetalous (monopetalous)..(320–536).
C. Incomplete flowers..................(537–773).
D. Diclinous flowers...................(774–903).
E. Composites........................(904–1000).

In the divisions, "Corolla polypetalous" and "Incomplete flowers," plants with polyandrous flowers will be generally missed, as they were done with in §§ 6–76; but when they are met with again, it is for good reasons.

Whenever a genus is represented by one species only (in our Flora), the specific name is given—for example, *Symplocos tinctoria;* and whenever the generic name alone —for example, *Lysimachia*—is found, it is to be understood that the genus is represented by more species than one. When species of a genus differ so widely as to be equivalent to sub-genera, they will be met with in the same or in separate paragraphs, as *Stylisma evolvulsides,* Choisy, and *Stylisma Pickeringii,* Gray, or *Phalaris arundinacea,* and *Phalaris Canariensis,* L.

Sometimes a genus (or a species) will be met with twice, or oftener, in distant divisions. Thus, we have *Penthorum sedoides* in the division "Corolla polypetalous," as well as in that of the "Incomplete flowers;" for the flowers of this species have sometimes a corolla, and oftener none. Again, the calyx is sometimes furnished with an obscurely toothed or obsolete limb, and the student may, therefore, look for the plant among those with incomplete flowers, and he will not miss it, as the case is provided for; still he finds the plant a second time among those with complete flowers. Finally, a genus may be identified, although the investigation may be made from the apparent relations of the floral parts, not from the true ones, as in *Euphorbia.*

The twelve plants we have analyzed and identified above, are either exogenous or endogenous. Thus it appears that the student may identify plants of both classes without understanding what the terms Exogens and Endogens signify. No one, however, is expected to consider the distinction between them of small importance.

In counting the parts of the floral circles, it is absolutely necessary to ascertain their *habitual* number. Thus, in *Acer* (Maple) the calyx-lobes (and the petals, if present) are usually 5, though sometimes anywhere between 4 and 12. We must always examine several flowers before we can be safely guided by the key. Looking in it for *Acer*, we shall miss it if we have not previously known that the calyx-lobes are usually five in number.

In the first part of the book, treating of Structural Botany, the plan has been to exclude all detail which is not absolutely necessary to the beginner.

When in the second part a figure in parenthesis is found at the right of a paragraphic figure, it indicates the paragraph giving the next previous step in the identification; see, for instance, **189** (175). When the name of a plant found in a paragraph has in parenthesis a figure at the left, this figure denotes a previous paragraph, in which it was met with already; and, again, a figure standing at the right gives reference to one of the subsequent paragraphs, in which, for some reason, the plant's name will again be found.

Terms and symbols, the meaning of which some readers are likely not to know, are explained in the Glossary.

In the sections of the key which treat of sedges and grasses, the student will find references to plates, crowded with engravings, appended to Gray's and Wood's manuals. Thus, in § 565, the notations, Gr. VIII., 62, and W. V., 63, refer to the 8th table of illustrations and the 62d genus of the order in Gray's, and 5th table and 63d genus in Wood's manual.

In preparing this work, I have freely availed myself of the labors of my predecessors—Chapman, Gray, the Hookers, Torrey, Wood, and others. Much of what they have laid down is stored up in my memory, or converted, as it were, into my own mind; so I may have used, here and there, their very words, without realizing the fact. Again, in botany, the descriptive part particularly, certain modes of writing have become so general that originality would be deemed absurd.

Having said thus much, I may be permitted to state, that of the 800 genera laid down in the key, I had myself analyzed and represented by drawings over 500 before I ventured to prepare this book; and most of the rest I have examined also, either in the fresh or dry state.

To any professional botanist who may suggest improvements and corrections, I shall be indebted.

In calling the book "Practical Botany," I hoped to indicate that it is planned to gratify practical needs.

May this book lead the young botanist through the intricacies of our flowery wildernesses, as Ariadne's ball of twine led Theseus through Dædalus's labyrinth!

PART I.
STRUCTURAL BOTANY.

CONTENTS OF PART I.

I. INTRODUCTION, ... §§ 1–63
II. GENERAL VEGETABLE MORPHOGRAPHY, ... §§ 64–68
III. ORGANOGRAPHY, ... §§ 69–153
IV. ARRANGEMENT OF LEAVES AND FLOWERS ON THE STEM AND BRANCHES, ... §§ 154–179
V. GLOSSARY OF BOTANICAL TERMS, ... Page 95

STRUCTURAL BOTANY.

I.

INTRODUCTION.

A. DEFINITION OF BOTANY.

1. BOTANY is the science which treats of the structure of plants, the functions of their organs, their places of growth, their classification, and the terms which are employed in their description and denomination.

B. BRANCHES OF BOTANY.

2. BOTANY is divided in *two cardinal branches*—THEORETICAL and PRACTICAL.

3. THEORETICAL BOTANY treats of plants without regard to their utility or hurtfulness. It is divided into, 1. *Vegetable Organography;* 2. *Vegetable Physiology;* and, 3. *Special or Systematic Botany.*

4. ORGANOGRAPHY, also called TERMINOLOGY, or GENERAL BOTANY, treats of the terms employed by the botanist to designate the several parts of plants, or to describe their form.

Organography may be preceded by General Morphography, which considers the forms of plants, without taking into account the particular organs. While Organography proper tells what *ovate* leaves, pods, seeds are, General Morphography states what the term *ovate*, etc., means.

5. VEGETABLE PHYSIOLOGY and ANATOMY treat of the life of plants and the structure of their parts. It is superfluous to mention *Vegetable Anatomy* as a particular branch of Botany, since Physiology can not be studied without anatomizing the parts of a plant. A special department of vegetable physiology is *Morphology*, which must not be confounded with Morphography, and relates to the typical transformations undergone by the organs in the course of their development.

6. SPECIAL or SYSTEMATIC BOTANY teaches how the different plants are to be identified according to certain rules, derived from *Special Organography;* it gives the method of *analyzing* a plant—that is, of noting its constituent parts, all or some, in order to ascertain its name and place in a botanical system, or to identify it. *Analysis* of a plant is a methodical operation, by which we separate a whole plant, or parts of it, into its components, for the purpose of an exact inspection of each. *Identification* or *Diagnosis* of a plant means the result of this analysis. To *determine* or *identify* is to make the diagnosis of a plant.

7. PRACTICAL BOTANY considers plants with respect to their utility and hurtfulness. It has different branches, such as *Agricultural, Technical, Toxicological, Medicinal,* etc., *Botany*.

(But as Practical Botany can not exist without the study of Systematic Botany, the latter may also be called Practical Botany.)

C. METHOD OF STUDY.

8. The best method of becoming speedily interested in the study of Botany is to consider plants in their natural relations to each other. This is assisted by classifying vegetable INDIVIDUALS into a combination, which we call a SPECIES; again a number of such species into a higher combination, called GENUS; and, finally, an assemblage of genera into a still more comprehensive assemblage or category—namely, an ORDER. The orders are finally distributed into CLASSES.

9. Nature offers a great many facilities for such methodical study. In the nearest bit of country, wherever we may turn our eyes, we notice VEGETABLE INDIVIDUALS, of which many agree in general aspect, or in having a certain number of properties in common.

10. Any cornfield, any tract of land covered with buckwheat, clover, cauliflower, etc., will answer the question, *What is a vegetable individual?*

11. The several individuals of the cornfield—the straws with their parts—may differ in size and color; still, at a glance, we conclude that *they all belong to the same stock.* The formation of root, culm, leaves, spikes, flowers, and fruits presents everywhere the same characters. All these straws or *individuals are,* therefore, *numbers of the same community, which we call a* SPECIES.

A SPECIES IS A COLLECTION OF INDIVIDUALS PRESENTING THE SAME PECULIARITIES.

12. Viewing another field, bearing what one, who is not a botanist, would briefly call CLOVER, we, on closer examination, ascertain that one part of the ground is covered with stone-clover, another with red-clover, and a third with white-clover. These different sorts are allied members of a community, which we call *Genus* (kind). *Clover* (Trifolium) is the Genus, which is represented by several species—namely, *Stone-clover*, *Red-clover*, *Zigzag-clover*, etc.

A GENUS IS A COLLECTION OF SIMILAR SPECIES.

13. Perchance not far away from the clover-field there are two pieces of ground, one covered with MELILOT, the other with MEDICK. Seen from afar, these sorts of vegetables may be taken for clover; but, examined closely, they prove to be distinct, but similar kinds or genera—that is, MELILOT and MEDICK represent a genus each, just as a genus was represented by CLOVER.

14. Finally, the genera CLOVER, MELILOT, and MEDICK are members of a higher community—namely, of an ORDER (or Family); and this particular order is called that of the LEGUMINOUS PLANTS. It happens that, during our botanical excursion, we soon meet with other members of the order. We stop short in front of a copse of plants, which bear a striking resemblance to those we have just examined. There is a plant climbing over bushes, the *Wild Bean* or *Ground-nut*, bearing short racemes of butterfly-like flowers, similar to those of Clover, Melilot, and Medick. And not far off, on the very border of a wood, there are some trees with flowers of the same sort—namely, some specimens of the beautiful Locust-tree.

Reviewing the members of the order we have thus become acquainted with, we learn that their affinity is founded on peculiarities which they all have in common—namely, the compound leaves, the butterfly-like flowers, and the form of the fruit, resembling a pea-shell.

AN ORDER IS A COLLECTION OF SIMILAR GENERA.

15. In the Northern and Middle States, we have 134

orders of flowering plants, which comprehend 800 genera; and these genera are represented by 2350 species.

16. The classified plants have been designated by appropriate names and distinguished by concise descriptions in scientific language.

17. But this language can not be understood before one is somewhat acquainted with *Structural* or *Physiological Botany*.

D. VEGETABLE LIFE IN GENERAL.

18. *Physiological Botany* is a science which treats of the structure of plants, or of the organs, which enable them to live and reproduce themselves. The successive phases of vegetable life are GERMINATION, GROWTH, and REPRODUCTION.

19. Plants are organized or living beings, possessing the following attributes: 1st. The POWER OF SELF-SUPPORT OR ASSIMILATION—namely, the capacity to nourish themselves, by taking in surrounding mineral matter and converting it into their own substance; 2d. THE POWER OF SELF-DIVISION OR REPRODUCTION, by which they multiply themselves and perpetuate the species.

Plants alone are fitted to convert mineral into organic matter, while animals subsist solely on the organic matter, elaborated in vegetables.

20. There are only three chemical elements, which enter into the permanent structure of plants—*Carbon*, *Hydrogen*, and *Oxygen*, while the tissues of an animal contain an additional element—namely, *Nitrogen*.

21. The ANATOMICAL ELEMENTS of the vegetable organism are the CELLS.

When we properly examine by the microscope delicate slices of any of the parts of a plant, they will present a network, the meshes of which circumscribe little sacs or bladders, primarily of a spherical shape, but in time becoming very various, as polyhedral, oblong, cylindrical, or even tubular. The ordinary diameter of the cells is between $\frac{1}{20}$ to $\frac{1}{100}$ of a line.

22. A young and vitally active cell consists of the following parts: *a, a thin,* colorless, transparent *membrane,* which incloses and forms a cavity, and is lined by a soft and mucilaginous film, called the *primordial utricle; b, a jelly-like kernel* (nucleus) in the centre of the cavity; and, finally, *c, a liquid,* known under the name of *protoplasm,* which fills the rest of the cavity.

23. *All the cells of a plant constitute the elementary fabric, in which the products sustaining life are prepared.* The combined vital action of all the cells of the vegetable composes what we call vegetable life. The cell is the factotum, without which the existence of plants would be impossible. It manages the affairs of the whole economy of any vegetable individual.

24. Each cell is a chemical laboratory, which supports vegetable life. The cell is the true propagator of plants; it chains vegetable individuals to one another, thus establishing the constancy of the several species. It is also the Proteus, which creates genera, orders, and classes, and, finally, the parent of the whole vegetable kingdom.

25. *The lowest plants are single cells.* Such a plant is destitute of any external organs. It imbibes its food through its membrane, to assimilate it in its cavity, and produce new cells in the same. These (the progeny) escape by the decay of the mother-cell (or mother-plant). *Red Snow* is an instance of this. It occurs on damp rocks, in the form of dull crimson patches, which resemble blood-stains. These patches consist of numerous individuals of the species *Protococcus nivalis,* which can not be seen without the microscope.

E. PHÆNOGAMS.

How they Differ from Cryptogams—General Consideration of the Flower—Sorts of Flowers—The Phænogamous Orders of the Northern and Middle States East of the Mississippi, arranged according to a Natural System—The Artificial System of Linnæus.

26. The entire vegetable kingdom has two principal divisions: Flowerless Plants or Cryptogams, and Flowering Plants or Phænogams.

27. *The Flower is a system of organs which performs the reproduction of the Phænogams. The Cryptogams have no flowers.*

28. In Cryptogams, the organs of reproduction are apparatus, more or less analogous to flowers, which produce, instead of seeds, minute bodies called *Spores,* that do not contain any embryo prior to germination. The cryptograms are lower in grade than the phænogams. They are variously classified.

We may distinguish *Vascular and Cellular Cryptogams.*

Vascular cryptogams, also called *Acrogens,* are flowerless plants with a distinct axis, a branched stem, which grows from the summit only, containing woody fibres and vascular ducts, and commonly furnished with distinct foliage. The *Orders* of this class are: the *Horse-tails* (Equisetaceæ); the *Ferns* (Filices); the *Club-mosses* (Lycopodiaceæ); *Adder-tongues* (Ophioglosseæ), and *Quillworts* (Rhizocarpeæ or Hydropterides).

Cellular Cryptogams are flowerless plants, destitute of vascular ducts. We divide them into *leafy* and *leafless* cellular cryptogams.

The *Leafy division* comprises the *Chara family* (Characeæ), the *Liverworts* (Hepaticæ), and the *Mosses* (Musci frondosi).

The *leafless division*, that of the *Thallophytes*, consists of the *Lichens* (Lichenes); the *Mushrooms* and *Moulds* (Fungi), and the *Seaweeds* (Algæ).

29. PHÆNOGAMS, to which our ordinary herbs, shrubs, and trees belong, are the higher grade of plants. Their flowers produce seed. The seed contains *a ready-formed embryo*, or rudimentary *plantlet*. (Pl. III. 1*d*.)

30. The *Embryo* consists of the *Radicle* (stemlet, axis), 1 or 2, rarely more, *Cotyledons* (leaves), and the *Plumule*. The *Plumule* (a little bud) contains, in a rudimentary state, the future stem and leaves.

31. The organs of the flowers are, some of them, indispensable to the production of seeds, while others serve merely to protect and support them.

32. Thus we distinguish two kinds of floral organs: *a*, the ESSENTIAL ORGANS, which are the *Pistils* and *Stamens*, and *b*, the PROTECTING ORGANS, which are the *floral leaves*, also called *Perianth* or *Perigone*.

33. A *Pistil* is the seed-bearing organ of the flower, and distinguished into three parts—namely, beginning from below, the *ovary*, the *style*, and the *stigma*. (Plate IV., 6*c*.)

The *Ovary* is the hollow case, or initial fruit, which contains rudimentary seeds, called ovules. (Plate IV., 3*b*, 4*b*, 4*c*; Plate V., 7*a*, 8*b*.)

The *Style* is the tapering summit of the pistil, sometimes long and slender, sometimes short, and, not unfrequently, altogether wanting, it being not an essential part like the two other parts of the pistil. (Pl. IV., 1*a*, 6*c*.)

The *Stigma* is always that portion of the surface of the pistil which is not covered, like the rest of the plant, by a pellicle or epidermis, but a portion which consists of loose, projecting filaments or warts, destined to receive the fertilizing, powdery substance, shed by the anthers of the stamens. It is either a portion of the surface of the style, generally the tip, or, when there is no distinct style, of the top of the ovary. (Pl. III., 2*e*, 8*d*.)

The *Stamens* are the fertilizing organs of the flower. A stamen consists of two parts—namely, the filament, or stalk, and the anther. The *Anther* is a capsule or case, filled with a powdery substance, called *Pollen*. This powder, consisting of granules, is discharged through one or more openings of the anther. (Pl. III., 7*c*, 11*b*.)

34. The *Perianth* consists of *one* or *two* (rarely more) whorls of floral leaves. Most of the flowers have the perianth double—that is, two sets or whorls of it. The outer set or circle is called *Calyx*, and usually green; the inner is the *Corolla*, and commonly of some other color than green. The leaves of the calyx, if distinct, are named *sepals*; those of the corolla, if not united into one body, *petals*. (Pl. III., 1, 1, 10; Pl. V., 2.)

35. *All the organs of the flower, both essential and unessential, are situated on the top of the flower-stalk,* into which they are said, in botanical language, to be inserted, and which is called *Torus* or *Receptacle*. (Pl. III., 2*a*.)

36. The *Receptacle* is the axis of the flower, to which the floral organs are attached in a constant succession—namely, the *calyx* at its very base, forming the outermost set; the *corolla* just within (or above) the calyx; the *stamens* just within the corolla, and the (one or several) pistils within (or above) the stamens, occupying the centre of the flower.

The receptacle, usually short and inconspicuous, is sometimes considerably enlarged or elongated. Whenever the pistils are very numerous, the receptacle increases in size or modifies its shape, to give room for their insertion. In the *Strawberry* it assumes a conical form, bearing the pistils on its surface, and becoming at length the eatable part of the fruit, botanically considered a spurious fruit. In the flower of the *Rose* the receptacle is deeply excavated or urn-shaped, and invested by the adnate calyx-tube, while its cavity is lined by the pistils. (Pl. V., 5.)

37. As regards the essential (absolutely necessary) organs, we divide the flowers into *Perfect* and *Imperfect*. But each of these sorts presents important differences in reference to the unessential organs also.

a. Flowers, which have both kinds of essential organs, stamens and pistils, are called *Perfect* or *Hermaphrodite flowers*, whether they are furnished with a perianth or not. (Pl. XII., 2a, 5.)

b. Flowers, in which one set (the corolla) of the perianth is wanting, are *Incomplete flowers;* such flowers may be perfect. (Pl. XII., 5.)

c. Flowers which have both essential and unessential organs—namely, a complete perianth (calyx and corolla) and stamens, as well as pistils—are not only perfect, but also complete; we call them, therefore, *Complete flowers*. (Pl. III., 1, 1, 4, 6.)

d. Flowers which are barely furnished with the essential organs (stamens and pistils) are denominated *Achlamydeous* or *Naked flowers*. They are destitute of a garment (χλαμυς), perianthless. (Pl. XII., 2*a*.)

e. There are flowers designated as *Imperfect, Separate,* or *Diclinous flowers*—that is, flowers of two sorts, occurring either on the same specimen or on separate specimens of a species. One sort we call *Staminate* or *Sterile flowers*, because they have stamens, but no true pistils; the other sort *Pistillate* or *Fertile flowers*, since they are furnished with pistils, but not any (or at least no fertile) stamens. (Pl. XIII., 11, 11*b*; Pl. III., 8*b*, 8*c*.)

Diclinous flowers may be complete, incomplete, or naked, as the perfect flowers in their turn.

38. When sterile and fertile flowers grow on the same specimen or stem of a species, we call them *Monœcious* flowers (*monœcious:* in one household). When, on the other hand, some specimens of a species have sterile flowers, and others fertile ones exclusively, we say that they are furnished with *Diœcious* flowers (*diœcious:* in two households).

Therefore, we divide the plants with diclinous flowers into *monœcious* and *diœcious* plants.

39. Not unfrequently we meet with diclinous flowers, intermixed with perfect ones, both on monœcious and diœcious plants. In such a case, we speak of *polygamous*, or rather of monœciously and diœciously *polygamous plants*.

40. *The leaves of the Perianth, both sepals and petals, are either separate or united.*

If the *Sepals* are united so as to form a cup or tube,

having their tips distinct or obsolete, the calyx is said to be *gamosepalous* or *monosepalous* (one-sepalled). (Pl. V., 7.) If, on the other hand, they are not in any way united into one piece, the calyx is called *polysepalous* (many-sepalled.) (Pl. IV., 6.)

If the *Petals* are united into one body in the same manner, we call the corolla *monopetalous* (one-petalled) or *gamopetalous* (Pl. VIII., 2); and if the flower has its petals distinct, the corolla is said to be *polypetalous*. (Pl. IV., 4.)

41. The PHÆNOGAMOUS SERIES *divides into Two Classes.*

42. The *First Class of the Phænogams* (the class of the higher grade) comprises the *Plants with wood in a zone, or circle, or in concentric annual rings* (Pl. II., 29) *around a central pith; netted-veined leaves, parts of the flower mostly in fives or fours, and a dicotyledonous (two-leaved) embryo.* This is the *Class of the Exogenous or Dicotyledonous plants*, which we briefly call EXOGENS.

43. The *Second Class of the Phænogams* comprises those plants, which have their *wood disposed in separate threads, scattered through the diameter of the stem*, not in a circle (Pl. II., 28); the *floral parts usually in threes, never in fives;* the *leaves nearly always longitudinally veined;* and a *monocotyledonous (one-leaved) embryo.* This is the class of the endogenous or monocotyledonous plants, which we briefly call ENDOGENS.

44. The CLASS EXOGENS is divided into *two subclasses—the Angiosperms* and the *Gymnosperms.*

45. The GYMNOSPERMS are characterized by having their *ovules* (and seeds) *naked*—that is, not hidden in a closed pistil, but either *attached to the base of an open pistil* (an open *scale*, as in Pine, or a more evident *leaf*, as in Cycas), or, in the Yew, encircled at the base by an annular disk. The *Cycas* does not belong to our Flora. The number of cotyledons in Gymnosperms is often more than 2, in Pinus from 3 to 12. The Gymnosperms are represented in the Northern and Middle States by *Conifers* only, of the following genera: *Pinus, Abies, Larix, Thuja, Cupressus, Taxodium, Juniperus,* and *Taxus.* (See Pl. XIII., 1, 2, 3, and the description of the plate.) *Gymnosperm* from the Greek γυμνός, naked, and σπέρμα, seed.

46. The ANGIOSPERMS have *closed pistils*, which conceal the ovules in their cavity (*Angiosperm:* from Gr. ἀγγεῖον, a vessel, and σπέρμα).

47. The metamorphosing power of the plant is not exhausted by the production of a simple axis, but produces also secondary axes. In the lower orders of plants these secondary axes are not much different from the primary axis, but in plants of higher organization the difference is very great. The variety of forms displayed by these plants is astonishing. The primarily cylindrical shape of the leaf undergoes a variety of changes. The simple conico-cylindric axis usually develops leaves; and the leaf appearing first in the lower, then in the upper parts of the axis, and finally at its very top, there undergoes a series of protean transformations into sepal, petal, stamen, pistil, and fruit.

The parts of the flower, and so of the fruit, are nothing but whorls of leaves. A true simple pistil is a floral leaf, with its edges curved inward and united, forming a closed case, which is the ovary; and the ovules are borne on what answers to the united margins of the leaf. Several simple pistils may coalesce and consolidate into a compound pistil (as in Rose-Mallow, Flax, St. John's-wort, etc.) And now, whether we have to do with simple or compound pistils, every leaf employed for the formation of a closed pistil is called a *pistil-leaf*. But it is more convenient to use the term *carpellary leaf* instead of pistil-leaf, since this term can also be applied to open pistils. While the carpels of Angiosperms form closed pistils, those of Gymnosperms remain flat and open, not possessing the power to curve in and unite their edges. *Carpel* is a closed carpellary leaf.

As stated above, the several whorls of floral leaves—namely, calyx, corolla, stamens, and pistils—are inserted on the receptacle (the uppermost part of the axis) successively, beginning with the calyx, from below upward. It sometimes happens that the centre of the receptacle ceases to grow vertically, while the circumference rises higher; in such case the pistils (or the pistil) will be placed deeper than the other floral circles, or even in an excavation of the receptacle, as in the *Rose*. In theory, the pistils occupy the top, but the top is, as it were, depressed; in reality it is surpassed by parts of the receptacle, which stood deeper before.

48. The ANGIOSPERMS form *three Divisions:* the *Polypetalous,* the *Gamopetalous,* and the *Apetalous*. We here enumerate such of the orders belonging to these divisions, as occur in the Northern and Middle States.

I. Polypetalous Division.

49. This division has, as a rule, both calyx and corolla, the latter consisting of separate petals.

ORDERS:

1. *Crowfoots, Ranunculaceæ;* 2. *Magnoliads, Magnoliaceæ;* 3. *Custard-apples, Anonaceæ;* 4. *Moonseeds, Menispermaceæ;* 5. *Barberries, Berberidaceæ;* 6. *Waterbeans, Nelumbiaceæ;* 7. *Water-shields, Cabombaceæ;* 8. *Water-lilies, Nymphaceæ;* 9. *Water-pitchers, Sarraceniaceæ;* 10. *Poppyworts, Papaveraceæ;* 11. *Fumitories, Fumariaceæ;* 12. *Crucifers, Cruciferæ;* 13. *Capparids, Capparidaceæ;* 14. *Mignonettes, Resedaceæ;* 15. *Violets, Violaceæ;* 16. *Rockroses, Cistaceæ;* 17. *Sundews, Droseraceæ;* 18. *Parnassiads, Parnassiaceæ;* 19. *St. John's-worts, Hypericaceæ;* 20. *Water-peppers, Elatinaceæ;* 21. *Pinks, Caryophyllaceæ;* 22. *Purslanes, Portulaccaceæ;* 23. *Mallows, Malvaceæ;* 24. *Lindenblooms, Tiliaceæ;* 25. *Tea-worts, Camelliaceæ;* 26. *Flax-worts, Linaceæ;* 27. *Wood-sorrels, Oxalidaceæ;* 28. *Crane's-bills, Geraniaceæ;* 29. *Jewel-weeds, Balsaminaceæ;* 30. *Limnanths, Limnanthaceæ;* 31. *Rueworts, Rutaceæ;* 32. *Sumachs, Anacardiaceæ;* 33. *Vines, Vitaceæ;* 34. *Buckthorns, Rhamnaceæ;* 35. *Staff-trees, Celastraceæ;* 36. *Indian Soapworts, Sapindaceæ,* comprising the sub-orders, *Staphyllaceæ, Hippocastanaceæ,* and *Acerineæ;* 37. *Milkworts, Polygalaceæ;* 38. *Leguminous plants,* the *Pulse family, Leguminosæ;* 39. *Roseworts, Rosaceæ;* 40. *Calycanths, Calycanthaceæ;* 41. *Melastomes, Melastomaceæ;* 42. *Loosestrifes, Lythraceæ;* 43. *Onagrads,* or *Evening Primroses, Onagraceæ;* 44. *Loasads, Loasaceæ;* 45. *Indian Figs, Cactaceæ;* 46. *Currants, Grossulariaceæ;* 47. *Passionworts, Passifloraceæ;* 48. *Cucurbits, Cucurbitaceæ;* 49. *House-leeks, Crassulariaceæ;* 50. *Saxifrages, Saxifragaceæ;* 51. *Witch-hazelworts, Hamamelaceæ;* 52. *Umbelworts,* or *Parsleys, Umbelliferæ;* 53. *Araliads, Araliaceæ;* 54. *Cornels,* or *Dogwoods, Cornaceæ.*

II. Gamopetalous Division.

50. This division has both calyx and corolla, the latter more or less united; in exceptional cases not united.

ORDERS:

55. *Honeysuckles, Caprifoliaceæ;* 56. *Madderworts, Rubiaceæ* (*Stellatæ, Cinchoneæ,* and *Loganiaceæ*); 57. *Valerians, Valerianaceæ;* 58. *Teaselworts, Dipsaceæ;* 59. *Composites,* or *Asterworts, Compositæ;* 60. *Lobeliads, Lobeliaceæ;* 61. *Bellworts, Campanulaceæ;* 62. *Heathworts, Ericaceæ;* 63. *Beetle-weeds, Galacineæ;* 64. *Hollyworts, Aquifoliaceæ;* 65. *Storax-plants, Styracaceæ;* 66. *Ebonads, Ebenaceæ;* 67. *Soapworts, Sapotaceæ;* 68. *Ribworts, Plantaginaceæ;* 69. *Leadworts, Plumbaginaceæ;* 70. *Primroses, Primulaceæ;* 71. *Butterworts,* or *Bladderworts, Lentibulaceæ;* 72. *Trumpet-flowers, Bignoniaceæ;* 73. *Broomrapes, Orobanchaceæ;* 74. *Figworts, Scrophulariaceæ;* 75. *Acanthads, Acanthaceæ;* 76. *Vervains, Verbenaceæ;* 77. *Labiates, Labiatæ;* 78. *Borrageworts, Borraginaceæ;* 79. *Hydrophylls,* or *Water-leaves, Hydrophyllaceæ;* 80. *Phloxworts, Polemoniaceæ;* 81. *Bindweeds, Convolvulaceæ;* 82. *Nightshades, Solanaceæ;* 83. *Gentianworts, Gentianaceæ;* 84. *Dogbanes, Apocynaceæ;* 85. *Asclepiads, Asclepiadaceæ;* 86. *Olives, Oleaceæ.*

III. Apetalous Division.

51. This division has, as a rule, no corolla, sometimes also no calyx.

ORDERS:

87. *Birthworts, Aristolochiaceæ;* 88. *Marvelworts, Nyctaginaceæ;* 89. *Pokeweeds, Phytolaccaceæ;* 90. *Goosefoots, Chenopodiaceæ;* 91. *Amaranths, Amarant-*

aceæ; 92. *Buckwheats,* or *Sorrelworts, Polygonaceæ;* 93. *Laurels, Lauraceæ;* 94. *Daphnads, Thymelaceæ;* 95. *Oleasters, Elæagnaceæ;* 96. *Sandalworts, Santalaceæ;* 97. *Loranths,* or *Mistletoes, Loranthaceæ;* 98. *Saururads,* or *Lizards'-tails, Saururaceæ;* 99. *Hornworts, Ceratophyllaceæ;* 100. *Riverweeds, Podostemaceæ;* 101. *Waterstarworts, Callitrichaceæ;* 102. *Spurgeworts, Euphorbiaceæ;* 103. *Crowberries, Empetraceæ;* 104. *Nettleworts, Urticaceæ* (*Ulmaceæ, Artocarpaceæ,* and *Urticaceæ veræ*); 105. *Sycamores, Platanaceæ;* 106. *Walnuts, Juglandaceæ;* 107. *Mastworts, Cupuliferæ;* 108. *Galeworts, Myricaceæ;* 109. *Birchworts, Betulaceæ;* 110. *Willows, Salicaceæ.*

(The *Conifers,* spoken of above, make up the 111th order.)

52. The CLASS ENDOGENS is divided into *three Sub-classes:* the *Spadaciflorous,* the *Florideous,* and the *Glumaceous Endogens.* (The orders enumerated here occur in the Northern and Middle States.)

I. Spadaciflorous Sub-class.

53. The plants of this sub-class have flowers destitute of perianth, or furnished with a scaly perianth only, and massed on a thick, fleshy, and commonly club-shaped axis (*spadix*), the latter generally wrapped in a sheathing-leaf (*spatha*). To this spadiciflorous sub-class, however, we also refer plants whose flowers are not attached to a spadix, but scattered. In this case, the flowers are always periantheless.

Orders:

112. *Aroids, Araceæ;* 113. *Typhads, Typhaceæ;* 114. *Duckweeds,* or *Duckmeats, Lemnaceæ;* 115. *Najads, Najadaceæ.*

II. Florideous Sub-class.

54. The plants of this sub-class usually have perfect and complete flowers, with a perianth divisible into two 3-leaved whorls, both sometimes green, more often the outer one only.

Orders:

116. *Water-plantains, Alismaceæ;* 117. *Frog-bits, Hydrocharidaceæ;* 118. *Burmanniads, Burmanniaceæ;* 119. *Orchids, Orchidaceæ;* 120. *Amaryllids, Amaryllidaceæ;* 121. *Bloodworts, Hæmodoraceæ;* 122. *Bromeliads, Bromeliaceæ;* 123. *Irids, Iridaceæ;* 124. *Yam-roots, Dioscoreaceæ;* 125. *Sarsaparillas, Smilaceæ;* 126. *Lilyworts, Liliaceæ;* 127. *Melanths, Melanthaceæ;* 128. *Rushes, Juncaceæ;* 129. *Pontederiads, or Pickerel-weeds, Pontederiaceæ;* 130. *Spiderworts, Commelynaceæ;* 131. *Xyrids, Xyridaceæ;* 132. *Pipeworts, Eriocaulonaceæ.*

III. Glumaceous Sub-class.

55. The plants of this sub-class have flowers with an imbricated perianth of alternate glumes, instead of sepals and petals, and collected into spikelets, spikes or heads.

Orders:

133. *Sedges, Cyperaceæ;* 134. *Grasses, Gramineæ.*

(These 134 orders are enumerated in the same sequence, as in *A. Gray's Manual.*)

56. A *Natural System,* such as we have just displayed, has for its basis the whole plan of structure, so that each genus, tribe, and order is placed next to those genera, tribes, and orders which it most resembles in all, or nearly all, respects. In Artificial Systems, on the other hand, the real nature of the plants is disregarded, and merely some obvious external circumstance noted. An *artificial system* has no other aim than to serve as a convenient means of reference, as a contrivance for dentifying plants, and does not attempt to express fully their points of resemblance.

57. The best of all artificial systems is that of *Linnæus*. *Charles de Linné* (*Carolus Linnæus*) was born May 23d, 1707, in the hamlet Rashult, the province Smoland, in Sweden. His *system* consists of *twenty-four classes* and *a variable number of orders*. It was designed as a provisional substitute for the natural classes and orders, which Linnæus would have established himself, had such a thing been possible. In his day, when the list of known genera embraced a comparatively very small number of names, he divided the plants into *two cardinal series*, *Phænogams* and *Cryptogams*—a division which is accepted by the authors of natural systems. But in determining classes and orders, he proceeded differently. He divided the *Phænogams* into those with stamens and pistils in the same flower, and those with these organs in separate flowers. In the case of hermaphrodite flowers, he examined whether their stamens are united with the pistils or not; next, whether the stamens are united with each other, and, finally, whether they were of equal or unequal length, if numbering 4 or 6. *This system* is best understood from a tabular view.

I. SERIES: STAMENS AND PISTILS PRESENT.
PHÆNOGAMIA.

A. Stamens with the pistils in the same flower:

* Not united with the pistils,
** nor with one another;
*** of equal length, if 4 or 6.

(*Number of stamens in each flower.*)	CLASS.
1......	I. MONANDRIA.
2......	II. DIANDRIA.
3......	III. TRIANDRIA.
4......	IV. TETRANDRIA.
5......	V. PENTANDRIA.
6......	VI. HEXANDRIA.
7......	VII. HEPTANDRIA.
8......	VIII. OCTANDRIA.
9......	IX. ENNEANDRIA.
10......	X. DECANDRIA.
11–19......	XI. DODECANDRIA.
20, or more { on the calyx...	XII. ICOSANDRIA.
on the receptacle......	XIII. POLYANDRIA.

*** Of unequal length, if 4 or 6 — namely, if 4 : 2 longer and 2 shorter.....   XIV. DIDYNAMIA.

if 6 : 4 longer and 2 shorter   XV. TETRADYNAMIA.

** United with one another,

a, by their filaments:

  † into 1 set or tube....... XVI. MONADELPHIA.
  †† into 2 sets or bundles.... XVII. DIADELPHIA.
  ††† into 3 or more sets..... XVIII. POLYADELPHIA.

b, by their anthers into a ring. XIX. SYNGENESIA.

* United with the pistils.... XX. GYNANDRIA.

B. *Stamens and pistils in separate flowers:*

a, of the same individuals... XXI. MONŒCIA.
b, of different individuals... XXII. DIŒCIA.
c, some flowers perfect, others staminate or pistillate —in the same or in different individuals........XXIII. POLYGAMIA.

II. Series.   No Stamens nor Pistils.
CRYPTOGAMIA XXIV.

58. To denominate the first eleven classes, the Greek numerals μονος, δυο (δις), τρεις, τεσσαρες (τετρα), πεντε, ἑξ, ἑπτα, ὀκτω, ἐννεα, δεκα, δωδεκα, meaning 1, 2 (twice), 3, 4, 5, 6, 7, 8, 9, 10, 12, are combined with the word *andria* (from ἀνηρ, a man, used metaphorically for stamen). The numerals ἑικοσι, 20, and πολυς, many, were employed in forming the words *Icosandria* and *Polyandria*. *Didynamia* and *Tetradynamia* (compounded of δυναμις and δις, τετρα, respectively) mean 2, and 4 stamens more powerful. *Monadelphia*, *Diadelphia*, and *Polyadelphia* are made up of *adelphia*, brotherhood, and the numerals μονος, δις and πολυς (1, 2 and many). *Syngenesia* (συν, with, and γενεσις, generation) denotes a growing together, so as to form one body. *Gynandria* (γυνη, woman, and ἀνηρ, man) signifies that the (1 or more) stamens and the style, or the stigmas, are connate (united by growth). *Monœcia* and *Diœcia* (μονος, δις, and οἰκος, house). *Polygamia* and *Cryptogamia* (πολυς, many; κρυπτος, concealed; γαμος, marriage).

59. These *twenty-four Linnæan Classes* are each divided into *Orders*. In establishing orders, Linnæus noted either the number of pistils, styles, or sessile stigmas; the number of stamens; the fruit; or, finally, in the class *Syngenesia*, the more or less developed state of the florets.

The styles were chosen to distinguish the *Orders* of the *first* 13 *classes*.

60. *Synopsis of the Orders of the First* 13 *Classes.*
Order   I. Monogynia........ 1 style to each flower.
        II. Digynia.......... 2 styles to   "
        III. Trigynia.......... 3   "   "

Order	IV.	TETRAGYNIA.......	4 styles to each flower.
"	V.	PENTAGYNIA.......	5	" "
"	VI.	HEXAGYNIA........	6	" "
"	VII.	HEPTAGYNIA........	7	" "
"	VIII.	OCTOGYNIA........	8	" "
"	IX.	ENNEAGYNIA.......	9.	" "
"	X.	DECAGYNIA.........	10	" "
"	XI.	DODECAGYNIA......	12 or 11 "	"
"	XII.	POLYGYNIA........	more than 12	"

61. The number of stamens serves to establish the orders of the 16th, 17th, 18th, 20th, 21st and 22d classes. To find their orders we proceed as follows: If a plant belongs to *Monadelphia* (XVI.), *Diadelphia* (XVII.), or *Polyadelphia* (XVIII.), we examine how many stamens have gone to form the cluster, the tube, or the bundles of stamens. In the class *Monadelphia* we have the orders, *M. Pentandria, Decandria*, and *Polyandria;* in *Diadelphia* the orders, *D. Hexandria, Octandria, Decandria;* and in the class *Polyadelphia* only *one* order, *P. Polyandria*.

In the class *Gynandria* (XX.) we meet with plants, as *Spiranthes*, belonging to the *Orchids*, and *Aristolochia* to the *Birthworts*. *Spiranthes* has *one stamen*, and hence belongs to *Gynandria Monandria*. *Aristolochia*, with 6 stamens, belongs to *Gynandria Hexandria*. Of the *Orchids* there is only one genus, belonging to the second order of *Gynandria;* namely, *Cypripedium*.

In the classes *Monœcia* and *Diœcia* we find nearly all the orders represented, which are based on the number of stamens, and in *Monœcia* also the order *Syngenesia*, with the anthers united (in the *Gourd family*).

When a plant is *polygamous*, we ascertain whether it is

monœciously or diœciously polygamous, and then inquire for the *order* in *Monœcia* or *Diœcia*.

62. The quality of the fruit enables us to ascertain the *Orders* of the classes *Didynamia* (XIV.) and *Tetradynamia* (XV.)

63. The *Orders* of the 24th class, *Cryptogamia*, are, *Natural Orders*.

II.

GENERAL VEGETABLE MORPHOGRAPHY.

64. The FORMS occurring in the vegetable kingdom are not reducible to such strictly mathematical surfaces as are seen in the crystals of *rock-salt, common salt,* and sundry other minerals.

In describing the forms of plants, or their parts, we compare them either with simple geometrical figures or with familiar objects, as a bell, a cup, urn, top, etc.

Of strictly geometrical or mathematical forms, we find in plants perhaps only one—the spherical form of the cell.

In Botany, *regular* forms are those which may be divided into two equal parts, by more than one section coinciding with the axis; *symmetrical* forms, on the other hand, are those which can be divided only by one such section into true halves, each related to the other, like the right and left hand. Thus *malvaceous, rosaceous, campanulate* corollas are called *regular*, like a cube, a sphere, etc., in mathematics. A *papilionaceous flower*, on the other hand, or a *ringent corolla*, is said to be *symmetrical*.

65. ALL THE FORMS OCCURRING IN THE VEGETABLE KINGDOM ARE SOLID BODIES—that is to say, they present the three dimensions of space which all solid bodies present—namely, length, breadth, and depth (thickness). Accordingly, as one or other, or none, of these dimensions predominates, vegetable forms are said to be *linear*, as leaf-stalks, including cylindrical (terete), biangular (two-edged), triangular, quadrangular, etc., forms; *plane* or *flat*, orbicular, elliptic, or ovate, as leaves, etc.; or *solid* (stereometrical), as fruits, tubers, etc.

66. LINEAR FORMS *differ from one another in the shape of their cross-sections* (see cuts below, I. *a—h*).

All cross-sections are circumscribed by	curved lines, and exhibit different faces, which are	orbicular,	: *terete*, or cylindrical—Fig.(a)
		semilunar,	: *semi-terete* .... (b) (half-terete)
		elliptic or oval,	: *compressed* .... (c)
		biangular,	: *two-edged* .... (d)
	straight or slightly curved lines, and exhibit	triangles { with straight sides : *triangled*. with curved lines { curved inward : *triangular*. (e) curved outward : *three-sided*. (f) }	
		squares { with straight sides : *square*. with curved lines { curved inward : *quadrangular*. (g) curved outward : *four-sided*. (h) }	

(Usually the terms *triangular* and *three-sided*, as well as *quadrangular* and *four-sided*, are used indiscriminately.)

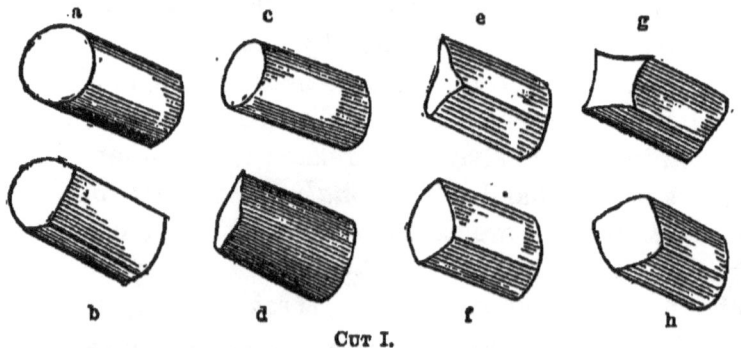

CUT I.

67. FLAT *or* PLANE FORMS *are named from their whole or partial outline.*

The following Key may serve to distinguish them:
1. *Names of flat forms derived from their whole outline, which, if it has sinuses, must be corrected by uniting mentally all its prominent points by a line* (see Cut II., Figs. a—k).................2
   Names of flat figures derived not only from the whole outline, but also from the shape of the base —that is, the part by which the organ is attached to a support, and the form of the apex—that is, the part opposite the base (see Cut III., Figs. a—o)...12
2. By comparing the longitudinal and transverse dimensions with each other......................3
   By imagining a centre or a middle line (x) on the surface of the figure, and then halving the distance thence to the corrected outline........11
3. Transverse dimensions, all of equal length, as in the cross-sections of terete bodies.....*orbicular* (II., Fig. a)
   Transverse dimensions of unequal length.........4
4. Greatest diameter in the middle of the figure......5
   Greatest diameter not in the middle.............6
5. Figure about twice as long as wide, forming a short ellipse..................*elliptic*, or *oval* (b)
   Figure three or four times as long as wide, forming an elongated ellipse................*oblong* (c)
6. Greatest diameter in the lower third of the figure..7
   Greatest diameter in the upper third of the figure, and the latter longer than wide............9
7. Figure longer than wide......................8
   Figure half as long (high) as wide, rounded above, and furnished with a roundish sinus below,
                reniform, or *kidney-shaped* (f)
8. Figure twice as long as wide..............*ovate* (d)

Figure three to four times longer than wide..... *lanceolate* (e)

9. Figure gradually narrowed toward the base.....10
Figure narrowing abruptly toward the base, with the summit rounded................*spatulate* (i)

10. Figure twice as long as wide...........*obovate* (g)
Figure about three times as long as wide, and rounded at the summit, *cuneate*, or *wedge-shaped* (h)

11 (2). Natural divisions extending down to the middle of the distance established above (in No. 2)................................*cleft* (k)
The same extending past the middle:..*parted* (l)
The same extending to the supposed line, or centre......................*divided* (m)

CUT II.

12 (1). Names derived from the outline of the apex..13
Names derived from the outline of the base..19
13. Apex with an angle........................14
Apex without an angle.......................18

14. Angle turned inward........................15
    Angle turned outward......................16
15. Forming a slight notch........*retuse* (III., Fig. a)
    Forming a deeper notch............*emarginate* (b)
    Forming a very deep notch ....*obcordate*
16. Legs of the angle curved outward.............17
    Legs of the angle curved inward, forming an angle of 90° and over..............*mucronate* (f)
17. Angle of 90°............................*acute* (e)
    Angle below 90°....................*acuminate* (g)
18 (13). Apex transversely cut off by a straight line
    truncate (c)
    Apex cut off by a convex line, rounded....
    obtuse, rounded (d)

Cut III.

19 (12). Base with an angle........................20
    Base without an angle, rounded......*obtuse* (m)
20. Angle turned inward........................22
    Angle turned outward......................21
21. Angle under 90°....................*attenuate* (o)
    Angle of 90°............................*acute* (n)
22. Angle acute..................................23
    Angle obtuse.................................24
23. Lobes rounded........*cordate,* or *heart-shaped* (h)
    Lobes pointed and turned downward............
    sagittate, or *arrow-shaped* (i)

24. Lobes rounded, surface wider than long (high).....
     reniform, or *kidney-shaped* (k)
 Lobes pointed and turned outward............
     hastate, or *halberd-shaped* (l)

68. *More compact* bodies—namely, organs in which the three dimensions of space are more obviously represented—are divided into hollow and not hollow (or solid) bodies.

Key to their forms:

1. Bodies hollow (Cut IV., a—g)................ 2
 Bodies not hollow........................... 9
2. Cavity of nearly equal width throughout....... 3
 Cavity of plainly unequal width.............. 4
3. A narrow tube of equal width, or nearly so, and open above......................*tubular* (a)
 A tube widening rapidly toward the rounded top...............*club-shaped,* or *clavate.*
4. Cavity abruptly narrowed above, near the limb, or orifice, so as to form a neck there........ 5
 Cavity enlarged above........................ 6
5. Neck short............*urn-shaped,*or *urceolate* (b)
 Neck long.....................*bottle-shaped* (c)
6. With a shorter or longer tube below.......... 7
 With no tube below........................... 8
7. Enlarging upward gradually, and representing an inverted, hollow cone........*funnel-shaped* (d)
 Abruptly enlarged above into a flat, spreading limb, and with a tube of rather even width.
    Salver-shaped, or *hypocrateriform* (e)
8. Forming a hemisphere, open above..*cup-shaped* (f)
 Resembling a bell................*bell-shaped* (g)
9 (1). Circumscribed by one curved surface only...10
 With one or several flat or plane surfaces.......13
10. Without a stalk...............................11

With a stalk..........................12
11. All sections presenting orbicular and equal surfaces........................*globose.*
Longitudinal (vertical) section, compressed orbicular in outline (see Cut I., c, the outline of the cross-section)..................*spheroidal.*
Longitudinal section, elliptic in outline,
ellipsoidal.
12. Bodies more or less spherical; the sections more or less orbicular.......*capitate* (Plate II., 24.)
Bodies resembling a pear,
pear-shaped, or *pyriform.*
13. With one plane, orbicular terminal, and one curved lateral surface; longitudinal section an equal-sided triangle................*conical.*
With two flat terminal faces and one curved lateral surface; longitudinal sections four-cornered; cross-sections orbicular..............14
With several plane surfaces.................16

CUT IV.

14. Terminal surfaces orbicular and much smaller than the lateral surface..........*cylindrical.*
Terminal surfaces orbicular and much larger than the lateral...............................15
15. Body resembling a thin slice of a cylinder,
cake-shaped, or *placentiform.*
Body resembling a far thinner slice of a cylinder,
disk-shaped, disciform.
16. Body resembling a cube.....*cubical,* or *tessular.*
Body resembling an angular column..*prismatic.*

III.
ORGANOGRAPHY.

69. *Plants* (organized bodies) *consist of organs.* These organs are distinguished into—1. *Simple*, or *elementary organs*, which are the most important of all, since they constitute all the solid parts of a plant: usually they are minute, and not perceptible without the aid of a microscope. 2. *Compound*, or *external organs*, which are formed of combinations of the elementary organs.

70. CONSPECTUS OF THE ORGANS OF PLANTS.

Simple, or elementary organs.	Solitary	cells, or primary elementary organs; vascular ducts, wood-cells, and other modified elementary organs.	
	Grouped	in the interior of the plant,	cellular tissue; bast-bundles (with various interstices).
		on the epidermis	pores, tubercles, hairs, bristles, spines, glands, warts.
Compound organs.	Organs of Vegetat'o	the root (primary and secondary roots, sorts); the stem with its branches (varieties of the stem, woody stem, herbaceous stem, culm, underground stems).	
		the leaves	leaves as foliage; leaves as something else than foliage.
	Organs of reproduction	floral organs	perianth; stamens and pistils.
		fruit	pericarp; sorts of fruits; seeds; seed-coats and kernel.

A. ELEMENTARY ORGANS.

71. The *Cell* (see §21) is the elementary organism, which in numbers constitutes the mass of all vegetation.

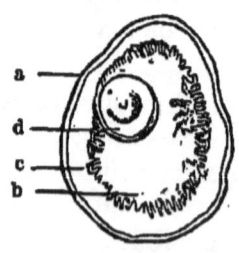

CUT V.

A magnified cell; *a*, membrane; *b*, protoplasm within *c*, the primordial utricle; *d* nucleus.

The cell-walls are readily permeated by fluids, which pass in and out through them incessantly; hence we must regard them as porous.

72. VARIETIES OF CELLS are: *a*, the *Wood-cells*—that is, elongated tubular cells or fibres with thickened walls, and grouped in bundles, with their tapering ends overlapping each other (very fine, long, and tough in the *bark*, where they are called *bast-cells*); *b*, the *Ducts*, more or less elongated tubes, either single or combined (they are combined, when formed of a row of cells placed end to end), larger than the wood-cells, and only in rare cases visible to the naked eye. There are different *sorts of Ducts*—namely, *Dotted Ducts*, the dots of which are not holes, but merely thin places in the cell-wall; *Spiral Ducts*, also called *spiral vessels*, in which the secondary layers consist of spiral, or ring-shaped fibres or bands, thickening the wall; finally, there are many *other forms* met with here and there.

73. A UNION OF SEVERAL CELLS, *forming a coherent mass, is called* CELLULAR TISSUE.

Owing to the various forms and arrangements of the cells, this tissue bears different names—namely:

a. Parenchyma—that is, ordinary cellular tissue, a system of rounded, lobed, or stellate cells, with frequent interstices; or of angular, prismatic, polyhedral cells, with but few, if any, intercellular spaces;

b. Pleurenchyma—that is, fibrous tissue formed of wood-fibres; and,

c. Trachenchyma, a tissue consisting of ducts.

(What is called *Cienchyma* is nothing but a system of canals and cavities between the cells. Only very rarely it happens that the cell-walls of a tissue come into actual contact. Between most of them there are intervening spaces known as *intercellular spaces*, or, in case they are large and regular, *intercellular passages*, or *air-passages*. They may contain special secretions, and in all probability the *milk-vessels* are, primarily, regular intercellular air-passages, instead of being composed of special cells. The *Pleurenchyma* and the *Parenchyma* together form the *main bulk of almost all plants*.)

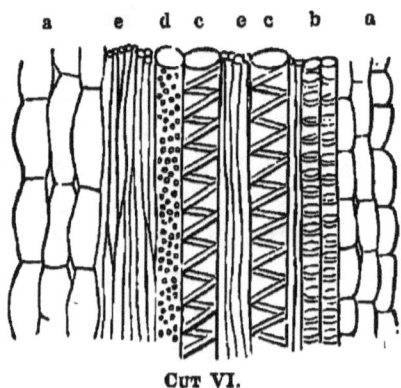

CUT VI.

Piece of a woody bundle, or compound fibre of an endogen ; *a*, Parenchyma cells ; *b*, ducts with rings, called annular cells ; *c*, spiral ducts ; *d*, a dotted duct ; *e*, wood-cells.

74. THE ARRANGEMENT OF THE ELEMENTARY ORGANS on the surface of plants constitutes the *Epidermial system*, that is to say, the *epidermis, stomata, hairs, glands, cuticle,* etc. (Cut VII.)

The EPIDERMIS is a membrane, formed of a layer of united, and commonly tabular, empty cells. The Mosses only excepted, it invests all plants, and all their parts, save the extremities, the stigma and the rootlets, and may be detached untorn from the underlying tissue. In certain places, especially on the lower surface of the leaves, it is pierced with a great many small openings, called *stomates* or *breathing pores*. Prominences are formed by elongated epidermal cells, and are called *hairs, glands, tubercles, warts, stings, bristles, prickles, scurf*.

75. By reason of certain peculiarities of their epidermis, plants and their parts are said to be:

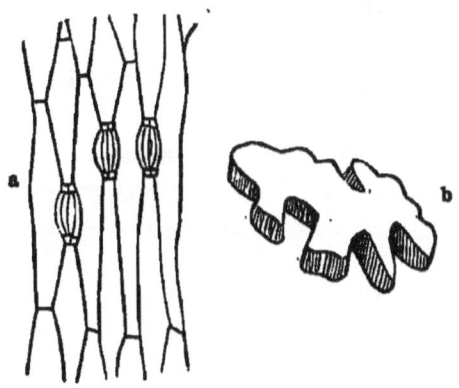

Cut VII.

a, A piece of epidermis of *Allium*, with 3 stomata; *b*, a tabular cell, detached from the epidermis of *Callitriche*.

Pilose: hairy, bearing soft, slender hairs, as the stem of *Stachys palustris;*

Pubescent or *downy:* bearing fine, soft hairs, or pubescence, as *Pentstemon pubescens;*

Puberulent: covered with fine, short, and almost imperceptible down, as the stem of *Lobelia puberula;*

Hispid: bristly, beset with stiff hairs, as *Lithospermum hirtum;*

Hirsute: hairy with rather stiff or beard-like hairs, as *Gonolobus hirsutus;*

Tomentose: covered with matted, woolly hairs, as the leaves of *Verbascum Thapsus;*

Villous: clothed with long and soft hairs, as the filaments of *Verbascum Thapsus;*

Sericeous: silken with usually appressed, shining hairs, *Potentilla anserina;*

Glabrous: smooth, without pubescence of any kind;

Lævigate: smooth, as if polished;

Scabrous: rough, or harsh to the touch, as the stem of *Equisetum hiemale;*

Hamose: bristle-prickly backward, as in *Galium Aparine;*

Aculeate: armed with thorns, as the Rose and common Greenbrier;

Echinate: prickly with rigid hairs;

Ciliate: bearing on the margin a fringe of cilia (hairs or bristles), somewhat resembling the eyelashes.

B. COMPOUND ORGANS.

AA. ORGANS OF VEGETATION.

76. The chief result of the nutrition of plants is the deposition in them of *carbon;* and their general and proper nutriments are *water, carbonic acid, ammonia,* and sulphur.

(The permanent fabric of the plant, or its real tissue, as distinct from the sap, consists of three elements—namely *carbon, hydrogen,* and *oxygen,* as we have stated in § 20. Other substances are sometimes deposited between the tissues, as, for example, silex; and other elements always enter into the sap—namely, with carbon, hydrogen, and oxygen, *nitrogen* and *sulphur.* These are indispensable to the protoplasm. Besides these, however, vegetables also take in potassium, calcium, magnesium, iron, phosphorus, chlorine, etc. Iron seems to be necessary for the formation of the green chlorophyl. Calcium, in the form of a salt, introduces sulphuric and phosphoric acid into the plant, and renders the oxalic acid harmless by combining with it. What physiological ends may be accomplished by phosphorus, chlorine, potassium, sodium, magnesium, etc., is not yet sufficiently ascertained. The plant's nourishment is received either in the gaseous or liquid form. In whatever mode imbibed, the main vehicle of alimentation is water, which, as fluid or as vapor, is in contact with the root, and as vapor continually surrounds the leaves.)

The *organs*, which imbibe and convey these aliments, are the *Root*, the *Leaves*, and the *Axis* (*stem* and *branches*). But it is the *root* which takes in the greatest quantity. From the root they are conveyed through stem and branches, and to every point where new organs are to be formed. The *leaves* absorb *carbonic acid* out of the air, and eliminate *oxygen* and *hydrogen*, at least by day.

I. THE ROOT.

77. The ROOT is an organ, which has its origin in the radicle of the embryo. It generally grows downward, and never produces leaves, but serves to fix the plant in the earth, whence it derives its nourishment. It lengthens by continued cell-multiplication, mainly at its lower extremity, the parts, once formed, scarcely elongating afterward. Still the structure of the root agrees essentially with that of the stem.

We distinguish two principal sorts of roots: 1. *Primary roots*, which start from the first joint of the stem of the plantlet, springing from the seed; and, 2. *Secondary roots*, which occasionally proceed from other parts of the stem. All stems, which creep on or under the surface of the soil, are apt to strike root from almost every joint. But even from erect stems we sometimes see roots springing—in Indian corn several inches above ground, and in some plants high in the air, as in Pandanus. When secondary roots spring from the upper parts of the stem, or even from branches, they are called *Aërial roots.*

CLASSIFICATION OF PRIMARY ROOTS.

I. *Simple primary roots,* which most plants send down from the root-end of the embryo; they are called

- from their direction
  - *perpendicular*..........*tap-roots* (Pl. I., 1–5).
  - *horizontal.*
- from their shape
  - very thin in proportion to their length, *thread-like,* or *filiform* (Pl. I., 1).
  - thick in proportion to their length, and with the
    - cross-sections orbicular, and all equal throughout......*cylindrical.*
    - cross-sections orbicular, gradually decreasing in diameter downward, and the vertical sections lanceolate, as in *Daucus Carota, spindle-shaped,* or *fusiform* (Pl. I., 5).
    - cross and vertical sections almost orbicular, and all nearly equal, *globular.*
    - cross-sections orbicular, and abruptly decreasing in size downward..*napiform,* or *turnip-shaped* (Pl. I., 4).
- from their division
  - not branching................*simple.*
  - branching.....................*ramose.*
- from their consistence
  - hard and woody................*lignose.*
  - soft and fleshy................*fleshy.*

II. *Multiple primary roots*—that is, several roots, which have sprung all at once from the root-end of the embryo. They are also called *Fasicled roots.* They are sometimes tuberous, as in Dahlia (see Pl. I., 10).

The absorbing surface of primary roots is sometimes greater than it appears to be, since there are root-hairs, or slender fibrils (*fibrillæ*), which abundantly cover the younger parts of the root. These fibrils, when examined with the microscope, are found to be slender tubes, which imbibe the moisture around them. Slender, thread-like, and freely branching roots are also called *fibrous roots.* Roots are often tuber-like (tuberous), whether simple or fascicled. Larger or smaller tuber-like excrescences form sometimes on the branches of the root, as in White Clover, in *Spiræa filipendula* (a garden plant), etc. Fascicled roots are also called *inaxial roots;* and secondary ones *adventitious*. Among the several peculiar forms of roots are also some, which we call *coralline root*, as that of *Corallorhiza*.

II. THE STEM.

78. The *stem* is that part of the plant which, originating in the plumule of the embryo, tends upward in its growth into the light and air, to produce, under their influence, leaves, flowers, and fruit.

The stem, generally called, with all it bears, the ascending axis of the plant, produces buds, resembling that from which it proceeded—namely, the plumule. All organs produced by the stem are merely repetitions of itself. The embryo is a primary stem (called radicle), with one or two—seldom more—leaves (called cotyledons) at its summit, which support a bud. The interstices between the successive (alternate, coupled, or whorled) leaves of the stem are merely new representations of the embryo-stem, and called *internodes* or *joints*. The points, where the internodes are united, or, as is sometimes the case, plainly articulated or jointed, are called *nodes* (popularly *knots* or *joints*). Since the internodes are repetitions of the stem-part of the radicle (the first internode), the apex of the stem, or of the uppermost internode, is always crowned with a bud. This bud contains the future continuation of the stem in miniature, just as the plumule contained the stem. These remarks may suffice to explain what the botanists mean, when they say: *a bud is a stem or branch in an undeveloped state*.

MODIFIED STEMS.

The stem is, wholly or in part, sometimes under ground (subterranean), and then it is distinguished from the root by having some sort of leaves, sheaths, scales, scars, and also by growing upward.

Rhizomes, bulbs, tubers, and *corms* are the types of underground stems.

Rhizomes or *rootstocks* are horizontal subterranean stems. *Acorus calamus* has a jointed, *Scrophularia nodosa* a knotty, and *Polygonatum* a scarred rhizome. (Pl. I., 9.)

Bulbs are abbreviated underground stems, furnished with an oval mass of thick, fleshy scales, closely packed together above, and with adventitious roots at the base, as the bulbs of the lily. (Pl. I., 12.)

Tubers are annual, thickened portions of a subterranean stem or branch, with minute scales, and often buds, sunk in small recesses, and called eyes. (Pl. I., 6.)

Corms are rounded or oval, fleshy, but compact underground stems, provided with more or less obvious buds. (Pl. I., 11.)

INTERIOR STRUCTURE OF THE STEM.

79. The internal structure of the stem exhibits the various forms of elementary tissue in certain combinations.

Among the *Cryptogams*, only the *Acrogens* have true stems—that is, stems containing woody fibres and ducts. The lower grades of the flowerless plants have either no stems, or stems consisting of parenchyma only.

The stems of the *Phænogams* have woody fibres and ducts.

The *parenchyma* of the stem *grows* equally fast *vertically and horizontally.* Into this both the *pleurenchyma* and the trachenchyma are imbedded vertically; they run longitudinally through the branches. The parenchyma we may call the *horizontal system* of the stem, and the pleurenchyma and trachenchyma together the *vertical* system.

80. *In* PHÆNOGAMS *the woody system is arranged in two widely different modes, the one mode characterizing the Exogens, and the other the Endogens.*

In EXOGENS the woody system is arranged, if their stems last one year, in one zone; if they last from year to year, in several (annual) concentric layers between a central pith and a separable bark, which forms a peculiar layer of tissue forming the circumference of the stem. (Pl. II., 29.)

In ENDOGENS, on the other hand, the woody tissue is never disposed in a zone, nor in concentric layers, but runs, in the form of separate and scattered bundles, through the ordinary cellular tissue, and not between a well-marked central pith and a separable bark. (Pl. II., 28.)

The stem of the *Exogens*, if it lasts longer than one year, increases in diameter by the annual formation of new pleurenchyma and trachenchyma around the prior circles—that is, grows on the outside (*Exogens:* outside-growers); but that of the *Endogens* increases by the deposition of new woody fibres toward the centre—that is, by a gradual distention of the whole system of woody bundles (*Endogens:* inside-growers).

81. A cross-section of the stem, or of a branch of any *exogenous plant* presents zones of different structure, known as *Pith, Medullary Sheath, Wood, and Bark.*

The *Pith*, occupying the centre of the stem, consists of parenchyma.

The *Medullary Sheath*, surrounding the pith, is a delicate tissue, consisting of spiral vessels.

The *Wood* consists of woody fibres and ducts, arranged in concentric circles.

The *Bark* is the outermost of the concentric layers, consists chiefly of parenchyma, and is lined by peculiar wood-cells, which we call *bast-cells*. Hence these bast-cells form the *inner bark* (*liber*, or *endophlœum*). The *outer layer* of the bark, consisting of parenchyma, is divided into a *green layer* (*green bark*, or *mesophlœum*) and a *corky layer* (*brown bark*, or *epiphlœum*), the latter being the outer stratum. The bark, as a whole, is invested by the *epidermis*.

Pith and *Bark* are brought in communication by the *medullary rays*—narrow plates of parenchyma (and as such they present themselves when seen on vertical sections of the stem). In cross-sections they appear merely as narrow lines. They run from the pith to the bark on all sides and make the *silver-grain* of the wood. (Pl. II., 29.)

82. Cross-sections of the stems of ENDOGENS exhibit no central pith, no distinct, separable bark (which, however, is present), and no layer or ring of wood between both, the latter being scattered throughout the whole in the form of threads or bundles. (Pl. II., 28.) In *Endogens* the bark is constantly found in the circumference of the woody bundles. The section of each thread or bundle exhibits wood within (in or near the axis), elongated cells without (in the circumference), and a parenchyma, which is intermixed with elongated and punctuated cells between. The lengthened cells constituting the outermost layer, are bast-cells, and the punctuated cells in the middle stratum a sort of liber. On the circumference of the stem the bark will accumulate, wherever leaves are formed. *Exogens* have a *cambium layer* —that is, a stratum of nascent wood and nascent bark between liber and wood; *Endogens* have none. The wood-bundles of *Endogens* are therefore limited, and the wood-wedges of the *Exogens* unlimited in growth.

While EXOGENOUS STEMS have the oldest and hardest wood at the centre, and the newest and softest at the circumference, in ENDOGENS the wood is softest toward the centre, and most compact at the circumference. Their stems increase in diameter in consequence of the continued deposition of new woody bundles so long as the more or less complete outer rind is distensible.

PRINCIPAL KINDS OF STEMS, NOT INCLUDING UNDERGROUND STEMS.

83. Stems are distinguished into *trunks, herbaceous stems, scapes* and *culms.*

Trunk ligneous (with harder wood)
- branching above .... *tree, arbor.*
- branching also below
  - the younger branches becoming ligneous and persistent .... *shrub, frutex.*
  - the younger branches dying every year; plant commonly not higher than 1 to 3 feet .... *suffruticose plant, undershrub.*

Stem herbaceous (caulis) with softer wood
- flowering and ripening its seed in the first year, and then dying, root and all .... *annual.*
- flowering in the second year, and dying, root and all, after ripening its seed .... *biennial.*
- blossoming year after year, but dying down to the ground annually .... *perennial.*

Scape: a peduncle springing from the root, leafless, or only with bracts, and bearing a solitary, or several to many flowers (a spathe is also a bract).

Culm: the straw of the sedges and the grasses, usually jointed, often hollow, and rarely becoming woody.

(Plants are called *caulescent*, when they have branching stems. Scape-bearing plants are said to be *acaulescent*, although they have some sort of underground stem).

Particular Names of Stems and Branches, chiefly derived from their Direction.

84. Stems bear different names in view of their direction and their simplicity or complexity. They are:

a. With regard to their direction and regular or irregular growth.

Erect: rising vertically from the ground; *ascending:* first bending, after having started from the ground, and then rising vertically; *procumbent:* prostrate or trailing, growing along the ground without rooting; *decumbent:* reclining on the ground, after having at the base risen somewhat above it; *repent:* creeping upon the ground and rooting; *cernuous:* bent over; *nutant:* having the top bent downward; *geniculate:* kneed, ascending by forming angles; *nodose* or *knotty:* furnished with hard, intumescences here and there (stems of grasses); *articulated* or *jointed:* provided with soft, intumescences, and, therefore, fragile at the joints; *scandent* or *climbing:* rising by laying hold, in some way, of other objects (as the stem of the grape); *voluble* or *twining:* winding spirally around a support (hop, bindweeds, etc.); *radicant* or *rooting:* climbing on other objects and striking root in them (as the stem of ivy).

b. With regard to the production of branches.

Most simple or *simple:* without branches, or nearly so; *ramose:* branching; *furcate:* forked; *dichotomous:* repeatedly furcate; *trichotomous:* divided into three branches.

85. *Branches* may be *opposite* (two at a node or knot, on opposite sides); *alternate* (one after another); *virgate* (wand-like, slender, weak, and ascendent); *verticillate* (whorled, several placed in a circle upon the nodes); *equal* (of equal length); *fastigiate* (springing from different heights of the stem, but terminating at one and the same level); *erect* (rising straight); *patulous* (spreading); *horizontal* (forming a right angle with the stem); *divaricate* (straddling widely); *pendulous* (drooping).

We must describe a few other sorts of branches called *Suckers*, *Stolons*, *Offsets*, and *Runners*.

The *Sucker* is a branch rising from some subterranean portion of the plant, bearing leaves above, sending out roots from its own base, separating at length from the mother-plant and becoming independent. The rose and the raspberry have suckers.

The *Stolon* is a branch which, having risen from some above-ground portion of the stem, becomes decumbent and trailing, strikes root from near its extremity, and becomes an independent plant. This sort of branches is seen in the hobble-bush.

Offsets, like those of the house-leek, are short stolons with a crown of leaves at the extremity.

Runners, like those of the strawberry, are long, filiform, tendril-like, prostrate and leafless branches.

III. THE LEAVES.

86. *Leaves* are peripheric organs of stems and branches, protruded from the mesophlœum of the bark in the form of thin, expanded bodies or laminæ, usually placed in a rather horizontal direction, and made more or less rigid by a framework of woody fibres called ribs and veins.

87. The *structure* of the leaf presents (at least in the higher organized plants) two parts—*woody tissue* and *parenchyma*. While the former, derived from the liber, and composing the framework, exists in small quantity, the latter, the green pulp, is predominant, and is nearly identical with the green layer of the bark. The cells of the green pulp exist in two layers, beneath the epidermis. The upper layer consists of oblong cells, placed perpendicularly to the surface of the leaf; the lower is made up of similar cells placed parallel to the same, and traversed by intercellular passages, so as to be less compact than the upper layer.

The green color of the leaf is owing to a green pigment, lying loose in its cells, and called *chlorophyll*.

Not only the *Phænogams*, but also the higher-grade *Cryptogams*, have leaves, consisting of both wood-fibres and parenchyma. The leaves of the cellular *Cryptogams*, however, consist of parenchyma alone.

88. As regards *their forms*, we divide the leaves into ORDINARY LEAVES on the one hand, and MODIFIED, TRANSFORMED, and DEGENERATED LEAVES on the other, such as *stipules, bracts, involucral leaves, spathes, scales, pales, cupules, hollow leaves*, etc. Modified, etc., leaves exhibit essentially the same structure as the ordinary sort.

* ORDINARY LEAVES.

89. *Ordinary leaves*—that is, those which serve particularly as foliage—are divided into, 1, *simple*, and 2, *compound leaves*.

† SIMPLE LEAVES.

90. In *simple leaves* we note their *parts*, their *consistency, position, insertion, outline* (general and partial), and *surface*.

1. PARTS, CONSISTENCY, POSITION, AND INSERTION OF THE LEAF.

91. PARTS OF THE LEAF: *a*, the *blade* or expanded part, with its nerves (and their ramifications); *b*, the *petiole* or leaf-stalk, which is often enlarged at its base, or wrapped, more or less, in the form of a sheath, around the stem or branch, or closely united with the stipules. The petiole is often wanting.

92. CONSISTENCY OF LEAVES.—The leaf is either: *a, carnose*, fleshy; *b, membranaceous*, very thin; *c, coriaceous*, leathery; or *d, herbaceous*, neither too thick, nor very thin.

93. POSITION OF THE LEAVES.—Leaves are said to be: *a, radical*, issuing from the root; *b, cauline*, springing from the stem; *c, fascicled*, when they grow in clusters, as those of the larch; *d, alternate*, one above the other, on nearly opposite sides; *e, opposite*, two against each other, at the same node or knot, as those of the Labiates; *f, decussate*, when the successive pairs of opposite leaves form a cross with each other at right angles; *g, cruciate*, four leaves in a whorl, placed crosswise, as those of some species of bedstraw; *h, distichous*, in two vertical ranks, as the leaves of the yew; *i, verticillate*, in whorls; *k, sparse* or *scattered*, irregularly spiral; *l, equitant*, riding astraddle, when conduplicate leaves alternately embrace. (Pl. II., 10.)

94. INSERTION OF THE LEAVES.—Leaves are: *a, sessile*, without a petiole (stalk), Pl. I., 22; *b, petiolate* or *petioled*, with a petiole (which is sometimes channelled, or vaginate, or winged), Pl. I., 13; *c, decurrent*, running down the stem, as in thistles (stem winged); *d, amplexicaule*, or clasping, Pl. I., 31; *e, perfoliate*, when the stem appears to pass through a leaf, as in *Bupleurum rotundifolium*, Pl. I., 30; *f, connate*, opposite leaves with the bases united so as to form one piece of the two, Pl. I., 33; *g, peltate*, with the petiole in the centre of the blade, or nearly so, Pl. I., 29.

2. Outline and Surface of the Leaf.

95. General Outline of the Leaf.—As regards the general outline, the leaf is said to be: *a, setaceous*, bristle-like, as the leaves, or rather, branches, of asparagus, and those of some grasses; *b, subulate*, or awl-shaped, as those of the common juniper, Pl. I., 15; *c, linear*, more than four times as long as wide, with nearly parallel margins, Pl. I., 14; *d, acerose*, or needle-shaped, as the leaves in pines, Pl. I., 14; *e, cuneate*, wedge-shaped, two to four times as long as wide, and gradually tapering from the broad and truncate apex toward the base (§ 67, Cut II., *i*); *f, spatulate*, rounded above, and long and narrow below, Pl. I., 17; *g, lanceolate*, or lance-shaped, when several times longer than wide, and tapering upward, or both upward and downward, Pl. I., 23; *h, oblong*, when nearly twice as long as broad (see § 67, Cut II., *o*.); *i, elliptical*, oblong with a flowing outline, the two ends alike in width (see above § 67); *k, oblong-ovate*, more ovate than oblong; *l, obovate*, ovate, with the petiole at the tapering end of the blade; inversely ovate, Pl. I., 20; *m, orbicular*, circular; *n, reniform*, or kidney-shaped, Pl. I., 28; *o, rhombic*, rhomb-shaped, Pl. I., 18; *p, ensiform*, sword-shaped, as the leaves of Iris.

96. Special Outline of the Leaf.—As regards the *apex* and *base* of the leaf, see above § 67, with figures.

STRUCTURAL BOTANY.

In view of the *margin*, the leaf is *entire*, when it is even-edged—that is, when the margin is destitute of incisions. If a *leaf* has deep incisions, it is called *lobed, cleft, parted, divided* (see § 67 with figures in III.) The segments may be entire, or furnished with small incisions.

If the margin of a leaf has small incisions, its outline is expressed as follows:

Compare the figures below (VIII.)

		Legs of each angle unequal... *serrate*, or *saw-toothed* (a)
Outer and inner angles of the incisions both pointed.	Legs of the angles equal.	the outer and inner angles obtuse, rarely right, or moderately acute... *dentate* or *toothed* (b)
		the outer and inner angles very acute and the divisions narrow and long..... *fringed* or *ciliate* (c)

Outer angles rounded, the inner ones pointed, the teeth broad and rounded......... .... *crenate* or *scalloped* (d)
Outer angles rather pointed, the inner ones rounded............. *repand, repand-toothed, wavy* or *undulate* (e)
Both outer and inner angles rounded.......... *sinuous* or *sinuate* (f)
Both outer and inner angles irregular and small..... *erose, jagged,* or *gnawed* (g)

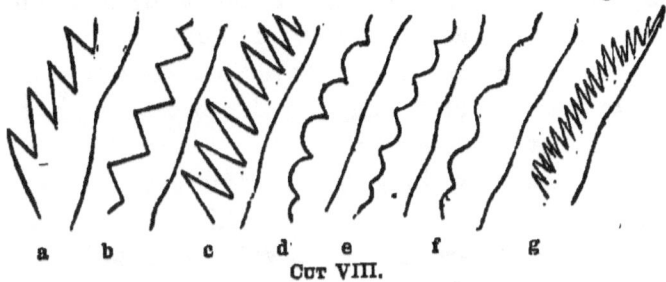

a　b　c　d　e　f　g
Cut VIII.

If the incisions of a serrate leaf (the serratures) are very deep, the leaf is said to be *cut* or *incised-serrate*, and if the indentations (teeth) of a dentate leaf are very deep, it is *cut* or *incised-toothed.* We employ such terms as *double-serrate, double-dentate,* or *double-crenate,* when the incisions are unequal.

97. The Surface of the blade of a leaf is said to be *undulate*, or wavy, if bent up and down; *crisp*, when bent irregularly up and down; *rugose*, wrinkled; *plicate*, or folded, gathered into longitudinal folds; *nervose*, or nerved, with strong vascular bundles, *nerves*, or *ribs*; *veined*, having thin and slender vascular bundles, especially branching ones, *veins*. We distinguish *netted-veined* (Pl. II., 12) and *parallel-veined leaves* (Pl. II., 10), and other sorts of venation, which will be considered presently in connection with certain divisions of the margin.

98. The division of the margin of leaves with reference to the framework of their surface.

Leaves are, as regards the principal bundles of their woody fibres, either *feather-veined* (*pinnately-veined*) or *radiate-veined* (the same as *palmately-veined*).

The first-named sort has only one longitudinal rib or nerve, extending from the top of the petiole to the apex of the leaf, and this rib, also called the *midrib*, sends out branches, or *veins*, which are divided into branchlets or *veinlets*. Sometimes the two lowest branches of the midrib are very strong and long, and the leaf is then called *triple-ribbed* or *triple-nerved*.

The other sort has three to five or more ribs or nerves, instead of a single one; hence we distinguish *three-ribbed* (triple-nerved), *five-ribbed*, *seven-ribbed*, etc., *leaves*.

Now, according to these two modes of venation, we distinguish two modes of division into segments, calling some leaves *pinnately-* and others *palmately-cleft, parted,* or *divided*. (See Pl. II., 3–6.) A few *modifications of pinnately-cleft or parted* leaves are designated by particular terms. *Pectinate* we call pinnately-parted leaves, with very close and narrow divisions, resembling the teeth of a comb; *lyrate, or lyre-shaped*, those with the segments decreasing in size toward the base (Pl. II., 30); *runcinate*, those with rather sharp lobes pointing toward the stem or branch, or downward. (Pl. II., 2.)

Palmately-cleft leaves are sometimes briefly called *palmate leaves*, and a palmate leaf with the lateral lobes cleft into two or more segments, is said to be *pedate* (since it resembles a bird's foot).

The segments of a lobed or divided leaf may be again divided, parted, or cleft, and then the leaves are said to be *bipinnatifid, tripinnatifid* (that is, bipinnately or tripinnately parted), or *twice-palmately, thrice-palmately cleft, parted,* etc.

†† COMPOUND LEAVES.

99. A COMPOUND LEAF is a leaf composed of two or more blades, called *leaflets*, borne on a common petiole, and usually supported by stalklets of their own, between which and the main petiole an articulation is formed. The leaflets we describe just as we do simple leaves, as entire, serrate, toothed, cleft, parted, etc.

100. COMPOUND LEAVES are of two principal sorts—namely, the *pinnate* and the *palmate* (or *digitate*).

The first sort are produced when a simple leaf of the pinnately-veined variety becomes compound; they have their leaflets, or pinnæ, along the sides of the common petiole; the second sort result from the palmately-veined kind of simple leaves becoming compound; they bear their leaflets, or folioles, on the apex of the common stalk.

When the division of a simple leaf reaches the midrib, or the top of the petiole, the leaf is said to be compound. (Pl. II., 6, 11.) Not unfrequently it is rather difficult to decide, whether a leaf is simple or compound; and we may even find in different parts of the same leaf different degrees of division, illustrating the gradual transition of leaves from simple to compound in all degrees.

Again, we distinguish *simply* and *repeatedly compound leaves*.

101. Leaves simply compound, both of the pinnate and the palmate sort, receive different names from the number of their leaflets.

In *pinnate* leaves the number of leaflets varies from three (or rarely one) to sixty, and upward. When a pinnate leaf has three leaflets, it is said to be *trifoliolate*, or *ternate;* when two, *binate*. When we find among three-foliolate leaves one with only one leaflet, as in *Desmodium, Rhynchosia*, and *Baptisia*, we must, in theory, regard such a leaf as a compound, since its single leaflet is articulated to the petiole.

When a pinnate leaf has an uneven number of leaflets, the odd one being borne on the very tip of the common petiole, it is said to be *odd-pinnate*, or *unequally-pinnate;* and when it has an even number, it is *equally-* or *abruptly-pinnate*. The latter may bear a *tendril* at the end, in place of an odd leaflet, as in the Vetches and in the Pea. (Pl. II., 8.)

A lyrate leaf, becoming compound, is said to be *lyrately-pinnate*.

Pinnate leaves with smaller leaflets irregularly intermixed with larger ones, as in *Agrimonia Eupatoria*, are said to be *interruptedly-pinnate*.

Palmate leaves may be *ternate, quinate, septinate*, etc. —that is, they may have three, five, seven, etc., leaflets springing together from the top of the common petiole. In Pl. II., 6, we have a septinate leaf. A palmately three-foliolate leaf must not be mistaken for a three-foliolate pinnate one, which has the odd leaflet raised above the other two on the prolonged common petiole, while the palmate has its three leaflets either sessile, or stalked alike. (Pl. II., 7.)

102. Leaves repeatedly compound are those pinnate or palmate leaves, in which the common petiole bears, instead of leaflets, secondary petioles, supporting more than one leaflet, which may have tertiary petioles, bearing a number of leaflets in their turn. When the common petiole of a leaf of the pinnate sort bears, instead of leaflets, petioles furnished with two or more (sessile or stalked) leaflets, the leaf is *twice-pinnate*, or *bipinnate* (Pl. II., 13); and when the common petiole of a leaf of the palmate sort gives off three petioles, each of them bearing three leaflets, the leaf is *twice-ternate* or *biternate*. (Pl. II., 12.)

If the division goes one step farther in this direction, a leaf becomes *thrice-pinnate* or *tripinnate* (Pl. II., 14*b*), or *thrice-ternate* (triternate)—Pl. II., 14*a*.

When a compound leaf becomes more than thrice-pinnate, or thrice-palmate, the leaf is said to be *decompound*.

It happens, sometimes, that the secondary division becomes pinnate, although the primary one was digitate; in that case, we have the two sorts of compound leaves combined in the same leaf.

** MODIFIED LEAVES.

103. Modified leaves, whether extraordinarily formed, or in any wise arrested in their development, are those which are something else than foliage, as *pitcher-shaped leaves* (*ascidia*), *fly-traps* (as in *Dionæa*), *scales*, *bracts*, *spathes*, *air-bladders* (in some species of *Utricularia*), *tendrils* (which, however, are sometimes transformed branches), and *spines*.

Pitcher-shaped leaves are those of the Side-saddle flower, or *Sarracenia*, with the edges curved inward and united, forming a cavity which is usually half full of water. (Pl. II., 9.)

Leaves serving as fly-traps are to be seen in *Dionæa*, or Venus's Fly-trap. These bear at the summit an appendage, something like a steel-trap, which opens and shuts, to catch flies.

Bracts, not always sufficiently distinguishable from scales or from leaves, are generally leaves of an inflorescence. They often bear a striking resemblance to ordinary leaves, as in the Linden (Pl. IV., 6), but are usually smaller. Properly speaking, the bract is the small leaf (or scale), from the axil of which a flower or its pedicel springs. Sometimes we call the upper leaves bracts, since they support flowers, as in *Melampyrum Americanum*. (Pl. IX., 8.) The glumes and pales of the grasses are bracts. The application of the terms bract and scale is oftentimes discretional. Thus the involucre of composites may be said to consist of empty bracts, or of scales, and the use of either of these terms may here be justified.

Scales are poorly developed, membranous or fleshy, leaf-like organs, and often of another color than green. In some cases they are transformed leaves (in buds and bulbs); in others they are modified bracts (in aments, cones, etc.), or, again, they are appendages to the petals. The small bracts (chaff) between the disk-flowers of Composites are scale-like, and, therefore, often called scales, as also are glumes and pales.

Spathe is a bract, usually large, leaf-like, which inwraps an inflorescence. (Pl. II., 25.)

Tendril is a thread-like body, used for climbing, and

either a transformed branch, as in Ampelopsis, or a transformed leaf (at least a considerable part of the compound leaf), as in Lathyrus, and finally a changed odd leaflet, as in Vetch and Pea.

Spines are abortive leaves, or altered stipules. When in the Barberry a leaf-bud is produced in their axil, the spine is held to be nothing but a reduced leaf. In other cases they are petioles, changed after the leaves fall off. In Robinia we find a pair of spines at the base of the petiole, in place of stipules. We must distinguish spines, on the one hand, from *thorns*, which, originating in axillary buds, are abortive branches, and, on the other hand, from *prickles*, such as those of the Rose and Blackberry, which belong to the epidermis, and may be stripped off with it.

Stipules are lateral appendages of leaves, always occurring in pairs, situated on each side of the base of the leaf-stalk. In a great number of plants they are absent, but their presence is often the characteristic of all the species of an order or tribe. They are various in form, sometimes membranous or scale-like; in other cases, as we have stated above, transformed into spines. They may develop on young shoots only, as in the Beech, or in Magnolia. Not unfrequently they cohere with each other, or with the base of the petiole. Sometimes a stipule adheres to each side of the base of the leaf-stalk, as in the Rose, Strawberry, and Clover. In the Plane-tree the two stipules are free from the base of the leaf-stalk, but cohere by their outer edges, apparently making a single stipule opposite to the leaf. And again it happens, that they are united by both margins, so as to form a sheath around the stem, just above the leaf; such stipules are said to be *interfoliaceous*. When interfoliaceous stipules are membranous, they bear the name of *ochreæ*, as in Knot-weeds.

When opposite leaves have stipules, the latter usually occupy the space between the petioles on opposite sides, and are then called *interpetiolar stipules*. The stipules of each of the opposite leaves becoming thus contiguous, or even continuous, each pair of leaves appears to have but a single pair of stipules, as in several Madderworts. The stipules of leaflets are called *stipels*, and leaflets with stipels *stipellate leaflets*, in contradistinction to *stipulate leaves*.

Leaves may differ from one another in the shape of their petioles. The latter, generally terete, or half-terete, and not unfrequently channelled on the upper side, may be winged (see § 94). Sometimes the petiole is flattened at right angles with the blade, as in the Aspen; or it is *dilated below into an inflated membranous sheath*, as in many Umbelworts. (Pl. I., 37.)

In many Endogens the leaf-like petiole consists entirely of a sheath, inwrapping the stem, which in grasses bears above a membranous appendage, to be regarded as a double axillary stipule and called the *ligule*. (Pl. I., 36.)

BB. ORGANS OF REPRODUCTION.

*THE FLOWER.

I. ESSENTIAL ORGANS OF THE FLOWER.

104. THE PISTILS vary in number. A flower with a solitary pistil is said to be *monogynous*, and one with 2, 3, 4, 5, 6 or many pistils, *digynous, trigynous, tetragynous, pentagynous, hexagynous, polygynous.* When in great number, they are arranged in spiral rows on an enlarged receptacle, as in most Crowfoots, the Magnolia, Strawberry, etc.

PISTILS are either SIMPLE or COMPOUND.

105. A SIMPLE PISTIL consists (as was stated in § 47) of the transformed blade of a carpellary leaf. The inner surface of the ovary answers to the upper surface of the leaf, and the style or stigma to its apex, rolled together. The line of the pistil, which was formed by the union of the edges of the leaf, is called the *inner* or *ventral suture*, a true seam, and always faces the axis of the flower. The opposite line, answering to the midrib of the leaf, is either obsolete (not at all, or scarcely perceptible), or conspicuous as a thickened stripe; this we call, though improperly, the *outer* or *dorsal suture*.

That part of the ventral suture within the ovary, to which the ovules are attached, and which commonly forms a projection, is called the *placenta*. The *placenta is the ovule-bearing part of the inside of the ovary*, and consists of two parts, one belonging to each edge of the transformed leaf. Hence, it is usually two-lobed or two-ridged. According as the union of the edges of the leaf is more or less perfect, the double nature of the placenta will be more or less obvious. When a pea-pod is laid open, the ovules or peas will be found in alternate order along each edge, so as to make but one row, when the pod is closed. In Aquilegia, on the other hand, the pods have their seeds arranged in two rows very conspicuously. Sometimes only one ridge of the placenta is fruitful, as in the one-seeded Cherry.

A simple pistil can have but one cell, one placenta, one style, one stigma. The fact, that a simple pistil is one-celled, does not imply that all single-celled pistils are simple, since many compound pistils are one-celled. The stigma may be two-crested, or even two-lobed. It is subject to great variations in form, but usually it is globular (capitate) and terminal, often linear and lateral.

106. A COMPOUND PISTIL consists of a pair or a whorl of carpellary leaves united more or less into one piece, and the degree of their union differs in proportion as the petals coalesce more or less perfectly. While the pair of carpels of Saxifrage make a compound pistil, their union proceeding from below upward, and extending only to about one half of their length, or less, the union of the two carpels of Pink leaves only the two styles separate. In some of our wild species of Hypericum, the 3-carpelled ovary has also the three styles united into one, its three stigmas remaining distinct; and in the three-celled pistil of Tradescantia, even the three stigmas are consolidated into one body. Neither the styles, therefore, nor the stigmas always indicate the number of carpels of which the pistil consists, although in most cases they do.

As was stated above, the ventral suture, which is the placenta, always faces the axis of the flower. Hence it is evident, that, when a pair or a whorl of *closed carpellary leaves* become consolidated, the placentæ will meet in the centre of the compound ovary, and that the latter has as many axile placentæ, more or less united into one, as there are pistil-leaves in its composition, and as many cells. (Pl. V., 7, 8; VI., 1c.) The partitions or *dissepiments* of a compound ovary are, of course, double, one of the two layers belonging to each carpel, and in ripe pods they often split into two layers. None of the carpels, forming the combination, can have a true dissepiment; if any ever occurs, it is a spurious one, an expansion of the dorsal suture, as in the 5-carpellary ovary of Flax.

On the other hand, the carpellary leaves may remain open and unite by their edges, as the petals in a gamopetalous corolla, and a *one-celled compound ovary*, with as many lateral placentæ, as there were carpellary leaves, will be the result. So it is in Helianthemum, in certain species of Hypericum, and many other plants. Each of the placentæ consists here of the contiguous margins of two carpellary leaves, grown together.

Finally we often meet with compound ovaries presenting one cell, a central placenta (placentæ united), and no dissepiments, as in Pinks, Primroses, etc. (Pl. IV., 4c.) In such cases we judge that the pistil was formed by the coalescence of several closed carpels (3 to 5 in Spergularia), and that the dissepiments vanished very early. In some or all pistils of this sort, the central placenta is regarded by many botanists, and sometimes perhaps correctly, as a growth from the axis or receptacle.

107. Ovules are modified buds. Their number varies from 1 to 100. While Buttercups, Composites, and Grasses have solitary ovules in their ovaries, in Verbascum and Papaver they are indefinite—that is, not readily numerable.

As regards the position of the ovule in the cell of the ovary (see Cut IX.), it is said to be *erect* when growing upward from the very bottom of the cell, as in the Composites; *ascending* (Fig. 1), when rising obliquely from its point of lateral attachment; *horizontal*, when projecting from the side of the cell, and not turning either upward or downward; *pendulous*, when turned obliquely downward; and *suspended*, when hanging vertically from the summit of the cell (Fig. 6).

An ovule consists of a *kernel, nucleus,* and usually one or two coats. *In the upper part of the nucleus is the embryo-sac, in which the embryo is formed after fertilization;* and the coats—an inner one, *tegmen,* or *secundine,* and an outer one, *testa*, or *primine*—are the initial integuments of the future seed. The ovule of the Mistletoe has no coats, and that of the Walnut one only. Both coats remain open at the summit of the ovule, leaving a small passage called the *micropyle*. The ovule is either sessile, or raised on a stalk, the *funiculus*. The part by which the ovule is attached to the cell-wall, the placenta, or the funiculus, the point where the ovule, when changed into a seed, breaks away, is called the *hilum*, and then forms the scar of the seed (*h* in the figures of Cut IX.). The nucleus and the coats are unconnected, save at the base; here they are firmly united with each other and with the funiculus, when there is any. This point, where coats and nucleus cohere, is called the *chalaza*.

108. Ovules occur under four principal forms. The ovule is said to be *orthotropous* (Fig. 2) when perfectly straight. But when it is more or less inverted or curved over upon its elongated funiculus, or upon itself, we have a few terms to express its conditions. It is *anatropous* (Fig. 3) when it is completely inverted on its funiculus, remaining straight; in this case, a portion of the funiculus adheres to the testa, and is called the raphe (*r*); the orifice or micropyle (*f*) is close to the point of attachment; and the chalaza (*c*) occupies the point directly opposite to the point of attachment. An ovule is *campylotropous* (Fig. 4), when it curves upon itself by growing unequally, so as to bring the orifice (*f*) near to the chalaza (*c*). It is *amphitropous* (Fig. 5), when it is half inverted, remaining straight, and furnished with a raphe (*r*), extending from the chalaza (*c*) about half way to the orifice (*f*). The amphitropous ovule differs from the anatropous merely by the shortness of its raphe. The orthotropous and campylotropous ovules have no raphe.

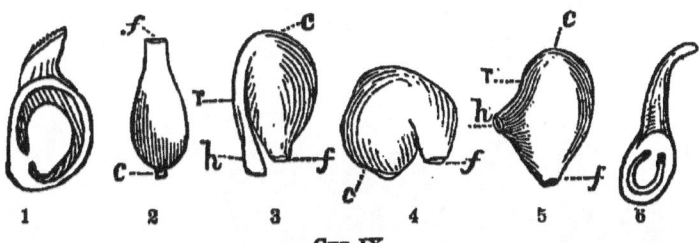

Cut IX.

The carpellary scales and naked ovules of the *Gymnosperms* have been already sufficiently described in § 45.

109. The stamens vary in number. A flower with a solitary stamen is *monandrous*. The terms *diandrous, triandrous, tetrandrous, pentandrous*, etc., and *polyandrous*, signify furnished with 2, 3, 4, 5, many stamens. Compare § 57. When the stamens are very numerous, they are arranged in several rows.

Stamens take their origin from floral leaves in the same way as pistils. The filament represents the petiole, and the anther the blade of a leaf. The blade curved in, until each of its edges unites with the midrib, forms a 2-celled anther. The usually well-marked stripe, which extends between the anther-cells, answering to the midrib of the leaf, is the *connective* or *connectile*. The *anther* is the essential part of the stamen (see § 33), and commonly it is 2-lobed and 2-celled—rarely 4- or 1-celled. It is the function of the anther to produce pollen, and discharge it at maturity.

With regard to their position, the stamens are *hypogynous*, when they spring from the receptacle below the ovary or ovaries (Cut X., Fig. 1); *perigynous*, when they are inserted on the calyx around the ovary (Fig. 2); *epipetalous*, when they are fixed on the corolla; *epigynous*, when they stand on a level, answering to the summit of the ovary so as apparently to spring from the top of this organ (Fig. 3), and *gynandrous*, when they cohere with the style. (Fig. XIII., 12a.)

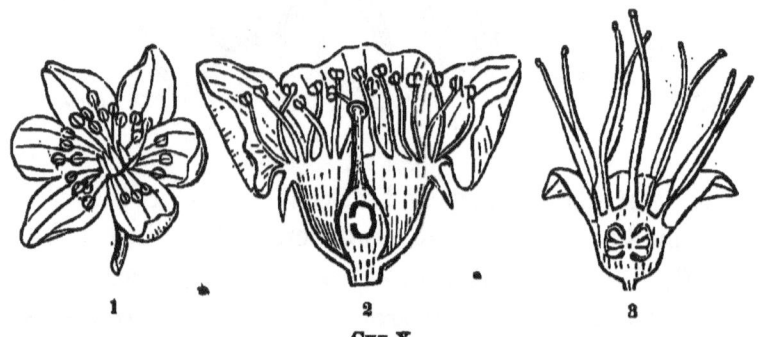

CUT X.

FIG. 1, a flower of Anemone, entire ; 2, of a Cherry ; 3, of a Vaccinium, both vertically divided.

110. The STAMENS, as we have stated in §§ 57 and 58, are either *distinct*, or *united* with each other, or with the style (§ 109). *Monadelphous stamens* have their filaments united either at the base only, as in *Stuartia*, or consolidated into a column or sheath, as in Mallows (Pl. IV., 5a) and some leguminous plants.

Diadelphous stamens are stamens united into two sets, or bundles, as in the *Fumariaceæ* (Pl. IV., 1a), and most *Leguminosæ* (Pl. V., 3b). *Triadelphous stamens* we see in several species of Hypericum, and *pentadelphous* ones in *Gordonia*. For triadelphous and pentadelphous we usually employ the term *polyadelphous*.

Syngenesious stamens are those which are *united by their anthers*, and mostly so as to leave the filaments distinct. But sometimes the filaments of syngenesious stamens are united. Stamens of this sort we find in the *Cucurbits*, etc. (§ 61). In this order they are also sometimes triadelphous (Pl. V., 9a). Syngenesious stamens with the filaments free, we have in the whole order of the Composites. (Pl. VII., 6c.)

Gynandrous stamens are seen in *Aristolochia* and in all Orchids (§ 61. Pl. XI., 4*b*; XIII., 12*a*).

111. The ANTHER is usually borne at the top of the filament. As has been stated in § 109, it usually consists of two lobes. Each of these lobes is generally 1-celled; in rare cases, however, 2-celled (whole anther, 4-celled), as in *Tetranthera*. Most anthers are 4-celled, when young, so that a slender partition runs longitudinally through each cell, which it divides into two portions, answering to the upper and lower layer of the green pulp of the leaf. Here and there we meet with 1-celled anthers. They become 1-celled either by *confluence*, the two cells of the kidney-shaped anther (Pl. IV., 5*c*) running together into one, as in the Mallows, in Veratrum, etc., or by the obliteration and disappearance of one half of the anther, as in the Globe Amaranth of the gardens. In Pentstemon (see Fig. *g* in Cut XI.) they are almost confluent.

112. There are three modes of attachment of the anther to the filament. The *anther* is said to be, *a, innate* when it is fixed by its base to the very summit of the filament, turning neither inward nor outward; the connective, in this case more or less conspicuous, is attached by two opposite sides to the anther-cells (see Cut XI., Fig. *a*); *b, adnate*, when both cells are placed on one side of a broad connective, leaving its opposite side free (Figs. *b, d, e*); *c, versatile*, when it is fixed by a point near its middle to the apex of the filament, so as to be freely movable—for instance, in *Œnothera*, and in all grasses (Fig. *c*).

The ADNATE ANTHER is either extrorse or introrse; *extrorse* (turned outward), when it occupies the outer side of the connectile, looking toward the floral envelopes, as in *Liriodendron* (Fig. *b*), *Asarum* (Fig. *e*), *Iris*, etc.; *in-*

trorse (turned inward), when it faces the pistils, as in *Magnolia* (Fig. *d*).

The VERSATILE ANTHER is usually *introrse* (turned toward the axis of the flower)—rarely extrorse.

The CONNECTIVE is often inconspicuous, or even wanting, so that the anther-lobes are in close contact. Sometimes it outruns the anther and tips it with an appendage, as in *Magnolia, Liriodendron, Violet, Asarum*, etc.

113. At maturity, the anther-cells open, or become *dehiscent*, to discharge the pollen. There are *three modes of dehiscence*—namely, *a, the valvular*—that is, the splitting open by two lateral, longitudinal lines, one on each lobe of the anther (Figs. *f, g*); *b, the porous*, in which the cells open by a pore or chink at the apex of the lobe (as in *Pyrola*, Fig. *h*), and each lobe is sometimes prolonged into a tube, as in *Vaccinium;* *c*, the *opercular*, in which each cell opens by a lid, which, as though attached to the apex by a hinge, turns upward like a trap-door, as in *Berberis* (Fig. *i*). The commonest mode of dehiscence is the valvular.

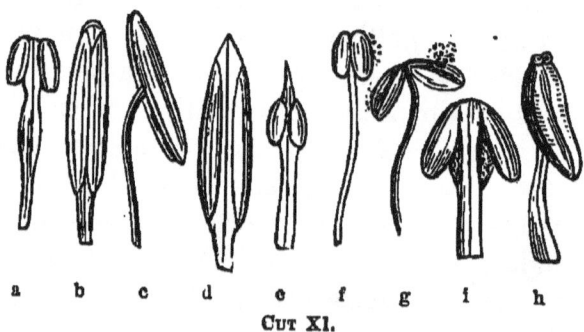

a   b   c   d   e   f   g   i   h
CUT XI.

114. The POLLEN, when examined under the microscope, is found to consist of grains, generally globular or ovoid, sometimes triangular or polyhedral, etc., but all alike in the same species. Each pollen-grain is a membranous sac, or cell, filled with a fluid. Its membrane, or coat, is double, the outer layer thick, but weak, the inner one very thin, but distensible, the former usually being striped or banded. The thickish fluid within contains minute molecular granules. When pollen-grains are exposed to moisture, they swell to such a degree that many kinds burst and discharge their contents. In the Asclepiads and Orchids the pollen-grains cohere in masses, called *pollinia*. (Pl. XIII., 12*b*.)

115. Of the REPRODUCTIVE ORGANS, the pistils and the stamens, the latter are adapted to fertilize, the former to be fertilized. The pollen-grains, after being brought in contact with the stigma, are retained there by its loose papillæ or projecting hairs. Having absorbed moisture from the stigma, they soon begin to grow or germinate. The inner coat of the grain expands and breaks through the outer one, which is weak and brittle, in the form of a tube, filled with the liquid and molecules, contained in the cavity of the grain. This tube penetrates like a radicle the soft tissue of the stigma, and, growing downward into the style, reaches the placenta, where, after entering the micropyle of an ovule (whether solitary, or one of many), it makes its way to the surface of the *embryo-sac*, but probably not farther (see figures of Cut XII.: Fig. 1, vertical section of the pistil of Buckwheat, enlarged; *o*, the ovule with its two coats; *e*, the embryo-sac; *v*, the embryonic vesicle; *p*, a pollen-tube, having entered the micropyle; *p'*, a pollen grain with its tube, separate).

It must be remembered that the embryo-sac at this stage as yet contains no embryo (§ 107).

The pollen-tube, having reached the surface of the embryo-sac, in some unexplained manner, causes the formation within it of the *embryonic vesicle*, which is attached to its wall next the micropyle (Fig. 2, the embryo-sac with the vesicle within, more enlarged). The materials of this vesicle are taken from the *cytoblast*, which consists of globular atoms, contained in the sac. The embryonic vesicle resembles all other new cells.

Let us trace its development after fertilization in one or more ovules of an exogen. After attaining a certain size, the vesicle divides by a cross-partition into two cells (Cut XII., Fig. 3), and the lower of these into another pair (Fig. 4). One cell of this pair continues the process of division in two directions (Fig. 5), and the resulting cells do the same, until the mass of cells assumes the outline of a rudimentary embryo, its upper extremity representing the radicle, in form nearly cylindrical, and its lower extremity the cotyledons under the form of a notch (Fig. 6). Gradual changes in the aspect of this embryo are shown in Figs. 7 and 8 and 9. The figures are, of course, all magnified. In the Asclepiads, the masses of cohering pollen-grains, dislodged in due time from their anther-cells and brought near the base of the stigma, produce a great many tubes, which penetrate the base of the stigma (Fig. 10). The fertilization of the naked ovules of the Gymnosperms usually leads to the production of several embryos, which, however, are commonly all but one abortive.

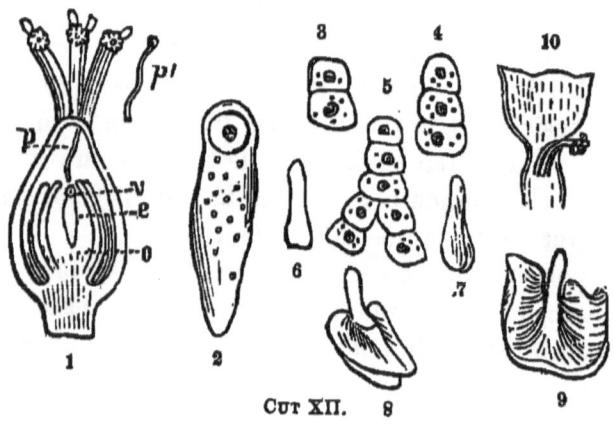

Cut XII.

The growth of the plantlet, after it springs from the seed, is simply a continuation of the same process of cell-division. But the plantlet, or embryo, must have grown and developed first in the seed, ere it can separate from it.

Of the seed we shall treat below, when we come to the fruit. Still we may at once state, that *the ovule, after fertilization, or after the embryo is formed, is called seed.*

II. Protecting Organs of the Flower.

116. The Protecting Organs of the flower are the *floral leaves*, called the *perianth*, which supports and protects the stamens and pistils. The perianth is, as was stated in § 34, either double or simple. But the term *perianth* is often used instead of the phrase simple perianth, for brevity's sake, or whenever we can not readily distinguish which leaves compose the calyx, and which the corolla.

117. The leaves of the perianth we consider to be modified ordinary leaves. In very many instances we find a gradual transition from ordinary leaves into sepals, and from sepals into petals; and the latter, when present in great number, grow smaller by degrees toward the axis of the flower, and are often stamen-like. Stamen-like petals we call *staminodia*, as in Nymphæa.

118. The Calyx, considered with respect to its form and duration, is simple, when consisting of one set of sepals only; double, when surrounded by an *involucel* (a whorl of bractlets), as in certain Mallows (we then have an involucellate calyx, in contradistinction to a naked calyx); persistent, not falling away after flowering, sometimes enlarging in fruit, as in Physalis, or becoming fleshy, as in the Rose; deciduous, falling away before flowering, as in the Poppy, or immediately after floration, as in the Cherry;

Gamopetalous, or Polypetalous (see § 38); parted, cleft, or toothed, from the deeper or shallower division of its limb; regular, with its lobes similar; irregular, with its lobes dissimilar, sometimes two-lipped, as in most Labiates; spurred, as in *Delphinium*; bisaccate in *Cheiranthus annuus* (a garden-plant).

119. The regular calyx receives different names from the diversities of its shape. It is called *wheel-shaped* or *rotate*, as in Potentilla (Pl. V., 4); *bell-shaped*, or *campanulate*, as in Hibiscus; *top-shaped*, as in Agrimonia; *urn-shaped*, or *urceolate* (in the Rose, Pl. V., 5), *funnel-shaped* (in Hyoscyamus), *tubular* (in the Pink), *prismatic* (in *Datura Stramonium*, Pl. X., 19), *inflated* (in *Silene inflata*), etc.

120. From its place of insertion the calyx is said to be *inferior*, or *hypogynous*, when inserted on the receptacle below the ovary (the *ovary* is then *superior*); *half-superior*, or *perigynous*, when adnate, by its tube, to the lower half of the ovary (and then the *ovary* is said to be *half-inferior*, or half-superior, as in Ceanothus, Mitella, Lophiola, and Aletris); *superior*, or *epigynous*, when its lobes or teeth spring from a level coinciding with the top of the ovary, its tube being adnate to the whole circumference of the ovary, as in the Pear, the Apple, the Blueberry, etc., or only constricted above it, as in Rhexia (the *ovary* is then said to be *inferior*).

(A calyx superior is also that of the Composites—its limb, if not wanting, represented by a crown, scales, or bristles.)

121. The Corolla, sometimes early deciduous, is, as we have seen in § 40, either gamopetalous, or polypetalous.

122. The forms of the Gamopetalous Corolla are to be viewed with reference to limb, tube, and mouth, and are designated as follows:

Tubular, having a cylindrical tube and a slightly spreading limb or border, as in the Primrose, or the tubular flowers of the Composites (Pl. VII., 6 *a, b*); *cup-shaped*, with segments cohering into a concave border (in Kalmia, Pl. VIII. 6); *urn-shaped*, or *urceolate*, as in the Whortleberry; *bell-shaped*, with the tube enlarging abruptly at the base and gradually into the limb, as in the Bell-flower (Pl. VIII., 2); *funnel-shaped*, narrow-tubular below and gradually widening into the border, as in Convolvulus, Quamoclit (Pl. X., 16), etc.; *wheel-shaped*, as in Veronica, Solanum (Pl. X., 17), etc.; *salver-form*, with the tube terminating abruptly in a horizontal limb, as in Phlox; *ligulate*, or *strap-shaped*, resulting from the splitting down of a tubular corolla (in the Liguliflorous Composites, Pl. VII., 7*b*, 9*b*); *labiate*, or *bilabiate*, two-lipped—that is, with two of the petals placed higher than the rest, forming the upper lip, and the three remaining ones joined together on the opposite side of the flower, to form the lower (usually three-lobed) lip. This sort of corolla belongs to most Labiates, Figworts, etc. (Pl. IX. and X., with numerous bilabiate corollas, etc.)

123. The forms of the Polypetalous Corolla are the *rosaceous*, with 5, more or less equal, roundish, spreading, short-clawed petals; *malvaceous*, with 5 petals, their short claws united with the staminal column, as in Mallows; *cruciform*, with 4 long-clawed, spreading petals, placed at right angles to each other, as in the Crucifers; *caryophyllaceous*, with 5 regular, long-clawed, spreading petals, within a tubular calyx, as in Dianthus, Saponaria, Silene, etc.; *papilionaceous*, with 5 irregular petals, whereof the uppermost one, which is usually the largest, is called *vexillum* (banner; standard), the two lateral ones *wings* (alæ), and the two lowest, which are often united at their lower margin, the *keel* (carina) (Pl. V., 3*a*); *orchidaceous*, having, within three sepals, three petals, whereof one, the *lip*, which, in fact, is uppermost, but apparently, by the twisting of the ovary, lowermost, is variously enlarged and deformed. (Pl. XIII., 12.)

124. Certain appendicular organs of the flower are called *spurs, scales, glands, crown.*

These are good marks not to be overlooked, when we want to identify certain plants. All these appendages were formerly comprehended under the general name of *Nectary.*

Spurs are more or less elongated tubular appendages of the flower, usually projecting from behind it. Aquilegia has all its petals spurred (Pl. III., 4), the Violet only one (Pl. IV., 2), and Delphinium two of its petals and one of its sepals, the spur of the sepal enclosing the spurs of the petals (Pl. III., 5).

Scales are attached either to the throat of the corolla, as in certain Borrageworts, or to the claws of the petals, as in Buttercups.

A Crown is an appendage at the summit of the claw of some petals, as in Silene and Soapwort.

Glandular bodies are often met with on the receptacle; they are abortive organs of one kind or another.

125. *The arrangement of the floral leaves (perianth) in the bud is called* ÆSTIVATION. In identifying plants, we must not overlook it.

We distinguish four principal modes of æstivation—namely, the *valvate, plaited, contorted,* and *imbricated.* They take place in the calyx as well as in the corolla.

In the contorted and imbricated sorts, the segments of the calyx, or the corolla, overlap each other, in the other two they do not.

1. VALVATE ÆSTIVATION is that in which the pieces of a perianth are brought in contact, edge to edge, throughout their whole length, as the sepals of the Linden (see Cut XIII., 1). Two varieties of valvate æstivation are sometimes more obviously presented—namely: *a*, the *reduplicate*, in which each piece of the circle has its two edges bent outward, forming salient ridges, as in the calyx of Mallows (Fig. 2), and *b*, the *induplicate*, in which each piece has its two edges bent inward, as in the calyx of Clematis and the corolla of *Solanum tuberosum* (Fig. 3).

2. PLAITED or PLICATE ÆSTIVATION, occurring in tubular or gamopetalous flowers, is that in which the lobes of the tube are folded lengthwise, so that the plaits turn either outward, as in the corolla of Campanula, or inward, as in that of Gentiana. The most remarkable variety of plaited æstivation is the *supervolute*, in which the plaits are wrapped round all in one direction, so as to cover each other in a convolute manner, as in the corolla of *Datura Stramonium* (Fig. 4 a cross-section, and Fig. 5 the upper part of the corolla).

3. CONTORTED, or CONVOLUTIVE ÆSTIVATION is that in which every piece covers its neighbor by one of its edges, all in the same direction, appearing, therefore, as if twisted together, as in the corolla of Geranium (Fig. 6).

4. IMBRICATED ÆSTIVATION is that in which one or more pieces of a floral circle are wholly outside, and therefore overlie and enclose the rest in the bud. There are two principal varieties of this æstivation—namely, the *triquetrous* and *quincuncial*. While the former has *one* piece wholly outside (Fig. 7), there are *two* pieces outside in the

latter (Fig. 8). The quincuncial sort is found in the calyx of Geranium, etc.

Imbricated æstivation results from a spiral arrangement of the floral leaves, and the valvate and contorted sorts are probably due to their opposite, or whorled arrangement. The figures, except No. 5, are cross-sections, or horizontal slices of a floral circle in the bud.

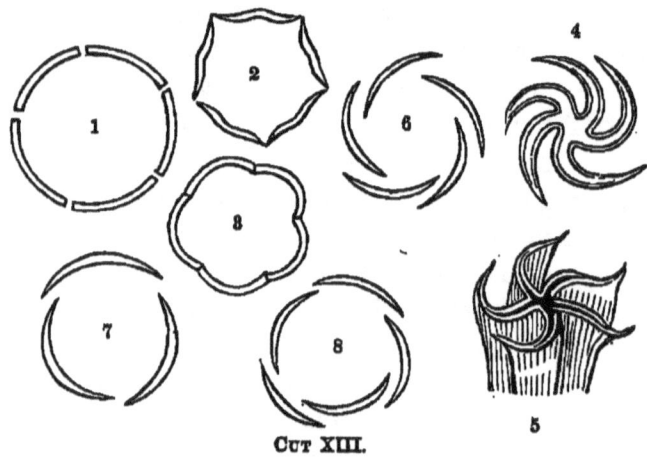

Cut XIII.

** FRUIT AND SEED.

126. A Fruit is a ripened ovary, with all its contents and appendages.

† ESSENTIAL PARTS OF THE FRUIT.

127. The fruit consists of two cardinal parts—namely, the *pericarp* and the *seed*, the former sometimes consolidated with other organs, which were adnate to the pistils.

128. The Pericarp is that part of the fruit which encloses the seed. It has its origin in one or more carpellary leaves. The pericarp often remains herbaceous in texture, like the pea-cod, or turns thin, dry, and membranaceous, as in Colutea. In other cases it thickens and becomes either hard and dry, or fleshy, and sometimes pulpy. Not unfrequently the pericarp consists of two or more layers of different texture. When there are two such layers, they may be distinguished as *Exocarp*, outer layer, and *Endocarp*, inner layer. In the drupe we call the exocarp *sarcocarp*, and the endocarp *putamen*. When there are three layers of the pericarp, we call them, beginning from the outside, *epicarp*, *mesocarp*, and *endocarp*. A drupe is indehiscent; and so are also many dry fruits, particularly one-seeded ones.

129. SEED is the fertilized ovule, and consists of two parts, the *seed-coat* and the *kernel*.

The names, given to the several parts of the ovule, are also applied to the corresponding parts of the seed. The *scar* left on the face of the seed by its separation from the seed-stalk at maturity, is termed the *hilum*, as was stated above. The *chalaza* and *raphe*, when present, are usually conspicuous in the ripe seed, as well as in the ovule. The *outer integument*, answering to the testa or primine of the ovule, is sometimes called *episperm*, or *spermoderm*, but more commonly *testa;* it varies considerably in texture (from papery to bony), and also in form (from its firmer or looser attachment to the nucleus). Not unfrequently this coat is expanded into a wing, as in Arabis Leavenworthia, Sullivantia, Tecoma, Chelone, etc., or provided with a tuft of hairs at one end (a *coma*), as in Epilobium and Asclepias, or even clothed all over with long wool, as in Gossypium (the Cotton-plant). The *inner integument* of the seed, answering to the tegmen or secundine of the ovule, often termed *endopleura*, but also *tegmen*, is frequently inconspicuous, and sometimes altogether wanting. The terms denoting the position of the seed in the pericarp and its inversions, are the same as those used to designate the same conditions in the ovule—namely, *erect, ascending, orthotropous*, etc. The seed is sometimes partially or wholly enclosed by a fleshy or pulpy enlargement of the seed-stalk or funiculus; and such enlargement we call *aril*, as the scarlet covering of the seed of Staff-tree.

The KERNEL enclosed in the seed-coat may consist of two parts, the *embryo* and the *albumen*, or of the embryo alone. The albumen has also been called *perisperm* or *endosperm*. Seeds with albumen are said to be *albuminous* (see Cut XIV., 1, 2), those without it *exalbuminous* (Fig. 3). When albumen is present, its quantity varies (see Figs. 1, 2). The *dicotyledonous embryo* has the plumule between two opposite minute leaves, or cotyledons—see Fig. 4 (an embryo with two cotyledons occurs in all Exogens, except the Gymnosperms, in which the embryo has 3–15 cotyledons). The *monocotyledonous embryo*, on the other hand, has only one cotyledon, and does not usually present a manifest radicle or plumule, but often consists merely of an undivided, club-shaped body, at least before germination. In Triglochin (an Endogen), we detect a vertical slit just above the radicular end of the embryo, through which the plumule protrudes in germination (Fig. 5). A horizontal section at this point reveals the plumule, enwrapped in the cotyledon (Fig. 6). The plumule is also conspicuous, under a lens, in grasses. Figures all magnified, the first three representing vertical sections of seeds, the other three embryos, the last in part: 1, seed of Delphinium, the minute embryo (e) at the base in abundant albumen (a); 2, seed of Violet, with the embryo in the axis of the albumen; 3, seed of Elodea, with the embryo, but no albumen; 4, embryo of the Bean; 5, embryo of Triglochin; 6, the same, transversely divided. The EMBRYO in the seed becomes a plant, provided it meets with the conditions for its development. Within the seed-coats we distinguish, as essentially permanent parts, the plumule and radicle, and as transient ones the cotyledons, with more or less albumen, which gives nutriment to the former, enabling them to grow. Their growth

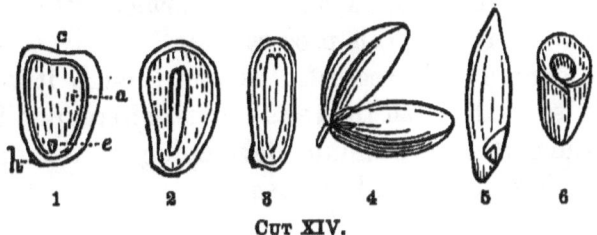

CUT XIV.

is what we call *germination* in general. The embryo may break through the seed-coats, as soon as its root-end is sufficiently developed, to draw food from the soil. (The seed is then usually one year old.) But it will not escape, until the seed is acted upon by moisture, with a certain amount of heat, varying from 50° to 60° Fahr., and free access of air. When separated from the plant and deprived of these conditions, its germinating power may remain dormant for some time, without becoming finally extinct. The seeds of certain leguminous plants have been known to germinate after sixty years. According to Lindley, raspberries were raised in the garden of the British Horticultural Society from seeds, found with the skeleton of a man, at the depth of 30 feet below the surface of the ground, near Dorchester; and as coins of the Emperor Hadrian were found together with the skeleton, it is supposed that the seeds were about 1600 years old. In a more restricted sense, the term germination denotes the continued or renewed growth of the embryo in the seed after the latter is severed from the plant, or placed under favorable conditions, which previously were either entirely or partially wanting. What we know about the process of germination is not much, and can be told in a few words. During germination, oxygen is taken up by the embryo and carbon eliminated, a portion of the food, stored up within or without the cotyledons, being burnt by the oxygen and transformed into carbonic acid and water. By this process heat is evolved, but the embryo generally decreases in weight, though it increases in bulk, so finally, that the testa bursts, whereupon its root enters the earth, the plumule developing in the opposite direction into stem and branches.

†† SORTS OF FRUIT.

130. When a flower has its ovary, or ovaries inferior, the *calyx* is apt to become part of the fruit, and sometimes forms its principal bulk, as in the Apple, Pear, etc. In the strictly botanical sense, it is the papery pods, arranged circularly in the core, which alone compose the true fruit; the other parts of the so-called fruit are accessory. The rose-hip, fleshy, and usually eatable at maturity, is a hollow, urn-shaped calyx-tube, lined with a concave disk or thin expansion of the receptacle, which within bears the ripened ovaries, in the shape of bony nutlets, called achenia. Again, instead of the calyx, it is the enlarged *receptacle*, which may become part of the fruit. In the Strawberry this becomes succulent and sweet, while it bears the real fruit—namely, numerous, hard achenia—on its surface.

Fruits like the Rose-hip and Strawberry are called *accessory*, or *anthocarpous fruits*—that is, fruits, the most conspicuous portion of which is superadded. Fruits of this sort are also seen in Gaultheria and Shepherdia. In these plants the fleshy part of the fruit is calyx; it encloses a dry pod in Gaultheria, and an achenium in Shepherdia; and this fleshy part is perfectly free from the ripened ovary.

131. We may divide fruits into *two principal classes*.—namely, 1st, *simple fruits*, and 2d, *multiple* (collective, or confluent) *fruits*.

Simple Fruits.

132. SIMPLE FRUITS are those which result from the ripening of a single pistil, whether with or without a calyx or other parts of the flower adherent to it. Most fruits belong to this class.

They may be divided into three sorts—namely, 1st, *fleshy fruits;* 2d, *stone-fruits*, and 3d, *dry fruits*.

FLESHY FRUITS are the berry, the hesperidium, the pepo, and the pome.

STONE-FRUITS are the drupe, the tryma, and the etærio.

DRY FRUITS are the achenium, the utricle, the caryopsis, the nut, the samara, the follicle, the legume, the loment, the pyxis, the capsule, the silique and silicle. The last seven of these dry fruits are dehiscent; the rest, the utricle of Amaranth excepted, indehiscent. The dehiscent dry fruits are all 1-celled, except the silique and silicle, which are 2-carpelled and 2-celled, and the capsule (which includes all other forms of dry dehiscent fruits from any compound pistil).

1. Fleshy Fruits.

133. The BERRY (bacca) is a fruit with a thin-skinned pericarp and a pulp within, which contains the seeds loosely imbedded. The fruits of the Gooseberry, Currant, Blueberry, Grape, Tomato, etc., are berries.

134. The HESPERIDIUM (orange-fruit), as the orange, lemon, and lime, is a succulent, many-carpelled fruit, with a thick and leathery rind, which is separable from the enclosed pulpy mass. It is merely a berry with a coriaceous pericarp.

135. The PEPO (gourd-fruit) is a compound fruit, usually formed of three carpels and the adnate calyx, having a thick, hardened rind and enclosing a pulpy mass. The pepo is a sort of berry.

136. The POME is a fleshy pericarp, formed of the calyx and receptacle, which together constitute the principal thickness of the fruit and enclose several papery or cartilaginous pods (bony carpels of the pistil). Pomes are apple, pear, and quince.

2. Stone-Fruits.

137. DRUPE is a 1-celled and 1- or 2-seeded fruit, with a double pericarp, or a pericarp of two layers, of which the outer, fleshy or pulpy layer is called the exocarp, epicarp, or sarcocarp, and the inner, hard and bony layer the endocarp, putamen, or nucleus. (Cut XV., 13.) (The term *drupe* is usually extended in a general way to fruits with two or more cells, as those of cornel, ginseng, etc.)

138. TRYMA is a sort of dry drupe, with the exocarp, or husk, fibrous-fleshy, or leathery, and the endocarp or nut-shell bony, and often rough and irregularly furrowed. Fruits of this sort are those of Juglans and Carya (Walnut and Hickory).

139. The ETÆRIO is an aggregation of (usually numerous) little stone-fruits or drupelets, in structure resembling cherries, and attached to an elongated receptacle. This sort of fruit belongs to the Blackberry and Raspberry.

3. Dry Fruits.

a, Indehiscent Dry Fruits.

140. Achenium, or akene, is the term applied to all 1-seeded, dry, hard, indehiscent and seed-like small pericarps (such as might be taken for naked seeds). The pericarp is tipped with the remains of the style. It is seen in the Buttercups, Anemone, Clematis, Geum, Strawberry, the Composites, and Umbelworts, the latter having a double achenium, called cremocarp, consisting of two mericarps (achenia), etc.

(Figures: Pl. VI., 4, cremocarps; Pl. VII., 6b, achenium with a pappus of bristles; Cut XV., Fig. 1, achenium of a Ranunculus, vertically divided, magnif.)

The achenia of the Composites are usually crowned with a pappus representing the calyx-limb, in the form of bristles, or (Cut XV., 2, 3, 4) they have a crown of scales or teeth. (Fig. 2, achenium of Cichorium; 3, of Helianthus; 4, of a Bidens.)

141. The utricle is an achenium with a thin and bladder-like loose pericarp, which commonly bursts irregularly, discharging the seed. We see it in the Goosefoot and Amaranth. In the latter plant it opens transversely all round, like a pyxis. (Fig. 5.)

142. Caryopsis, or grain, is a dry, hard pericarp, which firmly adheres to the seed, as in Wheat, Rye, Indian Corn, etc.

143. The nut (glans) is a 1-seeded, dry fruit, with a hard, crustaceous, or bony wall, commonly enclosed in a persistent involucre, which takes the form of a cup (cupula), as in the Oaks, or of a bur, as in Chestnut, or of a leafy covering (in the Hazelnut).

144. SAMARA, or key-fruit, is either an achenium, or any other indehiscent fruit, furnished with a wing. As instances, we may mention the fruits of the Ash, Maple, Elm. (Fig. 10 is the samara of Ulmus fulva, with the cell opened.)

b, DEHISCENT DRY FRUITS.

145. Dehiscent dry fruits are of two classes: those from a simple pistil and those from a compound one.

a, DEHISCENT DRY FRUITS FROM A SIMPLE PISTIL.

146. The FOLLICLE is a pod, resulting from a single carpel, and opening at the ventral or inner suture, as in Marsh-Marigold (Fig. 14), or in Spiræa, Asclepias, etc.

147. LEGUME is a dry dehiscent fruit, from a simple pistil, which splits open at both the ventral and dorsal sutures—that is, by two valves, as in most of the Leguminous plants. (Fig. 11.)

148. The LOMENT is a peculiar sort of pod, separating transversely into two or more 1-seeded joints, which commonly fall away at maturity. These joints usually remain closed, as in Desmodium (Fig. 12), but sometimes split into two valves, as in Mimosa.

149. The PYXIS, or PYXIDIUM, is a many-seeded dry fruit, which opens by a circular, transverse line, cutting off the upper part as a lid (circumscissile dehiscence). This sort of fruit belongs to Plantago, Portulaca (Fig. 6), Hyoscyamus, Anagallis, etc. (The fruit of Amaranth is a pyxis, as well as a utricle.)

β, DEHISCENT DRY FRUITS FROM A COMPOUND PISTIL.

150. The SILIQUE is a slender, usually linear, 2-carpelled, 2-valved, 2-celled capsule, with two parietal placentæ, between which a thin partition is drawn, from which, at maturity, the valves separate. The partition is a false one, an expansion of the placentæ (Fig. 7, silique of Cardamine dehiscent).

151. A SILICLE or POUCH is a very short silique, nearly as wide as long, or not over four times as long as wide. It belongs to the siliculous Crucifers. (Its valves are indehiscent in Senebiera.)

152. A CAPSULE is the pod of any compound pistil, whether regularly dehiscent, or opening by pores, or bursting irregularly. But usually it splits open lengthwise into equal pieces or valves.

The regular dehiscence of a capsule takes place in one of three ways. We distinguish a septicidal, loculicidal, and septifragal dehiscence.

SEPTICIDAL DEHISCENCE is that mode of dehiscence in which a capsule splits through the partitions, dividing each of them into halves (as in Hypericum, Fig. 8, lower part of the capsule, the upper being cut away).

LOCULICIDAL DEHISCENCE is a splitting open through the middle of the back of each cell of the pod (Fig. 9, lower half of the capsule), as in Cassandra, Cassiope, Oxydendron, Iris, etc.

SEPTIFRAGAL DEHISCENCE is that modification of either the foregoing two, in which the valves fall off, while the dissepiments remain united in the axis, as in Convolvulus.

MULTIPLE (COLLECTIVE, CONFLUENT) FRUITS.

153. MULTIPLE FRUITS result from the aggregation of a number of flowers (an inflorescence) in one mass. The fruits of Mitchella and of some Honeysuckles result from only two flowers. Their ovaries are united into a double or twin berry. Here, then, we have the simplest form of collective fruit. Collective fruits from a large number of flowers are the Mulberry, the Pineapple, and Fig. They are transformations of dense forms of inflorescence, the floral envelopes, coherent with each other, having become completely or in part succulent. In the *Mulberry*, which at first view resembles a blackberry, the grains are the ripened ovaries, not of a single flower, but of as many distinct, clustered flowers; and the pulp of the grains results from the transformation of the floral envelopes, and not of the ovary-walls. This sort of fruit, therefore, is not only confluent, but also anthocarpous. The fruit of the Fig issues from an inflorescence, enclosed in a hollow flower-stalk, which becomes pulpy. A collective fruit is also the *strobile*, or *cone*—a scaly, multiple fruit, which re-

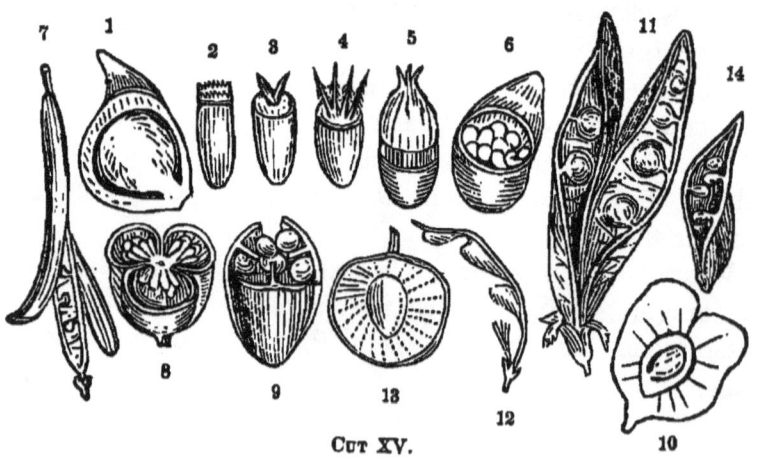

CUT XV.

sults from the ripening of a peculiar sort of ament. The terms *strobile* and *cone* are commonly applied to the fruit of Conifers; sometimes, however, also to that of the Hop, Hop-Hornbeam, etc.

IV.
ARRANGEMENT OF LEAVES AND FLOWERS ON STEM AND BRANCHES.

154. The several modes in which both leaves and flowers are arranged on stem and branches, have received different names.

A. PHYLLOTAXY.

155. The term *Phyllotaxy* (φυλλον, a leaf, and ταξις, order) denotes the arrangement of the leaves on stem and branches.

When they are found in pairs, threes, fours, etc., on the same level or node, it does not usually occur to us, to inquire into the reason of such arrangement. But when they are variously alternate, we are likely to ask, why they are disposed in such or such order, and which the physical law may be, according to which this succession takes place.

Here it may be observed, that the course of development in the growing plant is universally spiral.

156. As regards whorled leaves, including the opposite, it must first be noted that the leaves of a whorl are not placed directly above those of the whorl below, but at some point perpendicular to the intervals between them.

When the leaves are opposite, and each pair crosses at right angles the next above or below it, the leaves will occupy four imaginary, equidistant, longitudinal or vertical lines on the axis (stem or branch). Leaves so arranged are said to be *decussate*. In like manner, whorls of 3 or 4 leaves will occupy 6 or 8 lines of this sort.

157. But most plants have *alternate leaves;* and the arrangement of these will be best understood by reference to certain ingenious distinctions, drawn by botanists.

In the Oak, five leaves are spirally arranged around the axis, so that the sixth, succeeding the fifth, is placed in the same vertical or longitudinal line above the first. Suppose we examine a branch of some length: the seventh leaf will be placed above the second, and the eighth above the third. The same arrangement is seen in Prunus, Cratægus, Rhus, Salix, Populus, etc.

158. If we wind a thread around a branch of Hazel or Elm, so as to touch the insertion of the several leaves, the turns or revolutions of this thread will make a regular spiral. Taking one of the leaves as a starting-point, and counting those above it, we find that the third leaf is placed directly above the first, the fourth above the second, etc., and that all these leaves are arranged in two longitudinal lines or rows, separated by half a circumference of the branch. Such leaves are said to be *distichous* or *two-ranked*. This mode of arrangement occurs in all grasses and many other endogens, but also in certain exogens, as Hazel, Elm (as was mentioned above), Linden, some Leguminosæ, etc.

159. If, in the same way, we wind a thread around a branch of Birch, the fourth leaf is found to be directly above the first, the fifth above the second, etc. They are all on three equidistant vertical lines, and separated by one third of the circumference of the axis. In this case we say that three leaves are in one cycle, and that the fourth leaf commences a second one. This is the *tristichous*, or *3-ranked arrangement*; it is seen in Carex, Scirpus, Alnus, Betula, etc.

160. A CYCLE is a system of leaves, in which, after one or more turns of the spiral, we come to a leaf situated directly above the one from which the spiral started.

161. If on a branch of the Oak we form a spiral, so that it shall touch the insertion of the leaves, the sixth leaf, as stated above, will be found to be directly above the first. In this case the leaves are disposed on five equidistant verticals, and the circumference of the axis (the branch) is then divided into five equal portions, each of them an arc representing one fifth of the circumference. We notice, however, that the leaf which follows the first is placed not on the second but on the third vertical. Hence we say, that this vertical is two fifths the circumference distant from the first. While in the tristichous arrangement, the spiral, rising from the first leaf below, described one turn only, to arrive at a leaf (the fourth) directly above the first, in the present case, it must describe two turns or revolutions around the axis, to reach a leaf (the sixth), situated vertically above the first (lowest) of the five lower leaves. This arrangement, in which the distance between the verticals is two fifths of the circumference of the axis, is the *pentastichous, quincuncial*, or *5-ranked*, and by far the most usual in alternate-leaved exogens.

162. The *8-ranked arrangement* is also frequently met with, as in Aconite, Flax, Plantain, Holly, etc. Here the ninth leaf is situated vertically over the first; and the arc between the verticals of two successive leaves is $\frac{3}{8}$ of the circumference of the axis.

163. The arc interposed between the insertion of two successive leaves is called the *angle of divergence*. While the fraction $\frac{1}{3}$ expresses the angle of divergence of tristichous leaves, the fraction $\frac{2}{5}$ designates that of the quincuncial.

As regards distichous leaves, the term angle can not apply to their divergence, since the vertical lines are separated by half the circumference of the axis; it is, however, represented by the fraction $\frac{1}{2}$.

The fractions, $\frac{1}{2}$, $\frac{1}{3}$, $\frac{2}{5}$, $\frac{3}{8}$, $\frac{5}{13}$, etc., severally represent, not only the angle of divergence, but the whole plan of these modes. They have for their *numerator* the number of the spiral turns of which the cycle is composed, and for *denominator* the number of leaves in the cycle, or rather the number of intervals between the points of insertion of these leaves or their verticals.

It was *Bonnet*, who lived in the middle of the last century, that pointed out these modes of phyllotaxy, but they have recently been extended and generalized by Schimper, Braun, and others.

The 8-ranked arrangement is followed by the 13-*ranked*, 21-*ranked*, etc. The 21-*ranked arrangement* has 21 leaves in one cycle, with 8 turns of the spiral, therefore $\frac{8}{21}$ for an angle of divergence. The fractions, $\frac{13}{34}$, $\frac{21}{55}$, $\frac{34}{89}$, $\frac{55}{144}$, represent angles of divergence in cycles, consisting of 34, 55, 89, 144 leaves, and completed by 13, 21, 34, 55 revolutions or turns of the spiral.

Now if we arrange this series of fractions in a progression, thus, $\frac{1}{2}$, $\frac{1}{3}$, $\frac{2}{5}$, $\frac{3}{8}$, $\frac{5}{13}$, $\frac{8}{21}$, $\frac{13}{34}$, $\frac{21}{55}$, $\frac{34}{89}$, $\frac{55}{144}$, we shall readily perceive the relation that they bear to each other. Beginning with the third fraction, we notice, on comparing through the successive fractions, the numerators and denominators, that each fraction has for its numerator the sum of the two preceding numerators, and for denominator the sum of the two preceding denominators.

164. Sometimes it is very difficult to ascertain which leaf is vertical to the first, especially when the leaves are much crowded, as in Plantains, etc., or the cones of Pine, Fir, Larch. Suppose we have a plant under consideration, say *Sedum Telephium*, which has the leaves moderately distant below, but in a cluster at or near the summit of the axis. We see at once that below, the angle of divergence is $\frac{3}{8}$. At the top the spiral is considerably depressed. Compare it with a watch-spring, which may be compressed or more or less open. When the spiral (watch-spring) is compressed, the leaves nearest the circumference would have been the lowest in the more open spring, and those nearest the centre the nearest to the top. Now, to account for the upper crowded leaves (in *Sedum Telephium*), we plan three or four cycles, each of which contains eight leaves, and is completed by three revolutions of the spiral. If we draw a circle around the spiral, and then divide it into eight equal portions by as many radii, three of these portions will represent $\frac{3}{8}$ of the circumference of the axis. On the position of the first leaf, which we find, where the spiral touches the circumference, we place a number (1). From this point the coils of the spiral commence. Omitting the three first portions or arcs, which are together equal to $\frac{3}{8}$ of the circumference, we mark the position of the next leaf with a number (2), which will be placed at the intersection of the spiral and radius, which bounds the third arc; and so we go on marking a leaf-position at the intersection of every third radius, till we arrive at the centre. The plan will then represent all the leaves under consideration, numbered in order.

165. The place of every leaf on every plant is already fixed in the *bud* by unerring mathematical rule.

The mode, in which leaves are disposed in the bud, is usually called VERNATION. But the term also denotes the manner in which the leaf itself is folded, coiled, or packed up in the bud. Some botanists therefore use the term *præfoliation*, to denote the arrangement of the leaves in respect to one another, while by *vernation* they understand the manner in which the leaf itself is folded.

The leaf, considered in itself, is sometimes straight, flat, and open in the bud, but oftener bent, folded, or rolled. It is said to be *inflexed* or *reclined* in vernation, when its upper part is bent down upon the lower, as the young blade of Liriodendron is bent upon the petiole; *conduplicate*, when folded by the midrib, with the lateral halves brought together, face to face, as in the Magnolia, the Cherry, the Oak; *plicate*, or *plaited*, when folded like a fan, as in Maple, Currant, Vine, Birch. When the leaf is rolled, it is so in one of four modes. It is *circinnate*, when it is rolled downward from the tip, as in Sundew; *convolute*, when rolled up parallel with the axis from one of its sides, as in the Plum; *involute*, when it has both margins rolled inward, as in Water-Lily, Apple, and Violet; *revolute*, when it has both edges rolled outward, as in Azalea and Salix.

The arrangement of leaves in the bud with respect to one another, or *præfoliation*, is in descriptive botany briefly distinguished as *valvate* and *imbricate*. The valvate arrangement, in which the edges of the leaves meet with each other, is more common in plants with opposite leaves. In imbricate præfoliation the outer leaves successively overlap the inner, at least by their edges.

Cross-sections of buds show us, as it were, an epitome

of the phyllotaxy of the future stem, or branch, and the various modes of præfoliation are denominated the triquetrous, decussate, quincuncial, etc. The terms triquetrous and quincuncial denote in præfoliation the same arrangement of the parts as in æstivation, of which we have treated in the chapter on the flower.

166. Before we conclude this chapter on Phyllotaxy, we must treat of *buds* in general.

Leaf-buds contain the rudiments of a leafy stem or branch, while *flower-buds* enclose the same elements transformed, and destined to perform, directly or indirectly, the reproductive functions of the plant. *Leaf-buds* appear in the form of a tender-pointed cone, covered with scales. These buds regularly develop in the leaf-axils, and are then called *axillary* buds. Leaf-axil is the angle formed by the leaf with the supporting axis (stem or branch) on the upper side.

Buds become leafy branches, just as the plumule became a stem. From the axils of the leaves along the branches we see other buds making their appearance, which give rise to branches in their turn.

Some plants do not branch, but consist of a simple shaft. Plants of this sort, as palms, grow by the continued evolution of a bud, which crowns the summit of the stem, the *terminal bud*. But terminal buds occur in branching plants also, as on all the stems or shoots of Acer, Æsculus, Carya, etc. (Pl. I., 34, 35.) In growing, they only prolong the shoot, or stem. Yet down the sides of the same shoots other buds are produced, just over the scars, left by the leaves which fell the autumn previous—axillary buds.

A curious sort of buds are those known as *adventitious* or *accidental*. They are neither axillary, nor terminal, but appear anywhere on stem and branches, sometimes even on roots (which, as has been stated, are naturally unproductive of buds), where they have been wounded or in some way mutilated. We see them frequently on the trunks and roots of Poplars, Willows, Chestnuts, etc. From buds of this sort spring slender, beautiful twigs on the trunk and on stout branches of the Elm.

Accessory or supernumerary buds are those which spring from a leaf-axil, in pairs, threes, etc., in place of the single one. They are found in the Honey-Locust, the Walnut and Butternut, the Birthwort, etc.

Latent we call those axillary buds which suspend their activity from year to year, and sometimes, indeed, so long that their capacity for growth becomes extinct.

B. INFLORESCENCE.

167. *Inflorescence* is the term employed to designate the arrangement of flowers upon the stem or branch.

As has been intimated in § 166, a *flower-bud* is analogous to a leaf-bud, and a *flower* is a sort of leafy branch.

Flower-buds occupy either the extremity of the stem or branch, or the axils of leaves.

A flower with a stalk is called a peduncled or pedicelled flower; that without a stalk is said to be sessile.

168. When a stalk bears a cluster of flowers, it is known as a *general or main peduncle*, while the stalk of each particular flower of the cluster is called *pedicel*—that is, a partial peduncle. A stalked solitary flower is said to be peduncled, and a stalked flower of a cluster, pedicelled.

That part of the general peduncle which gives rise to the flowers is called the *axis of inflorescence*. The axis of inflorescence is termed *rhachis* when the flowers attached to it are sessile. When it is thick and covered with crowded flowers, it takes the name of *receptacle*.

The leaves of a flower-cluster are usually called *bracts* and *bractlets*—the former being the leaves of the main peduncle, at its base, and the latter, smaller ones, those on the points of attachment, and on the pedicels (if there are any) of the several flowers. The bracts are often so minute as to escape detection, and sometimes early deciduous.

169. In order fully to understand the various sorts of inflorescence, it is necessary to first consider the *solitary flower*, which, in fact, represents the simplest form of inflorescence. Such a single flower occurs either on the summit of the stem, or in the axil of a leaf.

170. Solitary terminal and solitary axillary flowers represent the two plans of inflorescence, the indefinite and definite, in the simplest possible form.

We have *indefinite*, or *indeterminate inflorescence*, when all the flowers spring from axillary buds; and the *definite*, or *determinate*, when they spring from terminal buds.

AA INDEFINITE INFLORESCENCE.

171. *Indefinite* or *indeterminate inflorescence* is that sort of inflorescence in which the flowers all spring from axillary buds, while the terminal bud, developing as an ordinary branch, continues the axis indefinitely.

The various modes of indefinite inflorescence run into each other through intermediate gradations.

172. *The principal sorts of* INDEFINITE INFLORESCENCE are, 1, *clusters with pedicelled flowers*—namely, the raceme, panicle, umbel, corymb, and thyrsus; and 2, *clusters* *with sessile flowers*—namely, the spike, spadix, catkin, and head.

1. CLUSTERS WITH PEDICELLED FLOWERS.

173. A *raceme* is that sort of flower-cluster in which the flowers, arranged along the main axis, and borne on distinct, simple pedicels, are oldest at the base of the axis and youngest at its top. (Pl. II., 16.)

A *panicle* is a compound raceme, in which the pedicels branch in an irregular manner, to support flowers. (Pl. II., 18.)

Umbel is an inflorescence, in which the pedicels rise, spreading apparently from the same point—that is, the top of the main axis, to the same level, like the rays of an umbrella, the outermost pedicels bearing the oldest, and the inner the youngest flowers. The flower-stalks of this sort of inflorescence are called *rays*. When the rays of an umbel are each terminated by a flower, we have a *simple umbel*, but when they branch again to form secondary umbels (umbellets), we have a *compound umbel*. Primrose, Milkweed, Crantzia, etc., have simple umbels. (Pl. II., 19.) Compound umbels are borne by most Umbelworts. (Pl. II., 20.)

A *corymb* is a flower-cluster in which the pedicels, springing from different points of the main axis, rise to the same level, or nearly so, thus forming a level-topped, or else a slightly convex, inverted cone, the lowest pedicels bearing the oldest flowers. (Pl. II., 22.) It is nothing but a modified raceme, a raceme with elongated pedicels.

Thyrsus is a compact panicle of pyramidal or oblong outline, as, for example, the flower-cluster of Horse-chestnut.

2. Clusters with Sessile Flowers.

174. A *spike* is that form of flower-cluster in which the flowers, all sessile, or nearly so, are arranged along a common axis of inflorescence. It differs from a raceme by having its flowers not pedicelled. (Pl. II., 15.)

The flowers are spiked in a great many plants. The so-called spikes of *Grasses* are *compound spikes*, bearing little spikes, or spikelets, instead of single or simple flowers (the spikelets of this order compose also racemes and panicles).

A *spikelet*, in grasses, is an axis, or a rhachis, subtended by a pair of scales, which we call *glumes* (the latter rarely solitary, or altogether wanting), and supporting one or more naked flowers, either perfect or imperfect, the one flower or each of the several flowers, as the case may be, between another pair of scales, called *pales* (rarely embraced by a solitary pale). In descriptive botany, a pair of pales, and even a single pale, are usually called flowers, whether with or without essential organs. The scales are distichously arranged. (Grasses have commonly 3 stamens, rarely 1, 2, or 6, and a simple ovary, with 1 ascending ovule, 2 styles, and 2 feathery stigmas.)

A *spadix* is a fleshy spike enveloped by a large leaf-like bract, which is called a *spathe*, as in Indian Turnip, Skunk-Cabbage, Calla, etc. (Pl. II., 25.)

A *catkin* or *ament* is a slender, often pendulous spike, with scaly, deciduous bracts beneath the sessile flowers. This sort of inflorescence is seen in Oak, Beech, Birch, Willow, etc. (Pl. II., 21.)

A *head* is a round or roundish cluster of sessile flowers, or a short spike. This sort of inflorescence belongs to the Button-bush, Red Clover, etc. (Pl. II., 24; VII., 2.) The head may be surrounded by empty bracts, forming an involucre, as in Composites. (Pl. VII., 7.)

175. The *order of flowering* in indefinite inflorescence is *centripetal*, or *ascending*, which is the same thing. In its several forms we notice that either the lowest or the outermost flowers are the oldest. If we suppose the main axis to be contracted, as is really the case in the umbel, the lower, lateral axes will be the outer, while the upper will be the inner. The lower lateral axes, regarded as the outer, will represent the periphery, and the upper, regarded as the inner, the centre of a circle. The youngest flowers are in the centre.

BB. DEFINITE INFLORESCENCE.

176. *Definite* or *determinate inflorescence* is that sort of inflorescence in which the flowers grow from terminal buds. If it be represented by three flowers (which, by-the-by, resemble a simple umbel), the middle and largest, or oldest flower arrests the growth of the axis; and further growth can not take place but by the development of other axes from axillary buds. These secondary axes, if they are leafy shoots, as in our inflorescence of three flowers, will sooner or later be terminated each by a flower in their turn (Pl. II., 23); and when other (tertiary) axes spring from axillary buds of these secondary axes, they will also be checked by a terminal blossom. (Pl. II., 26.)

Definite inflorescence occurs chiefly in opposite-leaved plants; but sometimes in alternate-leaved, as the Rose and Potentilla. This sort of inflorescence assumes forms, which closely resemble those of the definite. Both sorts have often been confounded, and not unfrequently they are combined.

177. There are *three principal forms* of definite inflorescence—namely, the *cyme*, the *fascicle*, and the *glomerule*.

The *cyme* is a flat-topped, rounded, or expanded, definite inflorescence, either simple or compound, in which the order of flowering is centrifugal, beginning at the centre or top of the cluster. This sort of inflorescence occurs in the Elder, Arrow-wood, Dogwood, etc. Very curious forms of cyme are seen in the Borrage family, and are often, for convenience' sake, described as racemes or spikes.

The *fascicle* is a cyme with the flowers much crowded —that is, a compact, bundle-like flower-cluster, in which the order of flowering is centrifugal, as in Dianthus.

The *glomerule* is an axillary, condensed cluster of flowers, exhibiting a centrifugal evolution. When occurring in the axils of opposite leaves, so as to meet around the stem, they constitute a *verticilaster*, or *verticil*, as in Marrubium and Nepeta cataria.

178. The order of flowering is in all forms of definite inflorescence, centrifugal or descending, which is the same thing. In Buttercups it is obviously descending, the upper flowers expanding first, the lowest last. In the Elder it proceeds from the centre toward the circumference of the cluster, or flies the centre. If we imagine the flowering axis of a Buttercup to be contracted, the lower, younger flowers will become the outer, and the upper, older ones, the inner.

179. We conclude the chapter on inflorescence by briefly recapitulating the principal differences between its two modes.

The *indefinite* mode of inflorescence is that in which the primary or leading axes elongate indefinitely, or merely cease to grow from lacking nourishment; the *definite* is that in which the primary or leading axes are arrested in their growth by solitary flowers at their extremity, or checked definitely.

GLOSSARY

OF BOTANICAL TERMS.

A, prefixed to words of Greek derivation, signifies negation, as in apetalous.

ABORTION: non-development.

ABORTIVE: scarcely or poorly developed.

ABRUPT: suddenly terminating. *Leaves abruptly pinnate:* leaves pinnate, without an odd leaflet.

ACAULESCENT: without a conspicuous stem, which, however, is usually hidden under ground. Acaulescent plants are usually furnished with obvious scapes.

ACCUMBENT; see *Embryo.*

ACEROSE: needle-shaped.

ACHENIUM (plural: achenia): a 1-seeded, seed-like fruit.

ACINES: the separate grains of a fruit, such as the raspberry.

ACORN: the nut of the Oak.

ACUMINATE: taper-pointed.

ACUTE: sharp-pointed.

ADELPHOUS: united into one or more brotherhoods—a ring, column, clusters, or bundles; a term, combined with *monos, dis, polys,* to form the words *monadelphous, diadelphous, polyadelphous,* which, applied to stamens, signify: one, two, many brotherhoods—that is, stamens united in one, two, or more sets.

ADHERENT: growing fast to.

ADNATE: born adherent to, originally adherent.

APPRESSED: closely applied, but not adherent to.

ASCENDENT: rising gradually upward.

ÆQUILATERAL: equal-sided, opposed to oblique.

ÆSTIVATION: the arrangement of the floral envelopes in the bud.

ALTERNATE: one after another, opposed to opposite.
ALVEOLATE: honeycomb-like.
AMENT, or CATKIN: a slender, often pendent, scaly sort of spike, as in birch, beech, willow, and usually deciduous.
AMPLEXICAUL: clasping the stem by a broad base.
ANGIOSPERMS: plants with their seeds formed in a closed pistil, ovary, pericarp.
ANNULAR: ring-shaped.
ANTHER: the part of a stamen which contains the fertilizing, powdery substance called pollen.
ANTHERIFEROUS: Anther-bearing.
APETALOUS: without petals.
APHYLLOUS: without leaves.
APPENDAGE: any superadded part borne by another one.
APPENDICULATE: furnished with an appendage.
AQUATIC: growing in water.
ARBORESCENT: tree-like, or with a woody stem.
ARIL: a fleshy coat to a seed, as in *Celastrus*.
ARTICULATED: jointed, consisting of joints.
AURICULATE, or AURICLED: furnished with ear-like appendages at the base.
AWL-SHAPED: sharp-pointed from a broader base.
AWN: a slender, bristle-form appendage, as on the glumes, or pales of grasses.
AWNED: furnished with an awn.
AXIL: the upper space between the attachment of a leaf and the stem, forming an angle.
AXILLARY: placed in an axil.
AXIS: the central line of a body (the stem of a plant, also the general peduncle of an inflorescence, etc.)

BACCATE: berry-like.
BARBED: bearded, beset with short, stiff hairs.
BEAKED: terminating into an elongated narrow tip.
BEARDED: barbed.
BERRY: a thoroughly juicy fruit, containing one to several seeds. (A drupe is more fleshy, and contains one to several nutlets.)
BI- (bis) in compound words: twice, as bipinnate (twice pinnate).
BIFID: two-cleft to about the middle.
BIFURCATE: twice-forked.

BILABIATE: two-lipped, as a flower with the border of the corolla or inner set of the perianth cleft into two principal divisions.

BITERNATE: twice ternate (that is, 3 partial petioles are borne on the summit of a main petiole, and each of them bears on its top 3 leaflets). The terms ternate, bi- and triternate are always applied to palmately compound leaves.

BLADE of a leaf, or of a petal, is the expanded portion, distinguished from the leaf-stalk, or the claw.

BOAT-SHAPED: concave within and ridged (or keeled) without.

BRACTS: leaves of an inflorescence, usually smaller than ordinary leaves.

BRACTEOLE: a little bract.

BRACTEOLATE, or BRACTEOLED: furnished with little bracts.

BRACTLET: the bract of a pedicel in an inflorescence.

BRISTLES: slender, stiff, sharp hairs.

BRISTLY: beset with bristles.

BUD: an undeveloped branch compacted in a conical body.

BULB: a leaf-bud with fleshy scales, usually under ground.

BULBIFEROUS: bearing bulbs.

BULBOUS: bulb-like.

BULBLETS: small bulbs.

CADUCOUS: falling off very early.

CALLOSE, or CALLOUS: hardened.

CALYCINE: calyx-like.

CALYX: the outer set of a 2-whorled perianth.

CAMPANULATE: bell-shaped.

CANESCENT: grayish-white, or hoary, usually from white hairs.

CAPILLARY: as fine as hair.

CAPITATE: with a globular summit like the head of a pin; a term usually applied to stigmas.

CAPSULE: any dry fruit, splitting (more or less completely) into pieces at maturity; a pod.

CARINATE: keeled, furnished with a longitudinal ridge on the outer side. *Carina*, keel, a ridge, a prominent sharp mid-nerve; or the ridge formed by the junction of two petals, as the keel-petals of leguminous plants.

CARPEL: a simple pistil, or one of the floral leaves, which have helped to form a compound pistil.

CARTILAGINOUS: like cartilage in texture, firm and tough.

CATKIN: ament.

CAULESCENT: furnished with a conspicuous stem.

CAULINE: borne on the stem, as cauline leaves.

CENTRIFUGAL INFLORESCENCE: that sort of inflorescence in which the flowering proceeds from the centre outward, or from the top downward.

CENTRIPETAL INFLORESCENCE: that sort of inflorescence in which the flowering proceeds from the periphery inward, or from below upward.

CERNUOUS: nodding.

CHAFF: thin, little scales or bracts on the receptacle of Composites.

CHAFFY: furnished with chaff (scales, pales).

CHANNELLED: hollowed out like a gutter.

CHARTACEOUS: papery.

CILIATE: beset on the margin with hairs like the eyelashes.

CIRCUMSCISSILE: dividing in 2 parts by a circular, transverse line, as the pod of *Anagallis*. We speak, therefore, of a circumscissile dehiscence.

CIRRHOSE: tendril-bearing.

CLAVATE: club-shaped.

CLAW: the stalk-like base of some petals.

COALESCENT: growing together.

COBWEBBY: bearing hairs like cobwebs.

COLUMN: a tube formed by the united stamens, as in Mallows; or a cylindrical body formed by the union of stamens and pistils, as in the Orchids.

COMOSE: tufted, bearing a tuft of hairs.

COMMISSURE: the line of junction of 2 carpels, as in the fruit of the Umbelworts.

COMPOUND LEAF: a leaf-stalk, bearing several sessile or stalked leaflets, or a leaf-stalk, bearing leaf-stalks, which give rise to several secondary leaflets, etc. There are palmately- and pinnately-compound leaves.

COMPLETE FLOWERS: flowers with calyx, corolla, stamens, and (1 or more) pistils.

CONDUPLICATE: see *Embryo*.

CONE: the multiple fruit of Conifers; a term rarely applied to other multiple fruits.

CONGERIES: a collection of parts or organs.

CONNATE: grown together primarily.

CONNECTILE or CONNECTIVE: the part of the stamens connecting the 2 cells of the anther.

CONTIGUOUS: touching; meeting at the surface or border.

GLOSSARY.

CONTINUOUS: the reverse of jointed or interrupted.
CONTORTED: twisted together.
CONVOLUTE: rolled up lengthwise.
CORDATE: heart-shaped.
CORIACEOUS: leathery.
CORM: a solid bulb.
CORNEOUS: horny.
COROLLA: the inner set of a 2-whorled perianth.
COROLLINE: corolla-like.
CORONA, crown: one or more appendages at the top of the claw of some petals, or in the throat of some corollas or perianth-tubes.
CORYMB: a sort of centripetal inflorescence, a convex flower-cluster, somewhat resembling an umbel, from which it differs by having pedicels, which spring not from the same point.
CORYMBOSE: corymb-like, or forming a corymb.
COTYLEDONS: the initial leaves of the embryo, or young plant in each seed.
CREMOCARP: the 2-carpelled fruit of Umbelworts.
CRENATE: the margin scalloped into rounded teeth.
CROWN: see *Corona*.
CRUSTACEOUS: hard and brittle.
CUCULLATE: hooded, hood-shaped, having the summit rolled up.
CULM: a straw.
CUNEATE: wedge-shaped.
CUPULE: a little cup.
CUSPIDATE: tipped with a sharp, stiff point.
CYLINDRICAL: terete, like a quill.
CYME: a cluster of centrifugal inflorescence, usually umbel-like, as in the Elder.
CYMOSE: forming cymes.

DECA- (in compounded Greek words): ten.
DECANDROUS: with 10 stamens.
DECIDUOUS: falling off after some time.
DECLINED: turned downward.
DECOMPOUND: several times compounded.
DECUMBENT: prostrate to the ground.
DECURRENT: continued more or less on the stem beneath the place of insertion, as the leaves on Thistles.
DEFINITE: a well-marked number, not above twelve, or nearly so.

DEFLEXED : bent downward.
DEHISCENCE: the mode of splitting at maturity, a term applied to pods and anthers.
DEHISCENT : opening by regular splitting.
DENTATE : toothed.
DENTICULATE : finely toothed.
DEPRESSED : pressed down.
DI- (in Greek compounds) : two.
DIANDROUS : with 2 stamens.
DICHOTOMOUS or DICHOTOMAL : two-forked.
DICLINOUS : having the stamens in one sort of flowers, the pistils in another. These diclinous or separate flowers are either monœcious or diœcious.
DICOTYLEDONS : Plants with a 2- (rarely several-) leaved embryo. The Dicotyledons are also called *Exogens*.
DIDYNAMOUS STAMENS : 4 stamens in two pairs, and one pair shorter.
DIFFUSE : spreading widely and irregularly.
DIGITATE LEAFLETS : borne on the apex of the common leaf-stalk. *Digitate* is the same as *fingered*—that is, resembling the fingers of the hand. A *digitately divided* leaf : a leaf not parted as far as to the summit of the petiole.
DIŒCIOUS : diclinous flowers on separate plants of a species, one bearing the staminate, the other the pistillate ones.
DISK : 1, a fleshy expansion of the receptacle of a distinct flower; 2, the central part of the receptacle of Composites.
DISSECTED : cut deeply into numerous segments.
DISSEPIMENT : partition.
DISTINCT : separate, not united by growth.
DIVARICATE : straddling, widely divergent.
DIVIDED : cut into segments about to the base, or the midrib.
DODECA (in Greek compounds) : twelve.
DODECANDROUS : with 12 stamens.
DORSAL and DORSALLY : on or from the back.
DOWNY : clothed with short, soft hairs.
DRUPE : stone-fruit, a fleshy, succulent fruit, containing one or several stones or nutlets (not seeds, like the berry).
DRUPACEOUS : drupe-like.

ECHINATE : awned with bristles or prickles.
EMARGINATE : notched at the summit.

EMBRYO: the rudimentary plantlet in the seed. It consists of a stemlet (radicle) and 1, 2—rarely more—small leaves (cotyledons). The terms *accumbent, incumbent,* and *conduplicate* serve to distinguish the different modes in which the embryo is curved in the *Crucifers*. *Cotyledons accumbent:* the margins of the cotyledons on one side applied to the radicle (O=). *Cotyledons incumbent:* the back of one cotyledon applied to the radicle (O|). *Cotyledons conduplicate:* cotyledons folded upon themselves (O≫).

ENDOCARP: the inner layer of a pericarp.

ENDOGENS: plants with the substance of the stem, the branches, petioles, etc., not arranged in annual rings. They are also called *Monocotyledons*.

ENSIFORM: sword-shaped.

ENTIRE LEAVES: leaves with the margins even, not toothed, nor notched or divided.

EPICARP: the outermost layer of a fruit.

EPIGYNOUS: upon the ovary.

EQUALLY PINNATE: pinnate without an odd leaflet.

EQUITANT: riding straddle, a term applied to leaves, the older partially covering the younger ones.

EVERGREEN: holding the leaves over winter.

EXOGENS: plants with the substance of the stem, branches, etc., disposed in annual circles. They are also called Dicotyledons.

EXSERTED: growing beyond; a term applied to stamens and pistils, extending beyond the free border of the corolla or perianth.

EXSTIPELLATE: without stipels.

EXSTIPULATE: without stipules.

EXTRA-AXILLARY: a little out of the axil.

EXTRORSE: turned outward; applied to anthers which open on the side facing the perianth.

FALCATE: scythe-shaped.

FASCICLE: a bundle or close cluster.

FASCICLED: growing in a bundle or tuft.

FEATHER-VEINED LEAVES: leaves with the veins springing from along the sides of a midrib.

FERRUGINEOUS: rusty, red-grayish.

FERTILE: producing or bearing fruit, or yielding good pollen. We speak of fertile flowers, fertile stems, fertile stamens, etc.

FILAMENT: the stalk of a stamen.

FILIFORM: thread-shaped.

FIMBRIATE: fringed.

FISTULAR: terete and hollow.

Flora (the goddess of flowers): a term signifying metaphorically all the plants, which grow spontaneously in a country or district.
Floral: belonging to the flower.
Floral envelopes: the flower-leaves, the perianth.
Flower: the system of organs, which are subservient to the production of seed.
Foliaceous: leafy or leaf-like.
Foliolate: consisting of leaflets; a term applied to compound leaves. A *trifoliolate leaf* is a leaf with 3 leaflets. Six- or seven-foliolate (compound) leaves have 6 or 7 leaflets.
Follicle: a simple pod, dehiscent down the inner suture.
Frond: stem and leaves fused into one body, as in Duckmeat; also what answers to leaves in Ferns.
Fructification: the state of fruiting.
Fruticose: shrubby.
Fugacious: falling away early.
Funnelform: expanding gradually upward.
Fusiform: spindle-shaped.
Gamopetalous: having the petals united more or less into one piece.
Gamosepalous: having the sepals united into one piece.
Geniculate: bent, like a knee.
Genus: a kind; a rank above species.
Gibbous: swollen at one place.
Glabrous: smooth, without hairs, bristles, etc.
Glands: small projections or roundish bodies, secreting an oily substance, sometimes raised on hairs.
Glandular: gland-like, or furnished with glands.
Glaucous: covered with a bloom, or white powder, that rubs off.
Globose: spherical.
Globular: nearly globose.
Glomerate: aggregated in a close cluster.
Glomerule of flowers: a dense, head-like flower-cluster.
Glume: a bract, like those of Sedges and Grasses, subtending one or several flowers.
Granular: composed of grains.
Gymnospermous: naked-seeded, having the seeds not enclosed in an ovary.
Gynandrous: having the stamen or the stamens borne on and united with the pistil.

GLOSSARY.

HABITUAL: having by habit, or usually.
HEMI- (in compounds from the Greek): half.
HEMISPHERICAL: forming half a sphere.
HEPTA: seven.
HEPTANDROUS: with 7 stamens.
HERMAPHRODITE FLOWER: a flower having both stamens and pistils, a perfect flower.
HETEROGAMOUS: a term applied to the flower-heads of Composites, importing that each head contains 2 or 3 sorts of flowers.
HEXA- (in compounds from the Greek): six.
HEXAMEROUS: with the parts in sixes.
HEXANDROUS: with 6 stamens.
HILUM: the scar of the seed.
HIRSUTE: hairy, with beard-like, or stiffish hairs.
HISPID: bristly.
HOARY: grayish-white; the same as canescent.
HOMOGAMOUS: a term applied to the flower-heads of many Composites, signifying that each has its flowers all of one sort.
HOODED: see *Cucullate*.
HORN: a spur or some similar appendage.
HYALINE: transparent, or nearly so.
HYPOGYNOUS: inserted under the pistil.

ICOSANDROUS: having 12 or more stamens attached to the calyx.
IMBRICATE, or IMBRICATED: overlapping each other like the shingles on a roof.
IMMERSED: growing wholly under water.
IMPERFECT FLOWERS: flowers destitute either of stamens or pistils.
INCISED: cut deeply and irregularly.
INCLUDED: a part or parts not projecting beyond another. *Stamens included:* stamens not projecting beyond the border of the perianth.
INCOMPLETE FLOWERS: flowers with a simple perianth, or none.
INCUMBENT: see *Embryo*.
INDEFINITE: too numerous to mention.
INDEHISCENT: not dehiscent, at least not splitting open regularly.
INFERIOR: lower.
INFLATED: turgid, or bladder-like.
INFLORESCENCE: the arrangement of flowers on the stem.
INFRA-AXILLARY: beneath the axil.

INNATE ANTHER: anther, fixed by its base to the very tip of the fi ment.

INSERTION: the place of attachment of an organ, of leaves, stamens, e

INSERTED: attached to.

INTERRUPTEDLY PINNATE: pinnate with smaller leaflets among larg ones, as in Agrimony.

INTRORSE: turned inward; a term applied to anthers, with their lin of dehiscence turned toward the axis of the flower.

INVERSE, or INVERTED: turned upside down.

INVOLUCEL: a partial, or small involucre, as the little involucre at tl base of each umbellet of some compound umbels; also a whorl bractlets outside the calyx, as in Mallows.

INVOLUCELLATE: furnished with an involucel.

INVOLUCRATE: furnished with an involucre.

INVOLUCRE: a set of bracts around a flower, umbel, or head.

LABELLUM: the odd piece of the 3 inner perianth-leaves of an Orchi

LABIATE: two-lipped, the same as bilabiate.

LACTIFEROUS: containing a milky juice. *Lactiferous plants* are, witl out exception, or in part, those of the Nymphaceæ (with milk juice in the root-stocks only), the Papaveraceæ, Anacardiaceæ, L guliflorous Composites, Campanulaceæ, Lobeliaceæ, Sapotaceæ, Co volvulaceæ, Apocynaceæ, Asclepiadaceæ, Euphorbiaceæ, and, fina ly, Morus and Sagittaria.

LAMINA: the blade of a leaf, petal, etc.; also a thin plate.

LANCEOLATE: lance-shaped.

LATERAL: on the side.

LEAFLET: one of the divisions of a compound leaf.

LEGUME: a simple pod, splitting open into two pieces; the fruit the Leguminosæ.

LIGNEOUS: woody.

LIGULATE: furnished with a ligule.

LIGULE: a strap-shaped corolla in many Composites; also a men branaceous appendage on the leaf-sheaths of most grasses.

LIMB: the spreading portion, or open border of a calyx- or coroll; tube.

LINEAR: narrow and flat, with the margins parallel.

LIPS: the principal lobes of a 2-lipped corolla.

LOBE: a segment of a leaf, corolla, etc., especially a rounded one.

LOCULICIDAL DEHISCENCE: a splitting open through the middle of th back of each cell of a pod.

LOMENT: a pod, separating transversely into joints.

LYRATE: lyre-shaped, a term applied to a pinnatifid leaf spatulate in shape, with the terminal lobe largest.

MEMBRANACEOUS: thin and usually translucent.
MERICARP: one half (one carpel) of the two-carpelled fruit of Umbelworts.
MIDRIB: the main-rib of a leaf.
MONADELPHOUS: stamens by their filaments united into one set (or tube).
MONOCOTYLEDONS: plants with a one-leafed embryo. All the Endogens are Monocotyledons.
MONŒCIOUS FLOWERS: flowers of two sorts—pistillate and staminate ones—borne on the same plant.
MUCRONATE: tipped abruptly by a short point; a term usually applied to leaves.

NASCENT: beginning to exist or grow.
NECTARIFEROUS: honey-bearing.
NERVE: a term designating simple and parallel veins of leaves, sepals, etc.
NETTED-VEINED: having branching veins, or a net-work of veins.
NORMAL: not deviating from rule.
NUT: a hard, usually one-seeded, indehiscent fruit, as an acorn or a chestnut.
NUTLET: a little nut, meaning also the stone of a drupe.

OB- (in compounds) signifies: inversion.
OBCORDATE: inversely heart-shaped.
OBLANCEOLATE: inversely lance-shaped.
OBLONG: elliptical, and from two to four times as long as broad.
OBTUSE: blunt.
OCTO: eight.
OCTAMEROUS: with the parts in eights.
OCTANDROUS: with eight stamens.
OPPOSITE: on opposite sides of an axis, a term applied to leaves, branches, stamens, etc.
ORGAN: any constituent part of the plant, a leaf, petal, stamen, etc.
ORBICULAR: round, like a circle.
OVARY: the part of the pistil which encloses the ovules.
OVATE: egg-shaped, or shaped like the longitudinal section of an egg.
OVOID: ovate or oval in a solid form.
OVAL: short-elliptical.

OVULE: a primitive seed.

PALES: 1) the inner husks of grasses, representing petals, an upp< and lower pale; 2) the bracts (chaff) on the receptacle of the hea< of many Composites.

PALMATE LEAF: a compound leaf with its leaflets spreading from tl< summit of the common petiole, like the outspread fingers of the han<

PALMATELY LOBED or VEINED: lobed or veined in a palmate manner.

PANICLE: a compound raceme branching irregularly.

PANICLED, PANICULATE: arranged in panicles.

PAPILIONACEOUS: butterfly-like; a term applied to the corolla of mo< Leguminous plants.

PAPPUS: thistle-down—that is, the hair-, scale-, or margin-like caly< of the flowers (florets) of Composites.

PARALLEL-NERVED or VEINED LEAVES: leaves furnished with n< branching veins, running lengthwise, side by side (as the leaves < most Endogens).

PARIETAL PLACENTÆ: the seed-bearing ridges of the ovary (or pericarp< attached to the walls of its cavity.

PARTED: separated into segments almost to the base.

PECTINATE: pinnately dissected into narrow and close segment< somewhat resembling the teeth of a comb.

PEDATE: palmate or palmately cleft into segments, which are cleft j< their turn.

PEDATELY CLEFT, LOBED, etc.: divided in a pedate manner.

PEDICEL: the stalk of each particular flower of an inflorescence.

PEDICELLED, PEDICELLATE: furnished with a pedicel.

PEDUNCLE: the stalk of a single flower, or the main stalk of an infl< rescence.

PEDUNCLED: furnished with a peduncle.

PELTATE: shield-shaped; a term usually applied to leaves.

PENTA (in compounds of Greek origin): five.

PENTAMEROUS: having the parts in fives.

PENTANDROUS: with five stamens.

PEPO: the fruit of the Gourd family, as that of the Cucumber.

PERENNIAL: lasting from year to year.

PERFECT FLOWER: a flower furnished with both stamens and pistil<.

PERFOLIATE LEAF: a leaf with the base of its blade apparently pierce< by the stem.

PERIANTH: the floral envelopes; a term generally used when thes< envelopes can not be readily distinguished into two sets—that i< calyx and corolla.

PERICARP: the ripe ovary, enclosing the seeds.

PERIGYNOUS: borne on the calyx; a term applied to the petals and stamens when they are inserted into the calyx.

PERSISTENT: remaining long in place.

PERSONATE: masked; a term applied to a 2-lipped corolla which bears a projection (palate) in the throat.

PETAL: a leaf of the corolla.

PETALOID: petal-like.

PETIOLE: leaf-stalk.

PETIOLED, PETIOLATE: furnished with a petiole.

PHÆNOGAMOUS: having flowers.

PILOSE: hairy.

PINNATE LEAF: a compound leaf with the leaflets arranged along the sides of a common petiole. The common petiole may bear stalks, having leaflets in their turn laterally; we then call the leaf twice-pinnate. A leaf may also be thrice-pinnate, etc.

PINNATELY LOBED, CLEFT, PARTED, etc.: divided more or less deeply in a pinnate manner.

PINNATIFID: pinnately cleft.

PISTIL: the ovule-bearing organ of a flower.

PLACENTA: a projection, column, or ridge (central or parietal) within the ovary, which bears the ovules.

PLUMOSE: feathery; a term applied to branching bristles when they bear hairs along their sides, like the beard of a feather.

POD: any sort of capsule, also a follicle, particularly a legume.

POLLEN: the fertilizing powdery substance of the anther.

POLLEN-MASS: a more or less coherent congeries of pollen-grains.

POLY- (in compounds from the Greek): many.

POLYADELPHOUS: stamens united by their filaments into three or more sets or bundles.

POLYANDROUS: with more than twenty stamens inserted on the receptacle.

POLYGAMOUS FLOWERS: some of the flowers perfect, and others diclinous, either on the same or on different specimens of a species.

POLYPETALOUS: with the petals distinct, not united into one piece.

POLYSEPALOUS: with the sepals distinct.

POUCH: silicle.

PROCUMBENT: trailing on the ground.

PRODUCED: extended or prolonged.

PROSTRATE: lying flat on the ground.

PUBERULENT: clothed with almost imperceptible down.

PUBESCENT: hairy or downy.

PYXIS: a pod (sometimes membranaceous) opening by a circumscissile—namely, horizontal—circular line.

QUADRANGULAR: four-angled.

QUINCUNCIAL ÆSTIVATION: that variety of imbricated æstivation in which 2 of 5 leaves of a floral circle are wholly without, 2 wholly within, and one, one margin out, the other within.

RACEME: a sort of centripetal inflorescence, a flower-cluster with pedicelled flowers, arranged along the sides of the axis of inflorescence.

RACEMED, RACEMOSE: arranged in a raceme.

RADIANT: outer flowers enlarged, and often neutral, as in Hydrangea, or in Viburnum Opulus, or the same as *radiate*.

RADIATE FLOWERS: ray-flowers, the ligulate flowers of Composites in the margin of each head.

RADICAL: proceeding from the root; *radical leaves*: root-leaves.

RADICLE: the stem-part of the embryo.

RAMOSE: branching.

RAY: a ligulate marginal flower in a head.

RECEPTACLE: 1, the summit of the peduncle or pedicel, into which the parts of a single flower are inserted; 2, an expansion of the top of the stalk, which bears a flower-head (in the Composites, etc.).

REDUPLICATE (in æstivation): valvate with the margins turned outward.

REFLEXED: bent outward and backward.

REGULAR: with the parts similar.

RENIFORM: kidney-shaped.

REPAND: wavy-margined.

RETICULATED: with a network of veins.

REVOLUTE: rolled backward.

RHACHIS: the axis of a spike.

RHIZOME: a root-stock.

RINGENT: gaping open; a term applied to some 2-lipped corollas.

ROOT-STOCK: a root-like portion of the stem under ground, usually horizontal, as in Polygonatum.

ROSETTE: a cluster of leaves at about the same height of the axis, usually near the ground, directed to all sides, and almost whorled.

ROSTRATE: furnished with a beak.

ROTATE: wheel-shaped.

RUDIMENTARY: poorly, imperfectly, or not yet properly developed.

GLOSSARY.

Rugose: wrinkled.
Runcinate: coarsely saw-toothed or incised, the pointed teeth turned toward the base of the leaf.
Saccate: furnished with a sac.
Sagittate: arrow-shaped.
Samara: a dry, 1-seeded, indehiscent fruit, with a membranaceous margin or wing.
Sarmentose: bearing or resembling runners, or elongated, flexible branches.
Scabrous: rough to the touch.
Scale: a sort of imperfect leaf, and more or less rigid.
Scaly: scale-like, or furnished with scales.
Scarious: thin and dry, applied to flat organs, such as sepals, etc.
Scurf: an aggregation of minute scales on the surface of certain leaves.
Secund: one-sided. A *secund raceme* has the flowers all turned to one side.
Segment: a part of a divided body.
Sepal: a leaf of the calyx.
Septicidal dehiscence: that mode of dehiscence in which a pod splits through the partitions, dividing each of them into two laminæ.
Septum: partition.
Sericeous: silky.
Serrate: saw-toothed, with the margin cut into teeth pointing forward.
Serrulate: finely saw-toothed.
Sessile: sitting; not petioled, not pedicelled.
Sheathing: wrapped round the stem or scape.
Silicle and **Silique:** the fruit of most Crucifers, pods, 2-carpelled, 2-valved, and 2-celled by a false partition, drawn between the 2 placentæ, and opening, at maturity, lengthwise by the valves.
Silicle: a short silique, from nearly as long as wide to not more than 4 times as long as wide.
Silique: a pod of Crucifers, elongated so much as to be at least 4 to 6 times as long as wide.
Simple: in one piece; opposed to compound.
Sinuate: strongly wavy.
Spadix: a fleshy spike of flowers.
Spathe: a bract (usually leaf-like) wrapped round an inflorescence, particularly round a spadix.

SPICATE, or SPIKED: arranged in a spike.
SPIKE: a sort of centripetal inflorescence, with the flowers arranged along the sides of the axis.
SPIKELET: a secondary or partial spike; the inflorescence of grasses.
SPINE: a thorn.
SPINESCENT: tipped by a thorn.
SPINOSE: thorny.
SPUR: a projecting, slender appendage.
STAMEN: a fertilizing organ of the flower.
STAMINATE: furnished with stamens, but destitute of pistils.
STANDARD: the upper petal of a papilionaceous corolla.
STELLATE: star-shaped.
STEMLESS: really, or apparently, without a stem.
STERILE: imperfect, barren.
STIGMA: the glandular part of the pistil, which receives the pollen.
STIGMATIC: bearing accommodations for receiving the pollen.
STIPEL: the stipule of leaflets.
STIPELLATE: furnished with stipels.
STIPULATE: furnished with stipules.
STIPULE: an appendage at the base of certain leaves on each side.
STRAP-SHAPED: long, flat, and narrow.
STROBILE: the conical or oval fertile spike or catkin of Conifers, and also of the Hop, Hop-Hornbeam, etc., in fruit; a sort of aggregate fruit.
STYLE: the stigma-bearing, usually slender and tapering part of the pistil.
SUB- (prefixed to a descriptive term): somewhat, slightly; as in *sub-cordate, subsessile*.
SUBMERSED: growing under water.
SUBULATE: awl-shaped.
SUCCULENT: juicy.
SUFFRUTICOSE PLANTS: undershrubs.
SUPRA-AXILLARY: above the axil.
SUSPENDED: hanging down.
SUTURE: seam, particularly applied to the pistil. The line or seam down its inner side, bearing the ovules, is named the *ventral* or *inner suture*. An opposite line, running down the back of the pistil, we call the *dorsal suture*. The ventral suture, forming a projection or ridge in the cavity of the ovary, bears the ovules, and is the *placenta*, which is double. A simple pistil is formed of a floral leaf, with its edges curved inward and united; the junction of the margins is the ventral suture.

TAWNY: dull-yellowish, or somewhat brownish.

TENDRIL: a thread-shaped, coiling appendage (a modification of a branch or a leaf) to the weak stems of certain plants, serving the purpose of climbing.

TERETE: cylindrical and tapering.

TERMINAL: borne at the summit.

TERNATE: in threes. A *ternate leaf*: a leaf with three leaflets.

TERRESTRIAL: existing on the earth.

TETRADYNAMOUS STAMENS: 6 stamens, 2 of them shorter than the rest.

TETRAMEROUS: having the parts or sets in fours.

TETRANDROUS: having 4 stamens.

THROAT: the orifice of a gamopetalous corolla at the junction of the tube and border.

TOMENTOSE: clothed with matted, woolly hairs.

TOOTHED: furnished with erect teeth.

TOROSE: knobby.

TORULOSE: cylindrical with swells and contractions.

TRI-, as a prefix: three. Trifid: 3-cleft.

TRIADELPHOUS: filaments of the stamens in 3 clusters.

TRIANDROUS: with 3 stamens.

TRIMEROUS: with the parts in threes.

TRICUSPIDATE: with 3 cusps, 3-pointed.

TRIFID: 3-cleft.

TRIFOLIOLATE: with 3 leaflets, as the leaves of Trefoil.

TRIPLE-RIBBED, or NERVED: having a mid-rib divided into 3 branches near the base of the blade of the leaf.

TRITERNATE: three times ternate.

TRUNCATE: as if cut off at the top.

TRYMA: a sort of dryish drupe, 2-coated, the epicarp fibro-fleshy or woody, and the nucleus bony. Such fruits are the Hickory and Butternut.

TUBER: a thickened portion of an underground stem or branch, furnished with buds (eyes).

TUBERCLE: a small excrescence, or wart-like protuberance.

TUBERCLED: with a tubercle.

TUBULAR: hollow and of an elongated shape.

TUMID: swollen.

UMBEL: an umbrella-like sort of centripetal inflorescence, in which the pedicels resemble the rays of an umbrella, springing all from one point.

UMBELLATE: arranged in umbels.

UMBELLET: a secondary or partial umbel, or smaller umbel, borne on each ray of the primary umbel, which, therefore, has become compound.

UMBILICATE: with a sharp depression in the centre or at the end.

UNDERSHRUB: a partially shrubby plant, or a very low shrub.

UNEQUALLY PINNATE: odd-pinnate.

UTRICLE: a small, 1-seeded, thin-walled fruit, usually bladder-like.

VALVE: one of the pieces into which a pod splits at maturity.

VEINS: the small branches, furrows, or ridges, forming the framework of leaves.

VEINLETS: the smaller branches of veins.

VENTRAL: see *Suture*.

VERRUCOSE: warty.

VERSATILE ANTHER: an anther attached to the stalk of the stamen by one point, so that it may swing to and fro, as in the Evening Primrose.

VERTICIL: a whorl.

VERTICILLATE: whorled.

VEXILLUM: standard. See *Papilionaceous flower*.

VILLOUS: shaggy with long and soft hairs.

VINE: a trailing or climbing stem.

VIRGATE: wand-like, as a long, straight, and slender stem, twig, peduncle, etc.

VISCOUS, or VISCID: glutinous.

VITTÆ: the minute oil-tubes in the fruit-coat of Umbelworts.

WING: any membranaceous expansion.

WINGED: furnished with a membranaceous margin.

WOOLLY: clothed with long and soft hairs.

PART II.

SYSTEMATIC BOTANY.

CONTENTS OF PART II.

I. Analytical Key to the Flowering Plants growing spontaneously in the Northern and Middle United States east of the Mississippi.
II. Conspectus of the Orders and Genera of the flowering plants in the Key.
III. Index to the Key and Conspectus.
IV. Explanation of the Plates.
V. Lithographic Plates.

LIST OF ABBREVIATIONS.

Authors' Names.

A. D. C., Alphonse De Candolle.
Adans., Adanson.
Ait., Aiton.
Bart., Barton.
Beauv., Beauvois.
Big., Bigelow.
Cambd., Cambessèdes.
Cass., Cassini.
Cav., Cavanilles.
Crtz., Crantz.
D. C., De Candolle.
Desf., Desfontaines.
Desv., Desvaux.
Don, Don.
Ell., Elliot.
Ehrh., Ehrhart.
Endl., Endlicher.
Fenzl, Fenzl.
Fisch. & M., Fischer & Meyer.
Gærtn., Gærtner.
Gaud., Gaudin.
Gmel., Gmelin.
Gr., Gray.
Griseb., Grisebach.
Gron., Gronovius.
Hall., Haller.
H. B. K., Humboldt, Bonpland, and Kunth.
Herb., Herbert.
Hook., Hooker.
Host, Host.
Jacq., Jacquin.
Juss., Adrien de Jussieu.
Lag., Lagasca.
Lam., De la Marck.
L'Herit., L'Heritier.
Listib., Listiboudois.
L., Linnæus, Chs. de Linné.

Meisn., Meisner.
Mart., Martius.
Mill., Miller.
Mnch., Mœnch.
Moq., Moquin.
Muhl., Muehlenberg.
Mx., Michaux.
Nees., Nees von Esenbeck.
Nutt., Nuttall.
Pers., Persoon.
Planch., Planchon.
Poir, Poiret.
Presl, Presl.
Pursh, Pursh.
R. Br., Robert Brown.
Raf., Rafinesque.
Reichenb, Reichenbach.
Rich., Richard.
Rostk., Rostkovius.
Rupp., Ruppius.
Salisb., Salisbory.
Schreb., Schreber.
Schw., Schweinitz.
Scop., Scopoli.
Spreng., Sprengel.
Torr., Torrey.
Torr. & Gr., Torrey & Gray.
Tourn., Tournefort.
Tratt., Trattinick.
Trin., Trinius.
Vahl, Vahl.
Wahl., Wahlenberg.
Wallr., Wallroth.
Walt., Walter.
Wang., Wangenheim.
Wd., Wood.
Willd., Willdenow.

KEY.

☞ All flowering plants fall under one of the divisions of paragraph 2. The reader is there referred by the figure at the right to another paragraph giving more minute specifications. Under the second paragraph consulted, he will find a similar reference to a third, and so on, until he reaches one containing the name of the plant under examination.

1. Plants bearing flowers, **PHÆNOGAMS**..........2

Plants destitute of flowers, **CRYPTOGAMS.***

2. Several or many flowers enclosed in an involucre (an apparent calyx)..3

No calyx-like involucre to several flowers...........4

3. The anthers cohering more or less in a tube or ring, which encloses the style. Flowers densely crowded in a head on a common receptacle, surrounded by an involucre of bracts (scales). *Composites*......................904

Anthers not forming a tube or ring................4

4. All the flowers on a plant, or at least several of them, furnished with stamens as well as (1 or more) pistils. *Perfect, or Hermaphrodite flowers*......................5

Some flowers having stamens only and no pistils; others a pistil or pistils, but no stamens; *Diclinous or Separate flowers*...774

5. Stamens more than 20...........................6

Stamens fewer than 20..............................77

6. Stamens distinct, and not in obvious clusters........7

Stamens more or less united by their filaments, the latter sometimes forming a tube, or else disposed in clusters without cohering by their filaments....................66

---

* They are not analyzed in this key.

7. Stamens inserted on the receptacle (13. *Cl. of L.*)...8
   Stamens inserted into the calyx (12. *Cl. of L.*).....48
   Stamens inserted on a one-sided disk, all turned to one side. Calyx 4-parted. Petals 4, greenish-yellow; the upper one 3 to 5-cleft, the two lateral trifid, the lowest entire, linear. Leaves lanceolate, with a tooth on each side, at the base. Flowers spiked. **Reseda luteola,** L.

8. Pistil one (sometimes compound and many-celled)...9
   Pistils more than one......................25

9. Flowers regular..................................10
   Flowers irregular; one of the sepals and two of the petals spurred...................................42

10. Calyx of two deciduous sepals. Plants with a milky or colored juice..11
   Calyx only apparently of two sepals; the 2 outer of the 4 being large and leaf-like. Low, shrubby plant......23
   Calyx of more than 2 sepals, sometimes early deciduous..16

11. Petals 8–12, spatulate-oblong, the inner ones narrower. Stigma 2-grooved. Leaf radical, rounded, palmately lobed. Naked scape 1-flowered. Corolla white. Juice red-orange-color. **Sanguinaria Canadensis,** L.
   Petals fewer, commonly 4. Juice white or yellow...12

12. Juice white. Capsule opening by 4–20 pores or chinks under the margin of the broad crown of stigmas. Flower-buds nodding. Petals 4. **Papaver.**
   Juice yellow..................................13

13. Leaves and pods prickly, the former pinnatifid. Flowers pale yellow or white, peduncled, axillary, and terminal. Sepals sometimes 3. **Argemone Mexicana,** L.
   Leaves not prickly............................14

14. Pods bristly, 3- to 4-valved. Style columnar. Leaves pinnately cleft or divided. Peduncles axillary, 1-flowered. Corolla yellow. **Stylophorum diphyllum** Nutt.
   (*Meconopsis diphylla, D. C.*)
   Pods naked, or only rough with tubercles........15

15. Seeds crested, borne on 2 thin placentæ between the two valves of the linear pod. Leaves pinnatifid, the lobes irregularly toothed. Flowers yellow, in axillary peduncled umbels. **Chelidonium majus, L.**

Seeds crestless, partly sunk into a spongy false partition between the valves of the very long and linear pod. Leaves pinnatifid, or sinuate-lobed and toothed, the upper ones clasping by their cordate base. Flowers solitary, yellow. Herb, glaucous. **Glaucium luteum,** Scop.

16. Trees. Peduncle of each flower-cluster adnate to the midrib of a large, leaf-like bract 75
   No trees 17

17. Leaves palmately compound, twice to thrice ternate. Flowers white, in oblong or much elongated racemes. Sepals deciduous 18
   Leaves simple, or in one genus pitcher-form 19

18. Petals spatulate, long-clawed, 4–8. Raceme short, oblong. Fruit a berry, white or red. **Actæa spicata, L.**

Petals stamen-like, 2-horned at the apex, 1-8. Fruit a pod. **Cimicifuga** (45).

19. Leaves hollow, pitcher-shaped. Scape naked, bearing a solitary purple or yellow flower. **Sarracenia.**
   Leaves not hollow 20

20. Sepals 3, with 2 bracts outside, or bractless 21
   Sepals 4–6 (or apparently 2). Petals either 4 or numerous 23

21. Sepals 3, with 2 bracts outside, and larger than they. Petals 5. Fruit a 3-valved capsule 22
   Sepals 3, with no bracts outside. Fruit a prickly pod 13

22. Sepals united at the base. The 1-celled pod 3- to 6-seeded. Low shrubs, much branched. Leaves awl-shaped or thread-form, minute. Flowers small, yellow, along the upper part of the branches. **Hudsonia.**

Sepals not united at the base. The 1-celled pod 3-valved, triangular. Flowers of 2 sorts; the primary, terminal or dichotomal ones with large yellow petals and numerous stamens ; the later, in axillary glomerules, with small, or no petals, and only 3–10 stamens. Leaves lance-oblong. Capsule with many seeds in the primary flowers, and only few in the secondary ones. **Helianthemum.**

23. Petals numerous. Aquatics. Flowers solitary, showy...24

Petals 4, oblique, fugacious. Sepals 4, the 2 outer very broad and leaf-like. Low, rather shrubby plants. Styles 2, or 3–4. **Ascyrum.**

24. Sepals 4, green outside, as long as the white petals, which are arranged in many rows at the base of the ovary. Stigma many-rayed. Leaves orbicular, or sometimes kidney-shaped, cordate to the centre, floating. Flower fragrant.
Nymphæa odorata, Ait.

Sepals 4–6, of the same color as, and larger than the yellow petals, which are arranged in 1 or 2 rows, and furrowed at the back. Stigma deeply umbilicate. Leaves oval or oblong, sagittate-cordate, floating or erect. **Nuphar.**

25. (8.) Pistils more than 1, distinct, or not firmly coherent...26

Pistils many, completely coherent on an elongated receptacle, forming a dry, or fleshy cone in fruit. Trees or shrubs with large, showy flowers. Anthers linear........47

26. Pistils several, separately half sunk into cavities of the large, top-shaped receptacle, forming acorn-shaped nuts in fruit. The numerous sepals and petals colored alike, in several rows. Leaves orbicular, centrally peltate and cup-shaped. Flowers solitary, very large, yellow, on upright scapes. A huge aquatic. **Nelumbium luteum,** Willd.

Pistils free, on the surface of the receptacle......27

27. Leaves centrally peltate, large. Flowers small, dark purple. Aquatics.......................................264

Leaves not peltate..........................28

28. Herbs..................................29

Small trees. Sepals 3. Petals 6. Stamens in a globular mass. Pistils 3–7, longer than the stamens. Fruits oblong, torulous, about 3 inches long and 1 inch thick, pulpy, often 3 together. Flowers dull purple, axillary, solitary.
Asimina triloba, Dunal.

29. Ovaries 1-ovuled, forming achenia in fruit.......30

Ovaries more than 1-ovuled. Fruit a follicle or a 1- to 2-seeded acine, or berry........................38

30. Ordinary petals present..........................31

Petals none, or staminodia in their place........32

31. Petals equal. Achenia in a head, pointed. A scale or pore at the base of each petal, inside. **Ranunculus.**

Petals unequal, several, and much smaller than the sepals, passing gradually into stamens. Achenia in a head, with plumose tails. Vines, climbing by the leaf-stalks. Leaves ternate, whorled in fours at the nodes. Flowers purple, large, solitary on both sides of each node. Sepals 4, colored, purple. **Atragene Americana,** Sims.
(*Clematis verticillaris, D. C.*)

32. Petals none...................................33

Staminodia unequal. A vine with whorled ternate leaves...31

Staminodia equal. Herb erect, clothed with silky hairs. Achenia (50–75) bearing long, plumose tails. Sepals 4–6. Proper leaves radical, long petioled, cleft into many wedge-shaped, or linear segments. A concave, rather cup-shaped involucre below the middle of the stem. Flowers solitary, large, pale purple, silky outside, preceding the leaves.
Pulsatilla Nuttalliana, Gr.
(*Anemone Nuttalliana, D. C.*)

33. Achenia bearing the persistent styles in the form of naked, hairy, or plumose tails. Perennial herbs or vines. Leaves opposite. Flowers single, or in panicled clusters.
Clematis.

Achenia not tailed...............................34

34. An involucre present, near the flower, or remote from it..35

Involucre none..................................37

35. Involucre very close to the flower, resembling a calyx. Leaves all radical, cordate and 3- (rarely 5-) lobed. Scape 1-flowered, hairy. **Hepatica.**

Involucre remote from the flower...............36

36. Involucre of 2 or 3, variously divided leaves. The radical leaves 3- to 7- parted. Achenia (not grooved) with a short, rarely elongated, beak, on a hemispherical, conical, or even cylindrical receptacle. **Anemone.**

Involucre of 2 or 3 ternate, sessile leaves (apparently a whorl of long-stalked simple leaves). Leaflets 3-lobed at the apex. The radical leaves twice or thrice ternate, with leaflets, like those of the involucre. Achenia grooved, with a short style. Flowers several, white, or pale purple.
Thalictrum anemonoides, Mx. (892).
(*Anemone thalictroides, L.*)

37. Flowers all perfect. Achenia numerous in a head, membranaceous, compressed, 4-angled, and inflated. Leaves alternate, palmately lobed. Flowers corymbed, white.
Trautvetteria palmata, Fischer & M.

Flowers polygamous (or diœcious), in compound panicles, greenish or white. Achenia 4–15 in a head, ribbed-angled, and with a long stigma. Stem-leaves scattered, 3 to 4 times compound; leaflets rounded, stalked, either 3 or 5- to 7-lobed. **Thalictrum** (892).

38. Petals none.................................39
Petals present............................41

39. Sepals 3, deciduous. Fruit a berry. Pistils 12 or more, becoming a raspberry-like head of crimson acines. Leaves palmately 3- to 5-lobed, one of them radical, and 2 near the summit of the stem. A solitary greenish-white flower, terminal. **Hydrastis Canadensis,** L.
Sepals 5–9. Fruit of few or several follicles.....40

40. Sepals white, 5, deciduous. Leaves twice- to thrice-ternately compound, the leaflets 2- or 3-lobed. Flowers axillary and terminal. Pistils 3–6, some of them, usually 4, becoming divaricate pods.
Isopyrum biternatum, Torr. & Gr.

Sepals golden yellow, 6–9. Glabrous, trailing aquatics. Stems hollow. Leaves orbicular-kidney-shaped, crenate or entire. Follicles 5–10. **Caltha palustris,** L

41. Petals spurred, all or some of them.............42
Petals not spurred (sometimes 2 only, and the sepals very irregular)..43

42. Petals all spurred and all alike. Flowers terminal, scarlet without, and yellow within, pendulous, the spurs pointing upward, the yellow stamens and styles exserted.

The flower-stalk becomes upright in fruit. Leaves twice- to thrice-ternately compound, the leaflets lobed.
Aquilegia Canadensis, L.

Petals 4, unequal; 2 of them spurred and included in the spur of the uppermost sepal. Pistils 1–5, becoming many-seeded follicles. Leaves palmately divided, or incised. Flowers blue or purplish, in terminal simple or compound racemes. **Delphinium.**

43. Petals 2, long-clawed, covered by the hooded uppermost of the 5 sepals, the 3 lower petals wanting or rudimentary. Pistils 3–5. Leaves palmately cleft or dissected. Flowers blue (or white), in simple or compound racemes.
Aconitum.

Petals several, small, or minute.................44

44. Petals 2-horned, or 2-lipped....................45

Petals linear or club-shaped, small, or stamen-like. 46

45. Leaves ternately decompound. Petals 2-horned. Pistils 1–8. Flowers white, in long racemes.
(18) **Cimicifuga.**

Leaves pedately divided or parted. Petals 8–10, very small, tubular, 2-lipped. Pods many-seeded. Flowers nodding, large, greenish. Peduncles often in pairs.
Helleborus viridis, L.

46. Petals hollowed near the base, linear, 15–25, much shorter than the stamens. Pods 9–15, sessile. Leaves palmately parted, or incised. Flowers solitary, terminal, large, pale greenish-yellow. **Trollius laxus,** Salisb.

Petals hollowed at the apex, club-shaped. Pods 3–7, stalked, divergent, and pointed with the style. Leaves all radical, 3-foliolate, obscurely 3-lobed and toothed. Scapes with a solitary, star-like, white flower.
Coptis trifolia, Salisb.

47. (25) Anthers opening outward. Carpels indehiscent, falling away whole and separately, when ripe, from the long axis. Trees. Leaves long-petioled, truncate at the end and laterally 2-lobed. Flowers bell-shaped, greenish-yellow, marked with orange. **Liriodendron tulipifera,** L.

Anthers opening inward, longer than the filaments.

Carpels, when ripe, opening on the back, from which 1 or 2 berry-like seeds hang down, suspended by a long funiculus. Shrubs or trees. Petals 6-9, white or whitish. **Magnolia.**

48. (7) Ovary 1, simple or compound..............49
Ovaries more than one.........................57
49. Corolla none. Ovary superior.................50
Corolla present..............................51
50. Styles 3-5. Calyx 5-parted, purplish within. Stamens 5-60. Pod 3- to 5-celled, many-seeded, opening transversely. Prostrate, maritime herbs. Leaves lance-oblong, flattish. Flowers sessile, or short-peduncled, axillary, solitary. **Sesuvium Portulacastrum,** L. (639).

Styles 2. Calyx 5- to 7-toothed. Filaments long, thickened above, white or pink. Pod 2-lobed, 2-celled, the cells 1-seeded. Shrub 2-4 feet high. Flowers white, in a terminal, ament-like spike. **Fothergilla alnifolia,** L.

51. Calyx 5-lobed. Petals 5-10....................52
Sepals and petals numerous, imbricated in several rows, adherent to the 1-celled ovary. Fruit a 1-celled, eatable, crimson berry. Stem fleshy, jointed, the joints broadly obovate and flat, with spirally arranged small, awl-shaped leaves, and fascicles of prickles, often with a few yellow spines. Flowers large, sulphur-yellow, solitary, sessile. **Opuntia vulgaris,** Mill.

52. Shrubs or trees. Calyx 5-cleft. Petals 5........53
An herb, rough, branched, with barbed hairs. Petals 5-10. Stamens 20 or more. Styles 3, connate, and often spirally twisted. Leaves alternate, obovate, or oblong, lobed, or cut-toothed, short-petioled. Flowers golden yellow, solitary, terminal. **Mentzelia oligosperma,** Nutt.

53. Ovary superior, 1-carpelled. Style terminal. Fruit a drupe. **Prunus** (with Cerasus).
Ovary inferior, 1 to 5-carpelled.................54
54. Leaves opposite. Fruit a 3- to 5-celled pod. Style usually 4- (or 3- to 5-) cleft. Shrubs.
Philadelphus inodorus, L.
Leaves alternate. Fruit a pome. Styles 1-5....55

55. Petals oblong-spatulate. Styles 5. Flowers racemed, white. Leaves simple, sharply serrate. Fruit globular, eatable, purplish. **Amelanchier.**

Petals roundish. Flowers corymbed. Styles 1–5...56

56. Carpels 1-seeded. Styles 1–5. Flowers white, rarely rose-color. Leaves simple, sometimes lobed. Armed with thorns. **Cratægus.**

Carpels 2-seeded. Styles 2–5. Flowers white, or rose-color. Leaves simple or pinnate. Unarmed. **Pyrus.**

57. Sepals and petals numerous, in several rows, all united at the base in a fleshy cup, beset with leaf-like bractlets. Pistils few or many, enclosed by the calyx-tube, which becomes in fruit a sort of huge rose-hip. Leaves oval, chiefly acuminate, tomentose beneath. Flowers lurid purple on short branches. Shrub exhaling the fragrance of strawberries. **Calycanthus.**

Segments or pieces of the perianth definite.......58

58. Fruit a follicle. Pistils 3–12. Calyx 5-cleft. Petals 5. Flowers (rarely 4-merous) white or rose-color, sometimes diœcious, corymbed or panicled. Shrubs or herbs.
Spiræa (882).

Fruit an achenium. Pistils 2–50, free, in an open, or closed calyx..59

59. Achenia, dry or pulpy, in an open calyx..........60

Achenia enclosed in the calyx-tube, which becomes fleshy in fruit. Plants shrubby and prickly. Leaves odd-pinnate. **Rosa.**

60. Styles deciduous...............................61
Styles persistent on the dry achenia.............65

61. Calyx bractless, or minutely bracteoled, deeply parted..62
Calyx very conspicuously bracteolate (apparently 10-lobed)..64

62. Fruit a congeries of pulpy achenia on a spongy or succulent receptacle. Herbaceous, or somewhat shrubby, with or without prickles. **Rubus** (880).

Achenia dry or dryish............................63

63. Sepals 5 or 6, unequal. Leaves radical, rounded, heart-shaped, long-petioled. Dry, seed-like drupes, 5–10. Flowers white, on scape-like peduncles.
Dalibarda repens, L.

Sepals 5, equal. Leaves radical, 3- to 5-lobed. Achenia 2–6, minutely hairy. Small, yellow flowers, several on bracted scapes. **Waldsteinia fragarioides**, Tratt.

64. Receptacle globular, at length pulpy and red, the dry achenia on its surface. Calyx 5-cleft, with 5 alternate bractlets. Petals white. Acaulescent, with runners. Leaves trifoliolate. Scapes cyme-bearing. **Fragaria.**

Receptacle dry. Calyx with 4 or 5 lobes, and as many alternate bractlets. Herbaceous, or shrubby. Leaves pinnately or palmately compound. Petals mostly yellow, sometimes white. **Potentilla.**

65. Petals 8 or 9. Leaves oblong-ovate, slightly heart-shaped. Dwarf. **Dryas integrifolia**, Vahl.

Petals 5. Achenia numerous, heaped on a dry, conical receptacle, the long persistent styles forming hairy, or naked and straight, or jointed tails. Leaves pinnate or lyrate. **Geum.**

---

66. (6.) Stamens united by their filaments into one set or tube..67

Stamens united by their filaments, or by petaloid scales, into several bundles or clusters..................74

67. Stamens with their filaments united into a column. Anthers kidney-shaped. Pistils several, distinct, or united. Fruit a several-celled pod, or a cluster of carpels, their ovaries cohering in a ring. Herbs or shrubs..............68

Stamens with the filaments barely united at the base. Anthers not kdney-shaped. Pistil one. Sepals distinct. Ovary 5-celled, entirely free. Stigma 5-toothed. Seeds 1 or 2 in each cell. Leaves oblong-ovate, downy underneath. Flowers solitary, short-peduncled, axillary. Petals white.
Stuartia Virginica, Cav.

68. Column of stamens anther-bearing at the top. Carpels compactly aggregated in a ring around a central axis, from which they separate, when ripe....................69

Column of stamens anther-bearing for much of its length, naked and 5-toothed at the summit. Pod 5-celled, the axis evanescent after its dehiscence................73

69. Stigmas lateral, namely, occupying the inner face of the styles. Carpels 1-seeded, falling away separately....70

Stigmas terminal, capitate. Carpels 1 to few-seeded, opening before falling away. Involucel none..........72

70. Involucel of 6–9 bractlets. Leaves ovate, or slightly cordate, toothed, sometimes 3-lobed, velvety downy. Corolla rose-color. **Althæa officinalis,** L.

Involucel of 3 bractlets, or none................71

71. Petals inversely heart-shaped. Carpels beakless, rounded, blunt. Leaves either roundish, heart-shaped, and crenate, or sharply 5- to 7-lobed. **Malva.**

Petals wedge-shaped, truncate. Carpels beaked. Calyx either naked or with a 3-bracteoled involucel at the base. Leaves triangular-cordate, or hastate. **Callirrhœ,** Nutt.
(*Malva & Sida*, L.)

72. Carpels or ovary-cells 1-seeded. Seed pendulous. **Sida.**

Carpels or ovary-cells 2- to 3-seeded, the carpels 12–15, hairy. Leaves roundish, heart-shaped, taper-pointed, velvety. Flowers yellow, on axillary peduncles.
Abutilon Avicennæ, Gærtn.

73. Involucel of several bractlets. Pod 5-seeded. Roughish, hairy, 2–4 feet high. Leaves halberd-shaped and cordate, the lower ones 3-lobed. Flowers peduncled, axillary. Corolla rose-color. **Kosteletzkya Virginica,** Presl.

Involucel of many bractlets. Pod many-seeded. Upper leaves either ovate, or halberd-shaped, or 3-parted; the lower ones nearly always 3-lobed. Flowers large, showy, purple, rose-color, or pale greenish-yellow. **Hibiscus.**

74. (66.) Leaves alternate. Stamens in 5 clusters....75

Leaves opposite. Filaments united into 3 or more clusters at the base. Sepals and petals 5, the latter oblique. Flowers cymose, yellow. Herbs or shrubs.
Hypericum (199).

75. Peduncle of the cyme adnate to the midrib of a large, leaf-like bract. Each cluster of stamens attached to the base of one of the petaloid bodies, which are placed opposite the petals. Leaves cordate, serrate. Trees.
Linden. Tilia.

Bracts of this sort none..........................76

76. Calyx of 5 roundish, concave sepals. Ovary free. Fruit a 5-celled, 5-valved pod, with 2–8 seeds in each cell. Flowers large, white, peduncled, axillary. Leaves lanceolate, oblong, minutely serrate, coriaceous.
Gordonia Lasianthus, L.

Calyx 5-cleft, its tube coherent with the lower part of the 3-celled ovary. Fruit a 1-celled, 1-seeded, dry drupe. Flowers small, yellow, 6–14 in close, bracted clusters, fragrant. Leaves elliptical, acuminate, obscurely dentate, thick.
Symplocos tinctoria, L'Her.

---

77. (5.) Flowers with a double set of perianth, calyx, and corolla.....................................78

Flowers with only one set of perianth, or with no perianth at all.....................................537

78. Corolla gamopetalous.......................320
Corolla polypetalous........................79

A. COROLLA POLYPETALOUS.
(79–319.)

79. Ovaries, 1 or several, free in the bottom of the calyx, or nearly so; ovaries superior........................80

Ovaries, 1 or several, with the calyx-tube either adherent to, or constricted over them; ovaries inferior......273

80. One ovary, simple or compound, with one or several styles or stigmas..................................81

Several ovaries, with as many styles or stigmas..262

81. Corolla regular (or nearly so) and not spurred....82
Corolla irregular (parts obviously unequal or dissimilar) or spurred...............................218

82. Stamens 1 to 12 83
Stamens more than 12........................212
83. I. Petals 5, or more........................135
II. Petals 4....................................98
III. Petals 3...................................86
IV. Petals 2....................................84
V. Petal solitary. Stamens 10, monadelphous at the base. Pod oblong, 1- or 2-seeded, tardily dehiscent. Shrubs, with odd pinnate leaves; the leaflets minutely dotted. Flowers violet, in clustered, terminal, dense spikes.
Amorpha.

84. Herbs with (mostly) hollow, jointed stems (culms) and alternate leaves on tubular, split, or open sheaths. A corolla of 2 calyx-like petals; a 2-valved calyx; stamens 3 (rarely 1, 2, or 6); styles mostly 2 (rarely 1). **Grasses.** 565.
Herbs with stems.............................85

85. Sepals, petals, stamens and stigmas 2 (or 3).
Elatine Americana, Arnott.
The 2 inner pieces of the 4-leaved perianth broader. Stamens 4. Leaves kidney-shaped.
Oxyria Digyna (669), Campd.
(*O. reniformis*, Hook.)

86. A scurfy epiphyte, growing on the branches of trees. The 3 sepals linear, membranaceous, as long as the 3 lance-oblong, yellow petals. Stem and leaves thread-form, the former branched. Peduncles short, 1-flowered.
Tillandsia usneoides, L.
No epiphytes..................................87

87. Leaves veined lengthwise, grass- or rush-like, contracted at the base into sheathing petioles. Sepals and petals 3. Perfect stamens 6 or 3......................88
Leaves netted-veined (or coriaceous, often short-linear) 90

88. Stamens (perfect ones) 3. Petals white or yellow. 89
Stamens 6, with bearded filaments. Petals blue or rose-color. Flowers in umbelled clusters, axillary and terminal.
Tradescantia.

89. Petals white. Flowers single, terminating a naked

peduncle. Style filiform, its stigma simple. Imperfect stamens none. Moss-like, creeping, densely-leafy aquatics.
Mayaca Michauxii, Schott & Endl.

Petals yellow. Flowers imbricated in a head, each in the axil of a coriaceous bract. Style 3-cleft. The 3 fertile stamens alternating with 3 plumose filaments. The 2 lateral sepals keeled, boat-shaped, the anterior one larger and enfolding the petals in the bud. **Xyris.**

90. A whorl of 3 ample (usually broadly ovate) leaves at the summit of the otherwise naked stem, and a terminal showy flower between them. Sepals, petals and styles 3. Stamens 6. Berry 6-angled, 3-celled. **Trillium.**

Leaves not whorled at the summit of the stem.....91

91. Trees and tall shrubs. Flowers polygamous......92

Herbs or low shrubs. Flowers perfect, or rarely diœcious..93

92. Tree thorny. Leaves once or twice abruptly pinnate. Petals 3 (5). Stamens as many, greenish, in small spikes. Fruit a 1- to many-seeded legume. **Gleditschia** (149).

Shrub unarmed. Leaves trifoliolate. Flowers greenish-white, in compound terminal cymes. Sepals, petals and stamens 3 (to 5). Stigmas 2. Fruit a 2-celled, 2-winged samara, winged all around. **Ptelea trifoliata**, L.

93. Fruit an achenium. Stamens 6................94

Fruit a drupe or a capsule......................95

94. Fruit a dry, 3-cornered achenium. Perianth of 6 sepals, the three outer slightly coherent, the 3 inner larger and somewhat colored. Stamens 6. Ovary 1-celled; styles 3; stigmas tufted. Flowers crowded, or whorled in panicled racemes. **Rumex** (714).

Fruit of 3 (or 2, rarely only 1) roughish, fleshy achenia. Sepals 3. Petals as many, shorter than the sepals. Stamens 6. Ovary 3-celled (rather 3 ovaries united at the base; the style rising from between them.) Stigmas 3. A small herb. Leaves pinnate. Flowers small, solitary, axillary. **Flœrkea proserpinacoides**, Willd.

95. Fruit a capsule. Perianth not scale-like........97

Fruit a drupe. Flowers diœciously polygamous.

Perianth of 6 scale-like sepals, in 2 sets, the 3 inner petaloid. Stamens 3. Leaves linear, revolute at the margin. Undershrubs..96

96. Stigma nearly sessile, 6- to 9-rayed. Drupe berry-like, with 6–9 nutlets. Stem procumbent, 1–4 feet long.
Empetrum nigrum, L. (851).

Style filiform, 3- to 4-cleft. Drupe with 3 or 4 nutlets. Stem diffusely branching, 1 foot high.
Corema Conradii, Torr. (851).

97. Sepals, petals, stamens, and stigmas 3............85

Sepals 3, with 2 bracts outside. Petals 3, lanceolate. Stamens 3–12. Stigmas 3, plumose. Herbs, often shrubby at the base, 1–2 feet high. Leaves elliptical, lanceolate, or linear. **Lechea.**

98. Style 1, simple, or cleft, or with several stigmas...99
Styles or sessile stigmas, more than 1...........128

99. I. Stamens 2 100
II. Stamens 4..................................101
III. Stamens 6. Sepals and petals 4...........105
IV. Stamens 7, 8, or 11 (8–12)................126

100. Herb. Sepals and petals 4. Fruit a silicle.
Lepidium (115).

Shrub or low tree. Calyx 4-parted, very small. Petals 4 (rarely 5 or 6), narrowly linear and acute, 1 inch long. Stigma notched. Drupe fleshy, globular, purple. Leaves oval, oblong, or lance-obovate, petioled. Flowers snow-white, in loose drooping, terminal and axillary racemes.
Chionanthus Virginica, L.

101. Shrubs or small trees......................102
Herbs 104

102. Leaves compound 92
Leaves simple............................103

103. Leaves alternate, serrate. Fruit a drupe, berry-like, with 2–4 small nutlets, which are grooved on the convex or concave back, black. A disk lining the bell-shaped calyx-tube. Petals small, short-clawed, notched, and enfolding the short stamens (wanting sometimes). Flowers in axillary clusters. **Rhamnus** (661).

Leaves opposite, serrate. Fruit a pod, 3- to 5-lobed, crimson, when ripe. Stamens very short, on a broad, 4-angled disk, which is stretched over the ovary. Branchlets 4-sided. Flowers small, dark purple, in loose cymes, on axillary peduncles. **Euonymus atropurpureus**, Jacq.

104. Calyx of 4 sepals. Petals long-clawed. Leaves pinnately 5- to 11-foliolate, the terminal leaflet largest. **Cardamine hirsuta**, L. (121).

Calyx globular, or bell-shaped, 4-angled, 4-toothed, with a small horn at each sinus. The petals small, purplish, deciduous (or wanting). Stamens 4, short. Leaves opposite, linear-lanceolate. **Ammannia** (669).

CRUCIFERS.

[The genera of this family can not be identified without a thorough examination of the fruits.]

105. (99.) Pods jointed transversely—that is, pods composed of 2 or several pieces, which at maturity separate across; loments..106

Pods not jointed across, but 2-carpelled, 2-valved and 2-celled by a false partition, stretched between the 2 placentæ, and opening, at maturity, lengthwise by the valves; rarely 1-celled from the incomplete partition....107

106. Loments short, 2-jointed, each joint 1-celled and 1-seeded, the lower sometimes seedless, the upper sword-shaped. Fleshy. Leaves obovate, sinuate and toothed. **Cakile Americana**, Nutt.

Loments elongated, tapering upward, linear or oblong, 2-jointed, the lower joint often seedless and stalk-like, the upper necklace-form, several-seeded; the seeds, by constriction, transversely intercepted. Leaves lyre-shaped. **Raphanus Raphanistrum**, L.

107. Pods from nearly as long as wide to not more than 4 times as long as wide; silicles. Valves in one genus indehiscent..108

Pods elongated so much as to be at least 4 to 6 times as long as wide; siliques......................116

108. Silicle turgid or flattened, with a broad partition (the latter rarely incomplete)...................109

Silicle flattened contrary to the narrow partition..114

109. Flowers white, one case excepted, in which, however, the seeds are 2-rowed in each cell, and the petals emarginate...110

Flowers yellow.............................112

110. Seeds 2-rowed in each cell. Cotyledons (O=). Petals entire or 2-cleft. Silicle oval, oblong, or even linear. Flowers only in one species yellow. Leaves simple, entire, or toothed. **Draba.**

Seeds not 2-rowed. Silicles ovoid or globular...111

111. Silicle 2-celled, several-seeded, oval; valves turgid. Cotyledons incumbently folded on themselves. Dwarf. Leaves awl-shaped, tufted. Flowers minute.
Subularia aquatica, L.

Silicle 1-celled, from the insufficiency of the partition, ovoid or globular. Cotyl. (O=). Leaves undivided, or the lower pinnatifid. **Armoracia,** Rupp.
(*Nasturtium,* Gr.)

112. Seeds 2-rowed in each cell. Cotyl. (O=). Silicles oblong. Flowers usually small or minute.
Nasturtium, R. Br. (117).

Seeds not 2-rowed........................113

113. Valves convex, 1-nerved. Silicle turgid, obovoid, or pear-shaped, surrounded by a flat, sharp margin, pointed, many-seeded. Cotyl. (O||). On dehiscence, the style is split, and its halves are left on the valves. Leaves lanceolate, sagittate at the base, nearly entire. Flowers small.
Camelina sativa, Crantz.

Valves nerveless. Silicle globular, inflated, 2- to 4-sceded. Cotyl. (O=). Leaves elliptical, entire, sessile.
Vesicaria Shortii, Torr.

114. Seeds several in each cell. Cotyl. (O||). Silicle cordate-triangular. Valves boat-shaped. Radical leaves clustered, pinnatifid, or toothed; stem-leaves sagittate, sessile. **Capsella Bursa-pastoris**, Mœnch.

Seeds only 2, one in each cell................115

115. Silicles oval-orbicular, emarginate at the top. Valves keeled, dehiscent. Seeds pendulous. Cotyl. (O||), or in some species (O=). Stamens sometimes 2 only. Flowers white, small, often incomplete. Petals rarely none.
(100) **Lepidium** (674).

Silicles emarginate at the top and at the base, almost 2-tubercled. Valves ventricose, indehiscent, separating from the partition, when ripe, as 2, closed, 1-seeded nutlets. Cotyledons incumbently folded on themselves.
Senebiera (689).

116. Seeds in a double row in each cell. Cotyl. (O=). 117

Seeds in a single row in each cell............118

117. Valves nerveless. Siliques nearly terete, generally curved upward; sometimes shortened, so as to resemble a silicle. Seed lenticular. Leaves pinnate or pinnatifid.
(112) **Nasturtium**.

Valves 1-nerved. Silique long, linear, 2-edged. Seeds in some species margined. Stem-leaves mostly sagittate-clasping. **Turritis.**

118. Corolla purple, rose-color, or white............119

Corolla yellow, or, in one genus, yellow toward the base only, the upper part being purplish...............122

119. Valves 1-nerved in the middle, or longitudinally veiny. Seeds mostly winged. Cotyledons accumbent or oblique. Stem-leaves simple, or pinnately parted. Siliques often curved. **Arabis.**

Valves nerveless. Seeds not winged.........120

120. Leaves palmately compound; leaflets 3. Seed-stalks broad and flat. Cotyled. (O=). Siliques lanceolate. Flowers in a terminal raceme. Root necklace-form, or toothed; pleasant to the taste. **Dentaria.**

Leaves not palmately compound, or at least not the upper ones. Seed-stalks not dilated. Cotyl. (O=). 121

121. Flowers violet-purple, in panicled racemes. Claws of the petals longer than the calyx. Silique linear, terete. Leaves ovate-oblong, pointed, toothed, the lowest sometimes lyrate-pinnatifid. **Iodanthus hesperidoides,** Torr & Gr. (*I. pinnatifida,* Wood.)

Flowers white or rose-color. Silique linear, the flattened or concave valves usually separating elastically from the base. Seeds on slender seed-stalks. Leaves either pinnate or simple, and only the lower ones sometimes 3-lobed or divided. Stamens sometimes 4 only.
(104) **Cardamine.**

122. Seeds globular, with the cotyledons conduplicate. Siliques tipped with a stout, either empty or 1-seeded beak, and nearly terete, either smooth and 4-angled, or smooth, many-angled and torulose, or hispid and torose. Lower leaves lyrate, incised, or pinnatifid. **Sinapis.**

Seeds oblong, or lens-shaped.................123

123. Seeds winged. Siliques terete and flattened, nerveless. Petals twice longer than the sepals, yellow, at least their broad claws. Leaves lyrate-pinnatifid.
Leavenworthia.

Seeds not winged. Siliques 4- to 6-sided, the valves 1- or 3-nerved...............................124

124. Cotyledons incumbent......................125

Cotyledons accumbent. Siliques convex-4-angled, about 9 lines long, tipped with a conspicuous style, the valves keeled by a midnerve. Leaves lyrate-pinnatifid.
Barbarea vulgaris, R. Br.

125. Siliques sharply 4-cornered, linear, the valves keeled with a strong midnerve. Cotyledons often obliquely incumbent. Calyx closed. Leaves lanceolate, slightly toothed. **Erysimum.**

Siliques terete, awl-shaped, flattish, or obscurely 4- to 6-sided, sometimes curved. Valves 1- to 3-nerved. Calyx open. Leaves twice pinnatifid or runcinate.
Sisymbrium.

126. (99.) Leafless herbs, with scales or bracts in place of leaves, tawny, reddish, or whitish. Style longer than the ovary, hollow. Flowers several in a scaly raceme, all 4-petalled and 8-androus, excepting the terminal one, which is usually 5-petalled and 10-androus.
Monotropa Hypopitys, L.

Leafy plants...................................127

127. Trees or shrubs, with opposite digitate leaves. Stamens habitually 7 (rarely 6 or 8). Flowers in a terminal thyrsus or dense panicle..........................230

An annual herb, with alternate, 3-foliolate leaves. Stamens usually 11. Sepals and petals 4, the latter with claws and notched at the apex. Receptacle bearing a gland at the base of the ovary. Pod linear, or oblong, turgid, many-seeded, glandular-pubescent. Fetid, clothed all over with a clammy pubescence. **Polanisia graveolens,** Raf.

128. (98.) Shrubs.................................129

Herbs....................................131

129. Styles or sessile stigmas 4 or 5..............130

Styles 2. Petals strap-shaped, yellow, curled and twisted, about 1 inch long. Calyx with 2 or 3 bractlets at the base, 4-parted, cohering with the base of the ovary. Pod woody, 2-carpelled, 2-beaked. Leaves oval, or obovate, acuminate, wavy-toothed. Flowers in axillary clusters on thick stalks, a cluster consisting of 3–4 involucrate, usually 3-flowered glomerules. Only 4 of the 8 stamens fertile.
Hamamelis Virginica, L.

130. Petals oval or obovate, obtuse. Flowers mostly perfect, the rest abortive. Drupe with 4 bony, ribbed nutlets. Flowers single, or clustered near the base of the young branchlets. Leaves serrate, crenate, or sometimes spinously toothed.
Ilex (863).

Petals oblong-linear. Not more perfect than abortive flowers. Drupe with 4 horny, smooth nutlets. Flowers on very slender, axillary peduncles, solitary or sparingly clustered. Leaves deciduous, oval, entire, mucronate-pointed.
Nemopanthes Canadensis, D. C. (863).

131. Apparent calyx cup-shaped, 8- to 10-toothed, the teeth alternately petaloid. Apparent ordinary stamens about 12. Ovary with 3 bifid styles, and stalked. Herbs with a milky juice..538

Calyx of 2, 4, or 5 segments or pieces........132

132. Calyx 2-cleft.................................206

Calyx of more lobes or sepals...............133

133. Leaves twice- or thrice-ternately compound. Flowers diœciously polygamous, the sterile mostly apetalous.190

Leaves simple. Flowers perfect............134

134. Styles opposite the sepals. Annuals. Stem simple, erect, 2-4 inches high, 1- or 2-flowered. Leaves linear.
Mœnchia, Ehrh.
(*Cerastium quaternellum*, Fenzl.)

Styles alternate with the sepals. Little, matted herbs, annual or biennial. Leaves thread-like, or awl-shaped. Flowers axillary. **Sagina**, L. (187 and 653).

135. (83.) Petals more than 5....................136

Petals 5..144

136. Petals 7 or 8.............................137

Petals habitually 6 (or, in one genus, 6 or 9)....138

137. Petals, stamens, and calyx-lobes habitually 7 (rarely 6 or 8). Leaves 4-8 in a whorl at the top of a simple stem. Flowers 1-4, peduncled, star-like, white. Fruit a pod.
Trientalis Americana, Pursh (372).

Petals and stamens 8. Sepals 4, fugacious. Ovary ovoid, at length gibbous-pointed; stigma 2-lobed. Pod pear-shaped, opening by a lid. Scape 1-flowered. The radical, long-petioled leaves parted into 2 half-ovate leaflets.
Jeffersonia diphylla, Pers.

138. Calyx gibbous or spurred near the base, on the upper side, and commonly with 6 little processes in the sinuses.

Stamens usually 12. Ovary with a curved gland at the base; the slender style with a 2-lobed stigma. Flower solitary, peduncled, purple. A very viscid-hairy, branching annual. Leaves ovate-lanceolate. **Cuphea viscosissima**, Jacq.

Calyx not gibbous nor spurred. Stamens usually 6 (or to 12)..................................139

139. Thorny shrubs. Petals with 2 glandular spots inside, above the short claw. Stamens at the base of the petals. Berry 1- to few-seeded. **Berberis vulgaris**, L.

Herbs..................................140

140. Calyx gamosepalous. Petals 6, sometimes 7. Stamens 6, 7, or twice as many..................................143

Calyx polysepalous, with 2 or 3, sometimes early deciduous bractlets at the base..................................141

141. Petals thick, reniform or hooded, gland-like bodies. A simple and naked stem terminated by a small, simple, or compound raceme of yellowish-green flowers, and a little below bearing a large triternately compound, sessile leaf. (There is sometimes another, smaller biternate leaf at the very base of the panicle).
Caulophyllum thalictroides, Mx.

Petals not gland-like, but flat and oval, or oblong. Leaves orbicular, roundish-kidney-form, or broadly cordate, incisely lobed. Stems 1-leaved, or 2-leaved, the leaf in the first case centrally peltate, the two leaves in the latter peltate near the base. Flowers either in a terminal cyme or solitary, white. Fruit a berry..................................142

142. Anthers opening by uplifted valves. Petals 6, much longer than the sepals. Stamens 6. Berry 2- to 4-seeded. Seeds with no aril. **Diphylleia cymosa**, Mx.

Anthers not opening by uplifted valves. Petals 6 or 9. Stamens 12 and more (to 18). Berry pulpy, many-seeded. Seeds enclosed in a pulpy aril.
Podophyllum peltatum, L. (217).

143. Calyx-tube cylindrical, habitually 6- (sometimes 5- or 7-) toothed. Petals as many as the calyx-teeth. Stamens as many, or twice the number. Pod 2-celled. Flowers usually purple, rarely white. Mostly perennial. **Lythrum.**

Calyx scarcely with a tube, deeply 6- (or 5-) parted. Petals and stamens as many as the calyx-lobes. Pod 1-celled; placenta globose, with few seeds. Perennials. Leaves lanceolate, punctate, sessile. Flowers small, in dense thyrsoid racemes. **Naumburgia thyrsiflora**, Mœnch.

144. (135.) Style 1, with 1 or several stigmas (sometimes deeply 2- to 5-parted)..........................145

Two or more styles or sessile stigmas.........175

145. Stamens 5, or else 10 filaments, only half of them, the alternate ones, anther-bearing....................146

Stamens 6-12..............................163

146. Filaments 10, sometimes united in a tube, the 5 alternate ones without anthers........................147

Sterile stamens none; 5 fertile stamens only, sometimes more or less united 148

147. The filaments united in a 10-toothed tube, the 5 alternate teeth anther-bearing, the rest naked. Sepals and petals 5. Pod 3-celled; seeds many. Root-stocks tufted, sending up long-petioled, round-heart-shaped, crenate-dentate leaves and a slender scape, bearing a dense raceme of white flowers. **Galax aphylla**, L.

Filaments not united into a tube. Sepals and petals 5. Style with 5 stigmas (5 styles united). Fruit rostrate, of 5 coherent membranous pods, tipped with the long, spiral style, bearded inside. Flowers umbellate. Hairy, commonly prostrate. Leaves pinnate; leaflets sessile, pinnatifid, incised, acute. **Erodium cicutarium**, L'Her.

148. Filaments united at the base. Style 5-cleft or parted—that is, consisting of 5 partly united styles. Sepals 3-nerved, with rough, glandular margins. Petals sulphur-yellow. Pod of five united carpels, imperfectly 10-celled. Leaves linear. Herb with a tough, fibrous bark. Flowers racemed on the corymbose branches.

Linum Boottii, Planch.

Filaments not united with each other.........149

149. Thorny trees, with abruptly once- or twice- pinnate leaves, and small, greenish flowers in short spikes. Thorns above the axils, 3- to several-branched.

(92) **Gleditschia** (844).

No thorny trees, but shrubs or herbs, the former often climbing, or with watery juice also............150

150. Herbs...151

Shrubs..153

151. Leaves bipinnate. Flowers perfect, or polyga-mous, axillary, white. Stamens 5 (or 10). Legume flat, 4- to 6-seeded. Petioles with one or more glands.
Desmanthus brachylobus, Benth.

Leaves simple...............................152

152. Calyx 5- (to 7-) toothed or lobed.............143

Calyx of 2 sepals. The 5 stamens adhering to the short claws of the 5 rose-colored petals. Style 3-lobed at the apex. Stems from small tubers, simple, with a pair of opposite lance-linear or spatulate-oblong leaves, and a loose raceme of flowers. Petals rose-color, or white, tinged with red. **Claytonia.**

153. Anthers opening by 2 terminal pores. Calyx-lobes, petals and stamens 5 (the latter sometimes 6 or 7). Pod 5-celled, 5-valved. Leaves alternate, elliptic, ferruginously downy underneath, and strongly replicate at the margin. Flowers white, in terminal, umbel-like clusters, from scaly buds. Shrub 2–5 feet high.
Ledum latifolium, Ait. (661).

Anthers opening lengthwise................154

154. Leaves trifoliolate...........................92

Leaves not trifoliolate.....................155

155. Stamens opposite the petals. Fruit a berry, drupe, or capsule...156

Stamens alternate with the petals. Fruit a pod, sometimes resembling a berry. Seeds enclosed in a red aril. A disk filling the bottom of the calyx................161

156. Calyx minute, truncate or obscurely 5-toothed. A disk in the calyx, or none. Shrubs, with a watery juice, with tendrils, or none. Flowers small, greenish. Petals 5, deciduous. Berry 2-celled, usually 4-seeded..........157

Calyx larger, 5-cleft or parted, with a fleshy disk. Fruit a drupe or capsule. Shrubs without tendrils, sometimes climbing...............................159

157. Calyx filled with a fleshy disk, which bears the petals and stamens..158

Calyx without a disk. Leaves quinate, digitate; leaflets oblong-lanceolate. Flower-clusters cymose; tendrils with a foot-like, adhesive expansion at the end. Petals concave, thick, expanding before they fall.
Ampelopsis Virginica, Mx.

158. The 5 petals cohering at the top, while they separate at the base, and falling away unexpanded, resembling a mitre. Flowers diœciously polygamous, in a compound thyrsus, the pedicels usually umbellate-clustered. Leaves simple, rounded and cordate, often angularly lobed.
Vitis, L. (857).

The 5 petals expanding before, or when they fall away. Flowers usually all perfect, in small panicles. Leaves either cordate, and coarsely toothed, or twice pinnate. Tendrils sometimes none. **Cissus**, L.

159. Calyx and disk adherent to the base of the ovary, tubular, separating transversely after flowering. Petals hood-shaped. (Ceanothus)........................288
Calyx and disk perfectly free from the ovary..160

160. Petals small, short-clawed, notched, wrapped around the short stamens............................103

Petals larger, as long as the calyx, entire. Drupe with thin flesh, and a bony, 2-celled nut, dark purple. Leaves ovate, repandly serrate. Flowers polygamous, in terminal panicles. A glabrous, climbing shrub.
Berchemia volubilis, D. C.

161. Leaves alternate............................162

Leaves opposite, ovate-oblong, or oblong-lanceolate. Disk flat. Pod 5-angled, 5-celled, 5-valved, rough-warty, depressed, crimson, when ripe. Seeds with a scarlet aril.
Euonymus Americanus, L.

162. Pod oblong, 2-grooved, 2-celled, tipped with the style (2 united styles), 2-parted, when ripe. Seeds several in each cell. Flowers all perfect, white, in a terminal raceme. Leaves oblong, pointed, minutely serrulate.
Itea Virginica, L.

Pod globular, 3-celled, 3-valved, berry-like, orange-colored. Seeds enclosed by a pulpy, scarlet aril, 1 or 2 in each cell. Flowers diœciously-polygamous, small, greenish-white, in raceme-like clusters at the end of the branches. Shrub sarmentose and climbing. **Celastrus scandens**, L.

163. (145.) Leafless plants, with scales in place of leaves, parasitic on roots, tawny, reddish, or white. Herbs, smooth or pubescent, and either with a solitary, 5-petalled and 10-androus flower at the summit of the stem (the herb in this case smooth and waxy-white), or with a raceme of flowers, the terminal one usually 5-petalled and 10-androus, while the rest are 4-petalled and 8-androus, the herb in this case reddish or whitish, and slightly pubescent. (126) **Monotropa.**

Plants with ordinary green leaves............164

164. Leaves digitate, or pinnate..................165
Leaves simple, sometimes lobed..............166

165. Leaves digitate. Shrubs or trees............230
Leaves twice-pinnate. Stamens 10 (or 5). An herb, see 151. A tree, see 847.

166. Calyx 2-cleft, its tube cohering with the base of the ovary. Style 3- to 8-cleft, oftener deeply parted........206
Calyx of more segments or sepals............167

167. I. Stamens usually 8, on a disk. Calyx colored, commonly 5- (rarely 4- to 9-) lobed. Ovary 2-lobed; style bifid. Fruit consisting of 2 samaras, united at the base. Flowers often polygamous. Petals as many as the calyx-lobes, sometimes wanting. Leaves palmately-lobed, opposite. Flowers small, in umbel-like clusters or racemes. Shrubs or trees. **Acer** (662).

II. Stamens 6 or 7. Anthers opening at the top. 153

III. Stamens 10. Ovary not 2-lobed..........168

168. Leaves lobed or parted, opposite. Sepals and petals 5. Stamens united at the base, the 5 alternate ones longer, and each with a gland at its base. Style with 5 stigmas—that is, of 5 partly united styles. At maturity, the styles separate from the base, curling back elastically and carrying with them the small 1-seeded pods, their points remaining fixed to the summit of the axis. **Geranium.**

Leaves not lobed, nor parted, but entire, or serrate. Stamens not monadelphous, and rarely unequal..........169

169. Plants (conspicuously) caulescent.............170
Plants acaulescent, or nearly so..............174

170. Anther-cells opening by chinks or pores at the apex; anthers often 2-horned...............................172

Anther-cells opening lengthwise, not appendiculate. Pod 2- to 3-, or 3- to 5-celled........................171

171. Leaves alternate, or crowded. A low, branched evergreen, 6–10 inches high. Leaves smooth, shining, oval or oblong. Flowers small, white, in terminal, umbel-like clusters. Pods 2- to 3-celled. **Leiophyllum buxifolium**, Ell.

Leaves opposite or whorled, lanceolate, nearly sessile. Calyx broadly bell-shaped, 5- to 7-toothed, with horn-like processes in the sinuses. Stems recurved, 2–8 feet long, 4- to 6-sided. Stamens 10, half of them shorter. Flowers rose-purple, axillary. Pods 3- to 5-celled.
Nesæa verticillata, H. B. K.

172. Leaves strongly replicate at the margin, elliptical or oblong, entire, with a rusty down underneath........153

Leaves not replicate.......................173

173. Shrubs or trees. Anthers inversely arrow-shaped. Style slender, 3-cleft at the apex. Pod 3-celled, many-seeded. Leaves serrate, either wedge-obovate, or oval, or oblong, deciduous. Flowers in hoary, terminal racemes, white, often fragrant. **Clethra.**

Almost herbaceous plants. Anther-cells produced upward into tubes, opening by a 2-lipped pore; filaments dilated and hairy in the middle. The anthers sometimes violet. The short, ascending stems 3–10 inches high. Leaves wedge-lanceolate or lanceolate, serrate or toothed, whorled or scattered. Peduncles 1- to 7-flowered. **Chimaphila.**

174. Flowers in a raceme. Petals not spreading. Anthers slightly 2-horned, if at all. Style generally turned to one side, the 5 stigmas more or less conspicuous, mostly tubercular. Valves of the pod cobwebby on the edges.
Pyrola.

Flower single. Petals widely spreading. Anthers conspicuously 2-horned; filaments 2-spurred at the base. Stigma of the style radiately 5-cleft. Valves of the pod smooth on the edges. Plant only 2–4 inches high. Leaves rounded, serrate, clustered. **Monesis grandiflora**, Salisb. (*Monesis uniflora*, Gray.)

175. (144.) Stamens 2–10. Stipules scarious. Styles and valves of the capsule 3 (very rarely 5). Leaves linear. Low, decumbent herb. **Spergularia rubra**, Pers. (201).

Stamens 5 (rarely 3 or 4).....................176
Stamens 6–12..................................189

176. Styles, or sessile stigmas 2 or 3..............177

Styles, or sessile stigmas 4 or 5, and more, often deeply 2-parted (with the branches sometimes cleft).....182

177. Erect shrubs. Leaves pinnate; leaflets 3 to 31. Styles 3..178

Vines, climbing by tendrils. The filaments of the 5 stamens united into a tube, which sheathes the long stalk of the ovary; anthers fixed by the middle. Throat of the calyx crowned with a complex fringe. Fruit a many-seeded berry. Leaves alternate, palmately lobed. Peduncles jointed, axillary. Flowers large, greenish-yellow, or almost white, the crown purple, or rose-color. **Passiflora.**

Herbs..179

178. Leaves opposite, 3-foliolate. Capsule membranaceous, large, inflated, 3-sided, 3-parted at the top, and 3-celled, containing 1 to 4 bony seeds. Flowers all perfect, white, in drooping racemose clusters. **Staphyllea trifolia**, L.

Leaves alternate; leaflets 3, or 7–31. Fruit a dry drupe, either whitish to dun-colored, or red, sometimes pubescent. Flowers polygamous, greenish-white, or yellowish, in terminal or axillary panicles, or in clustered, scaly bracted, catkin-like spikes. **Rhus** (847).

179. Calyx of 5 sepals. Styles 3 (rarely 4)........180

Calyx bell-shaped, 5-cleft, somewhat coherent with the base of the ovary. Styles 2. Pod 2-beaked, many-seeded, opening between the beaks. Perennials. Radical leaves roundish-cordate, lobed and toothed, long-petioled. Flowers panicled, on scape-like stems........................181

180. Pod splitting to the middle, or farther into valves. Petals 2-parted. Stamens 3-5 (or 10). Stems spreading and marked with an alternate, or 2 opposite hairy lines. Flowers white, axillary and terminal. Leaves ovate.
Stellaria (200).

Pod opening at the top only, by teeth. Petals jagged, or denticulate at the summit. Stamens 3-5. Flowers white, in an umbel, on a terminal, glandular-pubescent peduncle. Pedicels reflexed after flowering. Leaves oblong.
Holosteum umbellatum, L.

181. Pod 1-celled, with 2 parietal placentæ. Seeds oval, roughish. Petals spatulate, small. Calyx-segments obtuse. Petioles with dilated margins or adherent stipules at the base. Flowers greenish or purplish. **Heuchera** (288).

Pod 2-celled, 2-beaked. Seeds wing-margined. Petals acutish, unguiculate, twice as long as the acute calyx-lobes. Leaves (mostly radical) rounded-heart-shaped, lobed and toothed. Flowers loosely panicled, small, white.
Sullivantia Ohionis, Torr. & Gr.

182. (176.) Herbs..............................183
Shrubs....................................130

183. Stigmas 4, sessile. Sepals and petals 5. A cluster of slightly united, gland-tipped filaments at the base of each petal. Fertile, anther-bearing stamens 5. Pod 1-celled, with 4 projecting parietal placentæ, 4-valved, many-seeded. Smooth perennials. Leaves rounded-ovate, cordate or reniform, entire, chiefly radical. Flowers solitary at the end of the naked stems. Petals white, marked with green and yellow veins. **Parnassia.**

Styles or stigmas 5 or more..................184

184. Stem leafy 185
Scape naked, simple, or ramose. Calyx 5-parted or toothed..... 188

185. Pod 10-celled, of 5 united carpels, the cell of each divided into 2 cavities by a partition projecting from the back of the carpel. Seeds 10. Sepals, petals, styles and stamens 5, the latter united in a ring at the base. Herbs

with a tough, fibrous bark. Leaves oblong-lanceolate or linear. Flowers scattered, usually yellow, sometimes blue.
(148) **Linum**.

Pod 1-celled. Sepals, petals, stamens and styles 5. 186

186. Petals 2-cleft. Pedicels longer than the obtuse sepals. Stamens usually 10, rarely 5. Stem hairy, viscid, spreading, 6–15 inches long. Leaves oblong-lanceolate. Upper bracts scarious-margined. **Cerastium viscosum**, L.

Petals entire............................187

187. Minute stipules present. Leaves linear, thread-form, whorled. **Spergula arvensis**, L.

Stipules none. Leaves opposite, thread-like or awl-shaped. (134) **Sagina** (204).

188. Calyx deeply 5-parted. Styles apparently 6 or 10, really 3, or sometimes 5 styles, deeply 2-parted. Low perennials. Leaves clothed with reddish, gland-bearing bristles, all in a tuft at the base. Flowers in a 1-sided raceme on a scape, white, or purple-rose color. Flower-parts rarely in sixes. **Drosera**.

Calyx funnel-shaped, 5-toothed, membranaceous above, herbaceous below. Petals clawed. Styles 5 (rarely 3). Flowers lavender-color, scattered, or loosely spiked, and the 3-bracted, 1-, rarely 2-flowered spikelets remotely secund on the branches. Leaves oblong-spatulate, or lance-obovate, tipped with a deciduous bristly point.
Statice Limonium, L.

189. (175.) Styles or sessile stigmas 2..............190

Styles or s. st. habitually 3 (rarely 4 or 5).....196

Styles or s. st. 5, or more...................206

190. Leaves twice- or thrice-ternately compound. Flowers polygamous. Calyx-lobes and petals 5 (4), the latter spatulate, small. Stamens twice as many. Pod 2-celled, separating into 2 few-seeded follicles.
Astilbe decandra, Don. (882).

Leaves not compound. Flowers perfect. Stamens 10.. 191

191. Pod 2-beaked. Calyx adherent to the base of the ovary, or perfectly free from it........................192
Pod not beaked............................193

192. Calyx deeply parted. Petals entire. Pod opening between the beaks. Root-leaves clustered, those of the stem chiefly alternate. Flowers greenish-white, yellow, or purple, solitary, corymbed, or panicled. **Saxifraga** (319).

Calyx bell-shaped, 5-cleft. Petals pectinately pinnatifid, slender. Stamens rarely 5. Leaves cordate, or orbicular-reniform, mostly radical; those of the stem opposite, if any, sometimes lobed. Flowers small, white, in a slender raceme or spike. Pod 1-celled, 2-valved at the summit. **Mitella** (317).

193. Leaves all radical, heart-shaped, sharply lobed and toothed, sparsedly hairy above, downy underneath. Scape naked, 5-12 inches high. Raceme simple. Petals clawed, white. Pod membranaceous, 1-celled, 2-valved, one valve much larger. **Tiarella cordifolia**, L.

Leaves none of them radical. Ovary stalked..194

194. Scaly bractlets at the base of the cylindrical calyx. Petals crenate, rose-color, with white dots. Leaves linear, hairy. Flowers in dense clusters of 3 or more.
Dianthus Armeria, L.

Calyx naked............................195

195. Petals with 2 teeth at the top of the claw. Calyx cylindrical, about one inch long. Leaves oval-lanceolate. Flowers large, rose-color, in corymbed clusters.
Saponaria officinalis, L.

Petals without teeth. Calyx ovoid-pyramidal, 5-angled, 5-toothed, and in fruit winged. Leaves ovate-lanceolate. Flowers pale red in corymbed cymes.
Vaccaria vulgaris, Host.
(*Saponaria vulgaris*, Medik).

196. The 3 styles 2-cleft........................538
Styles not cleft. Leaves entire. Herbs........197

197. Calyx tubular, swelling, sometimes much inflated, 5-toothed. Styles habitually 3 (rarely 4). Stamens 10. Petals long-clawed, often with a scale at the base of the blade, and inserted, with the stamens, on the stalk of the ovary. Stem swollen at the joints. Leaves opposite or in whorls of 4. **Silene.**

Calyx not tubular.........................198

198. Stamens 9, rarely more, united in 3 sets, which are separated by as many glands. Pod 3-celled. Flowers flesh-color, in the axils of the leaves. Perennials, in marshes or shallow water. **Elodea.**

Stamens not in sets, and without intervening glands...199

199. Three parietal placentæ in the 1-celled ovary. Stamens 6–12, very rarely more, distinct or obscurely clustered. Petals oblong or linear, small. Sepals narrow, erect. Annuals. Branches 4-angled. Leaves ovate, oblong, or lance-linear, or awl-shaped, usually 1- to 5-nerved.
Hypericum (in 4 spec.)

A central or basilar placenta. Stems more or less swollen at the joints. Ovary 1-celled, or, in one case, imperfectly 3- to 5-celled....................................200

200. Petals 2-cleft or parted. Stamens usually 10, rarely fewer. (180) **Stellaria** (692).

Petals entire or only slightly emarginate......201

201. Stipules present, membranaceous, cleft. Styles 3, rarely 5. Stamens 2 to 10. Sepals lanceolate, with scarious edges. Leaves filiform-linear, slightly mucronate. Flowers small, rose-color, on hairy pedicels.
Spergularia rubra, Pers.

Stipules none..............................202

202. Valves of the capsule 3, each 2-toothed. Styles 3..203

Valves of the capsule entire................204

203. Seeds appendiculate at the hilum (strophiolate). The white petals more than twice as long as the sepals.
Mœhringia lateriflora, L.

Seeds not appendiculate. The white petals much shorter than the hairy, 3-nerved sepals.
Arenaria serpyllifolia, L.

204. Stamens on a disk, 10........................205

Stamens not on a disk. Styles 4- (or 5), always as many as the sepals. Petals rarely wanting. Pod 4- (or 5-) valved, the valves opposite the sepals. Little, matted herbs, with filiform leaves, the upper short and awl-shaped, their axils bearing clusters of minute leaves. Petals much longer than the sepals. Stamens 10. **Sagina nodosa,** Fenzl.

205. Disk conspicuous and glandular, 10-notched. Ovary imperfectly 3- to 5-celled. Styles 3 (to 5). Sepals and leaves very fleshy, the latter ovate or oblong. Flowers solitary, axillary, white, diœciously polygamous. Maritime perennials. **Honkenya peploides,** Ehrh.

Disk inconspicuous, like a small gland. Petals rarely slightly notched. Styles 3. Leaves rigid, awl-shaped, bristle-form, or soft and filiform-linear. Flowers mostly white, solitary and terminal, or cymose. **Alsine.**

206. (189.) Calyx 2-cleft, half superior. Petals 5, cohering at the base. Stamens 7–12. Style deeply 5- to 6-parted. Leaves fleshy, alternate, obovate or wedge-form. Flowers sessile, pale yellow. **Portulaca oleracea,** L.

Calyx 5-sepalled, or 5-toothed................207

207. Calyx tubular, 5-toothed, the teeth linear. Styles 5, alternate with the calyx-teeth. Stamens 10, those opposite the petals adhering to their claws. Pod 1-celled, opening at the top by 5 teeth. Annuals or biennials. Leaves linear. Flowers large, purple, on long peduncles. Seeds black.
Agrostemma Githago, L.

Calyx of 5 sepals. Stamens 10 (rarely 5). Leaves simple or trifoliolate................................208

208. Leaves trifoliolate, stipulate, alternate, or radical. Stamens cohering at the base. Pod deeply 5-lobed, 5-valved, 5-celled. Petals yellow, violet, or white with reddish veins. **Oxalis.**

Leaves simple, opposite, or fascicled..........209

209. A 10-notched disk, bearing the stamens. Sepals and leaves very fleshy. (205) **Honkenya peploides,** Ehrh.

Disk none......................................210

210. Petals deeply 2-cleft or sharply emarginate. Pod 10-valved at the top. Flowers white, in terminal cymes.
(186) **Cerastium.**
Petals entire, or only obtusely emarginate.....211

211. Leaves all fascicled, whorled, linear-filiform, with stipules. Stamens 10 (or 5). (187) **Spergula arvensis,** L.
Leaves opposite, or only the upper ones fascicled, thread-form. Stipules none. (134) **Sagina.**

---

212. (82.) Filaments all united in a very conspicuous tube or column, sheathing the compound style, which is several- or many-branched at the apex. Petals 5, united with each other and the staminal column at the base........213
Stamens distinct...........................215

213. Anthers usually more than 20...............68
Anthers 15-20............................214

214. Calyx with a 3-leaved involucel. Stigmas capitate. Carpels 14-20, kidney-shaped, pointed, and at length 2-valved at the hispid apex. Annuals or biennials, humble, procumbent, or creeping. Leaves 3- to 5-cleft and cut. Flowers small, solitary, axillary, purplish.
Modiola multifida, Mœnch.
Calyx naked at the base, 5-toothed. Flowers diœcious, the sterile ones without pistils, and with 15-20 anthers, the fertile with 6-10 united carpels and a short column of filaments without anthers. Styles stigmatic down the inner face. Perennials, tall, roughish. Lower leaves very large, 9- to 11-parted, the segments linear-lanceolate, coarsely-toothed. Flowers small, white, in leafy panicles........880

215. Calyx gamosepalous. Ovary superior.........53
Calyx polysepalous........................216

216. Sepals 2. Plants with milky or colored juice....10
Sepals 2. Plant without milky juice. Petals 5. Stamens 15-20. Style 3-lobed at the apex. Pod 1-celled (or 3-celled at the base, when young), 3-valved, with a many-seeded globular placenta. Stems simple or branched, short,

thick, with the linear leaves crowded toward the summit. Peduncle long and naked, bearing a dichotomous cyme of purple flowers. Perennial. **Talinum teretifolium**, L.

Sepals 3 or 6 217

217. Sepals and petals 6. Stamens 12-18. Fruit a pulpy berry. Seeds enclosed in a pulpy aril. Leaves round, 7- to 9-lobed, peltate. (142) **Podophyllum peltatum**, L.

Sepals 3, united at the base only, with 2 bracklets outside 22

---

218. (81.) Corolla spurred or saccate at the base. Leaves alternate. Flowers usually drooping 219

Corolla not spurred nor saccate 226

219. Stamens usually numerous. Delphinium 42

Filaments 2 (compound) or 5 220

220. Membranaceous filaments 2, each consisting of 3 united ones, and bearing 3 anthers. Petals 5. Leaves compound. Ovary 1-celled. Fruit either a many-seeded capsule, or a fleshy, 1-seeded, indehiscent nut. Flowers white, rose-color, purple, or yellowish 221

Filaments 5, each with an anther. Leaves simple, sometimes pedately divided. Ovary either 1-celled and 3-carpelled, or 5-celled. Flowers white, blue, violet or yellow 224

221. Corolla 2-spurred, or 2-saccate at the base. Flowers either in axillary cymes, or in simple or panicled racemes, white, rose-color, or purple 222

Corolla 1-spurred at the base. Capsule either indehiscent, nut-like, fleshy, 1-seeded, or slender, 2-valved-, many-seeded. Flowers in racemes, whitish, rose-color, or yellow.
223

222. Petals united into an ovate corolla, becoming dry and persistent. A climbing vine. Flowers pale rose-color, in axillary cymose clusters. Young petioles tendril-like.

Adlumia cirrhosa, Raf.

Petals slightly united into a heart-shaped corolla, the inner ones crested, deciduous or withering. Flowers in simple or paniculate racemes, on scapes, white or purple. Seeds crested. **Dicentra.**

223. Pod globular, fleshy, indehiscent, 1-seeded. Seeds crestless. Flowers small, rose-color, in dense racemes.
Fumaria officinalis.

Pod slender, 2-valved, and many-seeded, sometimes torulose. Seeds crested. Flowers yellow or yellowish-red in racemes. **Corydalis.**

224. Stems swollen at the joints, very succulent. Fruit capsular, 5-celled, bursting elastically, when ripe, by 5 valves. Sepals and petals colored alike. Sepals apparently 4 (really 5, the upper 2 united), the lowest gibbous and spurred. Petals apparently but 2 (really 4, united in pairs). The filaments of the stamens furnished with a scale, and the 5 scales connivent over the stigma. Leaves oblong-ovate, or rhombic-ovate, scarcely serrate. Flowers axillary, or panicled, yellow, or orange-colored, often with brown dots. (Sepals and petals readily confounded.) **Impatiens.**

Stems not swollen at the joints. Leaves stipulate. Fruit a 3-valved capsule, with 3 parietal placentæ. Stamens with their broad filaments continued beyond the anther-cells, the anthers coherent with each other. Petals 5, commonly unequal, the lowest spurred, or gibbous and saccate at the base. Perennial herbs.............................225

225. Sepals unequal, more or less auricled at the base. The lowest and broadest of the petals spurred at the base. Two of the anthers appendiculate at the back. Flowers axillary, white, violet, blue, or yellow. Leaves undivided or divided, often cordate. Caulescent or acaulescent. **Viola.**

Sepals nearly equal, not auricled. The lowest and broadest petal more notched at the apex than the rest, and saccate or gibbous at the base. Stamens completely united into a sheath, enclosing the ovary, and bearing a broad gland on the lower side. Style hooked at the summit. Stems always leafy. Flowers small, greenish white, 1 or 3 together in the axils. **Solea concolor,** Ging.

226. Style or stigma 1.........................227.

Styles 3. Calyx deeply 4-parted. Petals 4, the upper one 3- to 5-cleft, the 2 lateral 3-cleft, the lowest linear and entire. Ovary 3-lobed. A fleshy, one-sided disk beneath the pistil...7

227. Sepals, calyx-lobes, or teeth, more than 3......**228**

Sepals 3, green. Petals 3, blue. Stamens 6, 3 of them sterile, with cross-shaped glands in place of anthers; one of the 3 fertile ones bent inward. The 2 lateral sepals partly united. The 2 lateral petals orbicular or kidney-shaped, the odd one smaller. Pod 3-celled. Leaves longitudinally veined, contracted at the base into sheathing petioles; the floral ones forming a sort of spathe. **Commelyna.**

228. Sepals 5, namely, 3 smaller greenish, and 2 larger petaloid ones. Petals 3, placed between the 2 larger sepals, connected with each other and the split tube of 6-8 stamens, the middle one boat-shaped and often crested at the back. Pod 2-celled, 2-seeded, flat, often notched at the top. Low, bitter herbs. Leaves simple, entire, often dotted. Flowers white, purple, or yellow. **Polygala.**

Calyx herbaceous throughout, regular, or sometimes 2-lipped......................................**229**

229. Stamens 5-10, all free......................**230**

Stamens 10 (very rarely 5), monadelphous or diadelphous, mostly with 9 united in a tube, which is cleft on the upper side....................................**234**

230. Stamens 7 (rarely 6 or 8), often unequal, sometimes declined or curved. Calyx tubular, 5-lobed, often oblique or gibbous at the base. Petals 5, unequal, clawed. Ovary 3-celled, each cell 2-ovuled. Fruit coriaceous, or, when young, prickly. Flowers polygamous, or staminate, panicled terminal. Leaves digitate; leaflets 5-7, serrate. Trees or shrubs. **Æsculus.**

Stamens 10 (or 5). Fruit a legume. Trees or herbs.......................................**231**

231. Calyx tubular, 5- (rarely 4-) toothed..........**232**

Calyx of 5, scarcely united sepals. Petals unequal, but not at all papilionaceous, spreading. Stamens 10, or, by abortion, fewer, unequal. Anthers opening by 2 pores or chinks at the apex. Herbs, often 5 feet high, with abruptly pinnate leaves and yellow flowers, in short axillary racemes, or supra-axillary clusters. **Cassia.**

232. Trees, 20 to 40 feet high 233

Herbs. Corolla perfectly papilionaceous. Keel-petals nearly separate and, like the wings, straight. Leaves palmately 3-foliolate, rarely simple (generally blackening in drying). Flowers yellow, indigo-blue, or white, racemed. Calyx 4- or 5-cleft. **Baptisia.**

233. Corolla perfectly papilionaceous. Standard large, roundish, reflexed. Keel-petals distinct, straight. Pod very flat. Leaves pinnate; leaflets 7–11, oval or ovate. Flowers showy, white, in drooping, panicled racemes.
Cladrastis tinctoria, Raf.

Corolla imperfectly papilionaceous. Standard smaller than the wings, and enclosed by them in the bud. Keel-petals larger and distinct. Stamens rather unequal. Leaves rounded-cordate. Flowers red-purple, in small lateral clusters, preceding the leaves. **Cercis Canadensis,** L.

234. Stamens 10 (rarely 9 or 5), monadelphous 235

Stamens 10, diadelphous (or monadelphous below, and then the pods jointed, or not jointed and 1-seeded) .. 241

235. Anthers of 2 forms. Leaves simple, palmately compound, or pinnately 3-foliolate 236

Anthers all alike 239

236. Leaves simple 237

Leaves compound 238

237. Calyx 2-lipped. Keel straight. Pod flat, several-seeded. Flowers yellow, in spiked racemes. A low, shrubby plant. **Genista tinctoria,** L.

Calyx 5-lobed, scarcely 2-lipped. Keel scythe-shaped. Pod inflated, oblong, many-seeded. Sheath of filaments cleft on the upper side. Flowers racemed, yellow. Stipules united and decurrent on the stem. Leaves oval or oblong-lanceolate, almost sessile. **Crotalaria sagittalis,** L.

238. Leaves pinnately 3-foliolate. Calyx 2-lipped; upper lip 2-, the lower 3-cleft. Flowers of two sorts intermixed in the heads, some complete, but unfruitful, yellow;

the rest fertile, consisting merely of a pistil between 2 bractlets. Style hooked. Pod reticulated, 1- to 2-jointed.
Stylosanthes elatior, Swartz.

Leaves palmately 7- to 11-foliolate; leaflets oblanceolate. Flowers showy, purplish-blue, rarely white, in a long and loose raceme. **Lupinus perennis**, L.

239. Pod several-seeded, linear, 2-valved. Leaves pinnate; leaflets 5-29. Flowers racemed, white or purplish.
Tephrosia (261).

Pod 1-seeded and indehiscent. Corolla imperfectly or slightly papilionaceous. Petals all clawed. Standard, in the bottom of the calyx, heart-shaped or oblong. Leaves pinnate; leaflets linear. Flowers in terminal, peduncled heads or spikes 240

240. The cleft tube of filaments bearing 4 of the petals on its middle; anthers 9 or 10. Corolla small, whitish.
Dalea alopecuroides, Willd.

The cleft tube of filaments bearing 4 petals at its summit; anthers 5. Corolla rose-purple. **Petalostemon**.

241. Leaves tendril-bearing, abruptly pinnate, of 1-several pairs of leaflets. Style bent at a right angle with the ovary 242

Leaves not tendril-bearing 243

242. Style filiform, hairy most outside and all around the summit. Stipules usually half arrow-shaped. **Vicia**.

Style flattened, hairy most inside. Stipules half arrow-shaped, half cordate, or halberd-form. **Lathyrus**.

243. Legumes jointed (loments), sometimes of one joint only 244

Legumes not jointed, 1-, or 2- to many-seeded .. 247

244. Leaves odd-pinnate, with more than 3, or many leaflets. Pod several-jointed 245

Leaves pinnately 3-foliolate 246

245. Stamens equally diadelphous (5 and 5). Calyx 2-lipped. Leaves with 37-51 sensitive leaflets. Racemes 3- to 5-flowered. Corolla yellow, reddish externally. Pod of several square joints. **Æschyomene hispida**, Willd.

Stamens unequally diadelphous (9 and 1). Calyx 5-cleft. Pods of several roundish joints. Leaflets 13–21. Stipules scaly. Raceme of many deflexed, purple flowers.
Hedysarum boreale, Nutt.

246. Calyx 2-lipped. Stamens diadelphous (or monadelphous only below). Pod 2- to several-jointed. Flowers all of one sort, and complete. Leaflets stipellate. Flowers in axillary or terminal racemes, often panicled, purple or purplish. **Desmodium.**

Calyx nearly regularly 5-cleft, the lobes slender. Pod 1-jointed (sometimes 2-jointed, with the lower joint empty or stalk-like). Stipules and bracts minute. Flowers either of two sorts, the fertile ones apetalous, in panicles or clusters, or all alike and perfect, in close spikes or heads. Corolla violet-purple, or whitish, with a purple spot on the standard. **Lespedeza.**

247. Twining or trailing plants. Leaves pinnately 3-, rarely 5- to 7-foliolate, mostly stipellate, without tendrils. Cotyledons thick and fleshy, becoming leafy in germination, but not in Wistaria. Flowers often clustered in racemes.248

Erect, or rarely prostrate plants, the latter with palmately 3-foliolate leaves. Cotyledons thin, not becoming leafy in germination..................................255

248. Keel primarily, or at length spirally twisted. Herbs......................................249

Keel not twisted...........................250

249. Leaves 3-foliolate, stipellate. Keel primarily twisted with the stamens. Flowers often clustered on the knotty joints of the raceme. **Phaseolus.**

Leaves 5- to 7-foliolate, not stipellate. Keel at first incurved, at length twisted. Subterranean shoots bearing eatable tubers. Leaflets ovate-lanceolate. Flowers brown-purple, in dense and short, often branching racemes.
Apios tuberosa, Mœnch.

250. Leaves pinnately 9- to 13-foliolate, not stipellate. Keel scythe-shaped. Standard roundish, reflexed, large, with 2 callosities at the base. Wings with 2 auricles below.

Seeds kidney-shaped. Cotyledons thin. A shrubby vine. Flowers large, lilac-purple, in dense racemes.
Wistaria frutescens, D. C.

⸺eaves pinnately 3-foliolate (rarely of a single leaflet). Herbs..251

251. Flowers purplish or white. Ovary several-ovuled. Leaflets usually stipellate........................252

Flowers yellow, racemed or clustered. Ovary 1- to 2-ovuled. Leaflets not stipellate, mostly roundish, rarely a solitary one. Calyx slightly 2-lipped, or deeply 4- to 5-parted. Herb downy. **Rhynchosia tomentosa**, Torr. & Gr.

252. Calyx 2-bracteolate........................253

Calyx not bracteolate, 4- or rarely 5-toothed. Flowers of 2 sorts; the upper purplish, in clustered or branched axillary racemes, complete, but often barren; the lower near the base, and on creeping branches, apetalous, but fruitful. Leaflets rhombic-ovate, stipellate. Bracts persistent, round, partly clasping, and, like the stipules, striate.
Amphicarpæa monoica, Nutt.

253. Standard with a short, spur-like projection at the back, near the base. Calyx short, 5-cleft. Leaflets stipellate, oblong-ovate to oblong-linear. Peduncles 1- to 4-flowered. Corolla 1 inch long, violet. Pod thickened at the edges, pointed with the awl-shaped style.
Centrosema Virginianum, Benth.

Standard not appendiculate...............254

254. Calyx 4-cleft, the upper lobe broadest. Leaflets stipellate, oval, or ovate-oblong. Flowers in somewhat interrupted, or knotty racemes, purplish. **Galactia.**

Calyx 5-toothed. Standard much larger than the other petals. Pod linear-oblong, flattish, knotty. Peduncles 1- to 3-flowered. Corolla pale blue, 2 inches long.
Clitoria Mariana, L.

255. Leaves palmately compound, 3- to 5-foliolate...256

Leaves pinnately compound.................257

256. Calyx equally 5-cleft, the teeth bristle-form. Pod 1- to 6-seeded. Stamens more or less united with the corolla; anthers all perfect. Flowers capitate, whitish, purplish, rose-colored, or yellow. Leaves 3-foliolate. **Trifolium**

Calyx unequally 5-cleft, lower lobe longest. Pod 1-seeded. Stamens sometimes slightly monadelphous, the 5 alternate anthers often imperfect. Flowers spiked or racemed, white or mostly bluish-purple. Leaves 3- to 5-foliolate, often glandular-dotted. **Psoralea.**

257. Leaves 3-foliolate........................258
Leaves with more than 3 leaflets, several or many.......................................260

258. Pods curved or coiled. Flowers purple or yellow, racemed or spiked. Stipules often toothed or slashed. **Medicago.**
Pods straight............................259

259. Flowers white or yellow, fragrant, in racemes. Legume longer than the calyx, rugose. **Melilotus.**
Flowers blue or violet, in spikes or racemes. Legume as long as the calyx, or nearly so. Stipules cohering with the petiole. Plants glandular-dotted. Half of the anthers often somewhat imperfect......................256
Flowers yellow, in heads. Stipules ovate......256

260. Trees or shrubs, often with prickly spines for stipules. Calyx slightly 2-lipped. Standard large and rounded, turned back, scarcely longer than the other petals. Pod 1-celled, compressed, elongated, many-seeded. Style bearded inside. Flowers showy, white or rose-color, and sometimes fragrant, in pendulous axillary racemes. Leaflets stipellate. **Robinia.**
Herbaceous or suffruticose plants. Calyx equally 5-toothed or cleft. Flowers white or purplish, spiked or racemed 261

261. Pod longitudinally 2-celled, more or less completely, by the turning inward of the dorsal suture. Standard small, its sides reflexed or spreading. **Astragalus.**
Pod 1-celled. Standard large, roundish, usually silky outside. Plants villous, with long, silky, or rusty hairs. (239) **Tephrosia.**

---

262. (80.) Stamens not more than 20................263
Stamens usually more than 20................25

263. Leaves 2-ranked, equitant, linear. Calyx of 3 bracts. Corolla (perianth) of 6 petals (sepals). Ovaries 3, united at the base. Stamens 6. Stem scape-like, leafy only at the base, bearing a close raceme of white or greenish flowers. **Tofielda** (735).

Leaves not 2-ranked, nor equitant............264

264. Leaves centrally peltate, elliptical, entire, 2-3 inches in width, floating. Flowers axillary, small, dark purple, rising to the surface of the water on slender axillary peduncles. Stalks clothed with clear jelly. Stamens 12-24. Pistils 4-18. **Brasenia peltata**, Pursh.

Leaves not peltate...........................265

265. Pistils indefinite, in a dense whorl or head, or in a long slender spike. Acaulescent herbs...............266

Pistils definite 268

266. Pistils spiked on a slender receptacle, usually 1 inch long. Seed suspended. Leaves (radical) linear. Scapes 1-flowered. Petals long-clawed. Stamens 5-20. Sepals and petals 5. **Myosurus minimus**, L.

Pistils in a roundish, or triangular whorl or head. Sepals and petals 3. Leaves petioled, 5-7-nerved......267

267. Stamens 6. Achenia not beaked.
Alisma Plantago, L.

Stamens 9-20. Achenia beaked. **Echinodorus**.

268. Calyx of 3-5 sepals......................270

Calyx 5-, or (apparently) 10-cleft............269

269. Calyx 5-cleft. Follicles 3-12. Shrubs or herbs
58

Calyx apparently 10-, really 5-cleft, with 5 bractlets. Petals 5, linear-oblong, minute, yellow, inserted with the 5 stamens into the margin of a woolly disk. Achenia 5-10. Styles lateral. Leaves 3-foliolate; leaflets wedge-shaped, 3-toothed. **Sibbaldia procumbens**, L.

270. Leaves simple. Herbs succulent, and usually fleshy 271

Leaves pinnate, with 5, incisely lobed and toothed leaflets. Sepals and petals 5, the latter smaller, clawed, slightly 2-lobed. Stamens and pistils 5-15. A shrubby plant, with polygamous, dull purple flowers, in compound, drooping racemes. **Xanthorrhiza apiifolia**, L'Her. (650).

271. Pistils entirely separate. Sepals, petals, and pistils 3–5..272

Pistils 5, united below. Stamens 10. Fruit of 5, beaked pods. Sepals and petals 5, the latter sometimes wanting. Leaves lanceolate. Flowers yellowish-green, loosely spiked along the upper side of the naked branches of the cyme. **Penthorum sedoides,** L. (652).

272. Stamens 3 or 4. Sepals, petals, and pistils as many. Very small, tufted annuals, with opposite linear leaves, and solitary, greenish-white, axillary flowers.
Tillæa simplex, Nutt.

Stamens 10 or 8. Sepals, petals, and pistils 4 or 5. Thick-leaved herbs. Flowers cymose, or 1-sided. **Sedum.**

---

273. (79.) I. Corolla of 2, inversely heart-shaped petals. Calyx tubular, with a disk, 2-lobed. Stamens 2. Style 1. Pod 2-celled, with 1 seed in each cell, bristly with hooked hairs. **Circæa.**

II. Corolla of 3 petals.....................274
III. Corolla of 4 petals.....................275
IV. Corolla of 5 or more petals.............284

274. Petals equal. Aquatics. Flowers diœcious, arising from sessile spathes............................885

Petals unequal (Orchids)....................758

275. I. Stamens 3 or 4.........................276
II. Stamens 8 or 10........................278

III. Stamens rarely fewer than 20. Styles 3–5, united below. Pod 3- to 5-celled. Shrub with opposite leaves. Flowers single, or few, at the end of the branches white, showy...54

276. Leaves in several regular whorls. **Galium** (529).

Leaves alternate, or opposite, rarely clustered, or only apparently whorled at the summit................277

277. Fruit a drupe, with a two-celled nut. Calyx minutely 4-toothed. Stamens 4. Leaves mostly opposite, entire, in one species alternate, in another crowded and apparently whorled at the summit of the stem. Flowers small,

in open naked cymes (rarely in close heads, surrounded by a showy, corolla-like involucre). Trees, shrubs, or herbs.
Cornus (698).

Fruit a pod, globular, or short and 4-sided, cubic. Leaves mostly alternate, rarely opposite, lanceolate, rarely ovate. Flowers axillary, yellow, or reddish.
Ludwigia (700).

278. Style 1..279

Styles 2. Petals strap-shaped. Stamens 8, but only 4 of them perfect..............................129

Styles 2, divergent. Petals 4 (or 5), ovate. Calyx hemispherical, 4- (or 5-) toothed, somewhat adherent. Petals 4 (or 5). Stamens 8 (or 10). Flowers rarely all fertile, perfect, and complete. Often also marginal and radiant ones present, consisting merely of a broad, rotate, 4-lobed, colored calyx, lacking petals, stamens, and pistils. Pod opening by a hole between the styles. Shrubs. Leaves opposite, ovate, rarely cordate, pointed, serrate. Inflorescence cymose.
Hydrangea arborescens, L.

279. Ovary hemispherical or globular. Fruit a berry or a pod..280

Ovary oblong or linear. Fruit a pod, 4-celled (4-valved or indehiscent), rarely at length 1-celled.......281

280. Ovary hemispherical. Fruit a berry. (Oxycoccus).
514

Ovary globular. Fruit a 4-celled pod, with 4 central, prominent placentæ. Seeds many. Petals oblique, inserted with the stamens at the summit of the 4-cleft calyx-tube. Anthers oblong or linear, 1-celled, opening by terminal pores, and sometimes spurred at the back. Leaves opposite, 3- to 5-nerved, ovate-oblong, or lanceolate. Stem terete, or square, often winged along the angles. **Rhexia.**

281. Calyx-tube not prolonged beyond the ovary....282

Calyx-tube prolonged beyond the ovary. Leaves always alternate..283

282. Seeds with a large downy tuft at the apex. Leaves chiefly opposite, at least the lower ones, or scattered. Pod linear. Flowers violet, purple, or white. **Epilobium.**

Seeds not comose, but naked. Leaves alternate,

lanceolate, decurrent on the erect, smooth, branching stem. Pod oblong, club-shaped, wing-angled. Flowers axillary, yellow. **Jussæia decurrens,** D. C.

283. Seeds many. Anthers mostly linear. Petals obcordate or obovate. Calyx-lobes reflexed. Pod oblong. Flowers yellow. **Œnothera.**

Seeds 1-4. Stamens alternately shorter, often declined. Ovary oblong, 4-celled, becoming 1-celled in fruit. Pod indehiscent. Leaves alternate. Flowers small, white or red, rarely 3-merous, spiked, or racemed. **Gaura.**

284. (273.) Stamens 5............................285

Stamens 8-20..................................315

Stamens usually more than 20. Petals 5-10. Herb, 1 foot high, dichotomously branched, rough with barbed bristles. Flowers yellow...............................52

285. Fruit a berry, or a berry-like drupe...........286

Fruit not a berry. Styles 2, or in one genus, 1- (mostly 3-) cleft.....................................288

286. Styles 2 (separate or united into 1). Berry 1-celled, with 2 parietal placentæ, many-seeded, prickly or smooth, crowned with the remains of the calyx. Calyx bell-shaped, 5-lobed, bearing the 5, often very small petals in the sinuses. Stems usually prickly or thorny. Leaves rounded-heart-shaped, and variously lobed or parted. Flowers in racemes. **Ribes.**

Styles 2-5. Drupe 2- to 5-celled. Leaves compound or decompound. Flowers, white or greenish, in umbels. 287

287. Styles and cells of the black, or dark purple fruit 5. Stem herbaceous or woody, one species a prickly tree. Leaves very large, either quinately or pinnately decompound, or twice pinnate. Flowers monœciously polygamous, or perfect, in panicled or corymbed umbels. **Aralia.**

Styles and cells of the red or reddish fruit 2 or 3. Stem herbaceous, low, simple, with a whorl of 3 palmately 3- to 7-foliolate leaves at its summit. Flowers diœciously polygamous, in a solitary umbel. **Panax** (881).

288. I. Ovary 3-celled, 3-lobed; style 1, mostly 3-cleft. Petals saccate-arched, with long claws. Flowers in panicles or corymbs, consisting of little umbel-like clusters, white. Shrubby plants. **Ceanothus.**

  II. Ovary 2-celled, 2-carpelled............... 289

  III. Ovary 1-celled, with 2 parietal, many-seeded placentæ, 2-beaked. Calyx bell-shaped, the tube cohering at the base only with the ovary. Petals spatulate. Leaves chiefly radical, round-heart-shaped and lobed.
          (181) **Heuchera.**

289. Seeds several. Calyx-tube top-shaped, cohering with the 2-carpelled, 2-beaked pod. Leaves palmately 5- to 7-lobed, or cut, petioled, alternate. Flowers white, in cymes. Stem glandular, 6–20 inches high.
       Boykinia aconitifolia, Nutt.
  Seeds 2 only, namely, 1 in each carpel......... 290

UMBELWORTS.

290. Inner face of each seed flat, or nearly so, not hollowed out....................................... 291
  Inner face of each seed hollowed out, either lengthwise or only in the middle. Flowers white....... 311

291. I. Flowers in simple umbels, sometimes spicate. Leaves simple. Calyx-teeth obsolete................. 292

  II. Flowers in capitate umbels—that is, sessile, and forming dense heads................................ 293

  III. Flowers in regularly compound umbels, not sessile in heads....................................... 294

292. Fruit flat, orbicular. Marsh herbs, with round-peltate or kidney-shaped leaves. Calyx-teeth obsolete.
          Hydrocotyle.
  Fruit globular. Proper leaves none; in their place fleshy, hollow, cylindrical or awl-shaped petioles, marked with cross-divisions. Calyx-teeth obsolete.
        Crantzia lineata, Nutt.

293. Flowers partly sterile. Fruits globular, thickly clothed with hooked prickles, few in a head. Leaves 3- to 5-, or 5- to 7-parted.   **Sanicula.**

Flowers all fertile. Fruits top-shaped, covered with little scales or tubercles, many in each head. Leaves linear-lanceolate, bristly-fringed or spiny-toothed.
Eryngium.

294. Fruit beset with weak prickles in single rows along the ribs. Involucre of several pinnatifid bracts, nearly as long as the umbel. Umbels concave, dense in fruit. The central flower of each umbellet abortive and dark purple. **Daucus Carota,** L.

Fruit smooth..............................295
295. Fruit strongly flattened dorsally.............296
Fruit not flattened on the back..............303
296. Fruit single-winged or margined at the junction of the 2 carpels..297
Fruit double-winged or margined at the edge. Each carpel 3-ribbed, or not rarely 3-winged on the back. 301

297. Fruit surrounded with a broad and tumid, corky margin, thicker than the fruit. Carpels nearly ribless on the back. Flowers yellow. Involucre none. Involucels bristly. Leaves twice-pinnate; upper leaves 3-cleft.
Polytænia Nuttallii, D. C.

Margin not corky..........................298
298. Flowers yellow. Involucre and involucels wanting, or inconspicuous. Stem grooved, smooth. Leaves pinnately-compound; leaflets cut-toothed. Root spindle-shaped. Calyx-teeth obsolete. **Pastinaca sativa,** L.

Flowers white.............................299
299. Flowers of 2 sorts, the marginal ones radiant. Petals inversely heart-shaped, those of the outer flowers larger. Plant stout, woolly. Stem grooved. Leaflets slightly cordate. Calyx-teeth minute. **Heracleum lanatum,** Mx.

Flowers all alike..........................300
300. Leaves none; in their place simple, long, terete, hollow petioles, with some cross-partitions.
Tiedemannia teretifolia, D. C.

Leaves simply pinnate, 3- to 11-foliolate.
Archemora rigida, D. C.

301. Seed loose in the cell, covered with vittæ.
Archangelica.

Seed not loose............ 302

302. Carpels with 3 wings on the back, half as narrow as those of the margins. Fruit longer than the pedicels. Leaves twice- to thrice-pinnate; leaflets pinnately-cleft. Calyx-teeth obsolete.
Conioselinum Canadense, Torr. & Gr.

Marginal and dorsal wings alike. Involucre none or scanty. Involucels many-bracted, small. Petioles membranaceous at the base. Leaves twice ternate, or the divisions quinate; leaflets dentate. Calyx-teeth obsolete.
Angelica Curtisii, Buckley.

303. Fruit flattened laterally, or contracted at the sides, wingless 304

Fruit not flattened either dorsally or laterally, at least not remarkably so. Cross-section of the fruit nearly orbicular, or square. Each carpel with 5 wings or strong ribs 309

304. Flowers yellow. Involucre none.............305

Flowers white............................306

305. Fruit oval, somewhat twin. Leaves twice- to thrice-ternate; leaflets entire. Calyx-teeth obsolete.
Zizia integerrima, D. C.

Fruit ovoid-oblong. Leaves all simple, entire, broadly ovate, perfoliate. Calyx-teeth obsolete.
Bupleurum rotundifolium, L.

306. Involucre none, or almost none..............307

Involucre of 2 to 8 bracts..................308

307. Fruit broader than long. Involucels many-bracted. Leaves thrice-pinnate or ternate; veins of the leaflets terminating in the notches. **Cicuta.**

Fruit elliptic-oblong. Leaves large, 3-foliolate, serrate. Involucels few-leaved. Calyx-teeth obsolete.
Cryptotænia Canadensis, D. C.

308. Calyx-teeth obsolete. Leaves pinnate, with serrate leaflets. **Sium.**

Calyx-teeth subulate. The leaflets bristle-form.
Discopleura.

309. Fruit ovate-globose. Carpels with 5, sharply keeled ridges, and with single vittæ in the intervals. Involucels 1-sided, drooping, longer than the umbellets. Leaves twice- to thrice-ternate; leaflets pinnately divided. Calyx-teeth obsolete. **Æthusa Cynapium,** L.

Fruit elliptical or ovoid....................310

310. Flowers white. Leaflets broadly ovate or rhombic-ovate, dentate. Involucre many-bracted. Calyx-teeth small, or minute. **Ligusticum.**

Flowers yellow, or sometimes dark purple. Involucre none. Involucels few-bracted. Calyx-teeth obsolete, or short. **Thaspium.**

311. (290.) Inner face of the seed deeply hollowed out in the middle. Fruit twin, a double sphere. The carpels nearly kidney-form. Plant low, flowering early.
Erigenia bulbosa, Nutt.

Inner face of the seed deeply furrowed lengthwise. Calyx-teeth obsolete.........................312

312. Ribs bristly, the bristles pointing upward. Fruit linear-club-shaped, tapering below. Calyx-teeth obsolete.
Osmorrhiza.

Ribs not bristly............................313

313. Fruit ovate or ovoid.......................314

Fruit linear-oblong, narrowed at the apex. Ribs broad. Stems spreading, 6-18 inches long, somewhat pubescent. Calyx-teeth obsolete.
Chærophyllum procumbens, Lam.

314. Ribs prominent, undulate-crenulate. Involucre and involucels 3- to 5-bracted. Leaflets lanceolate, pinnatifid. Calyx-teeth obsolete. **Conium maculatum,** L.

Ribs almost none. Involucels short and bristle-form. Fruit somewhat twin. Leaflets narrowly linear. Root a cluster of small tubers.
Eulophus Americanus, Nutt.

---

315. (284.) Calyx 2-cleft, half superior. Style 3- to 6-cleft or parted. Fleshy herb, prostrate, very smooth. Leaves

obovate, or wedge-shaped. Flowers sessile. Sepals keeled. Petals pale yellow. Stamens 7–12, rarely more.
(206) **Portulaca.**
Calyx 5-toothed or cleft........................316

316. Thorny shrubs. Styles 1–5..................56
Herbs, or unarmed shrubs and trees..........317

317. I. Pistil one. Flowers diœciously polygamous (perfect flowers intermixed with pistillate and staminate ones). Trees. Fruit a drupe..............................859
 II. Pistil one. Flowers perfect. Herbs. Pod short, 2-beaked. Ovary half-inferior; styles 2. Stamens 10. Leaves heart-shaped and 3- to 5-lobed, or reniform and crenate.
(192) **Mitella.**
 III. Pistils 2, separate or partly united.......318
 IV. Pistils 5. Stamens 10–20. Calyx-tube narrow. Petals slender, slightly unequal. Flowers in panicled corymbs, pale rose-color or white. Leaves trifoliolate.
Gillenia.

318. Shrubs with simple, ovate, rarely heart-shaped, opposite, petioled leaves.............................278
Herbs with compound leaves 319

319. Calyx-tube beset with hooked bristles; top-shaped. Stamens 12–15. Leaves interruptedly pinnate. Flowers yellow, in slender spikes or racemes. **Agrimonia.**
Calyx-tube naked. Stamens 10. Pod 2-celled, 2-horned, or 2, almost separate follicles. Leaves simple, entire, toothed, lobed, or cleft, often radical. (192) **Saxifraga.**

B. COROLLA GAMOPETALOUS.
(320–536.)

320. (78.) Ovary, or ovaries superior..............321
Ovary inferior............................507

321. Corolla regular, or nearly so; its lobes or segments perfectly or almost equal (in one genus each, with a little spur at the base)......................................322

Corolla decidedly irregular, often 2-lipped, and sometimes also spurred..................................434

322. Stamens fewer than the corolla-lobes (rarely with an additional rudimentary one) 323

Stamens as many as the corolla- or calyx-lobes..341

Stamens more numerous than the corolla-segments...418

323. I. Fertile stamen 1. A leafless, dull purplish, branching herb, parasitic on the roots of the beech, flower-bearing its whole length. Stamens 4, still usually all but one imperfect. Only the lower flowers fertile.
Epiphegus Virginiana, Bart. (800).

II. Stamens 2..................................324

III. Stamens 4 (3 of them usually imperfect in *Epiphegus;* see 323 I.).............................330

IV. Stamens 5, one of them rudimentary and sterile (sometimes only 2 perfect stamens). Calyx 5-cleft, 2- to 3-bracteolate at the base. Corolla bell-shaped, unequally 5-lobed (or somewhat 2-lipped), gibbous at the base. A strong-scented herb, usually decumbent. Leaves cordate, nearly orbicular. Flowers large, racemed. Pods with an incurved beak, which splits into 2 hooked horns, 4-celled.
Martynia proboscidea, Glox (450).

324. Herbs 325

Trees and shrubs.........................327

325. Ovary deeply 4-lobed, the style rising from between the lobes. Corolla 4-cleft. Leaves opposite, oblong, or oval-lanceolate, serrate, or sinuate-toothed. Flowers in dense axillary whorls, white. Stem 4-angled. **Lycopus.**

Ovary not 4-lobed........................326

326. Corolla wheel-shaped, 4-, rarely 5-parted. Pod sometimes notched at the apex, 2-celled. Leaves opposite, or whorled. Flowers blue, flesh-color, or white. **Veronica.**

Corolla slightly bell-shaped, somewhat irregularly 4-cleft. Pod 2 (rarely 3-) celled, 2-grooved (rarely 3-lobed). Radical leaves round-ovate; stem-leaves bract-like, clasping. Flowers greenish-yellow, in a spiked raceme.
Synthyris Houghtoniana, Benth. (446).

327. Corolla of 4, long, linear, acute, snow-white petals, barely united at the base. Stamens 2 (rarely 3 or 4)....100
  Corolla bell-shaped, or funnel-form 328

328. Fruit a very long and slender pod (often 1 foot long). Flowers large, showy. Only 2 of the 4 or 5 stamens fertile. Calyx deeply 2-parted. Corolla bell-shaped, 4 to 5-cleft, slightly 2-lipped. Leaves ovate, cordate, nearly entire, pubescent beneath, large. Flowers in large, terminal panicles. Corolla white, tinged with violet; throat dotted with purple and yellow. **Catalpa bignonioides**, Walt.
  Fruit a berry or a drupe...................329

329. Corolla funnel-form, its limb 4-cleft, longer than the calyx. Fruit a berry, 2-celled, 2- (or 1-) seeded, black.
  Ligustrum vulgare, L.
  Corolla short-bell-shaped, or salver-form; its limb 4-parted. Fruit a drupe, violet-purple, with a bony 2- or 1-seeded stone. Flowers polygamous (or diœcious), fragrant. Leaves coriaceous. **Olea Americana**, L. (853).

330. Ovary entire............................331
  Ovary deeply 4-lobed, the style rising from between the lobes. Herbs clammy-pubescent..................340

331. Ovary 2-celled. Style short, long, or thread-form. Pod many-seeded...................................332
  Ovary 4-celled. Style slender..............339

332. Woody vines, climbing either by tendrils or by rootlets. Ovary 2-celled. Pod 6–8 inches long. Seeds winged. Corolla 5-lobed, funnel-form, or bell-shaped. Flowers large, showy. Leaves compound..................333
  No vines..................................334

333. Climbing by tendrils. Corolla bell-shaped, slightly 2-lipped. Leaflets 2....................................460
  Climbing by rootlets. Leaves pinnately 5- to 11-foliolate; leaflets ovate, pointed, toothed. Corolla funnel-form, slightly irregular, orange and scarlet (2–3 inches long).
  Tecoma radicans, Juss.

334. A stemless herb. Leaves radical, fleshy, awl-shaped, or thread-form, long-petioled, clustered around the 1-flowered

scape, and longer than it. Flower small, whitish-flesh-color. Style short, club-shaped. Anthers confluently 1-celled.
Limosella aquatica, L.

Herbs with stems........................335

335. Leaves entire or only toothed. Calyx 5-toothed. Corolla nearly equally 5-lobed, purple, or blue..........337

Leaves pinnatifid, all of them, or the lower ones. Corolla nearly equally 5-lobed, yellow; throat woolly, as are the stamens..336

336. Corolla-tube short. Anther-cells not pointed at the base. **Seymeria macrophylla,** Nutt.

Corolla-tube elongated; the 2 upper lobes smaller and more united. Anther-cells awn-pointed at the base.
Dasystoma, Raf.
(*Gerardia*, L.)

337. Anthers 1-celled. Corolla salver-form, pubescent, its tube curved. Pod many-seeded. Leaves partly entire, partly sparingly and coarsely dentate, ovate-oblong, or lanceolate. Spike interrupted. **Buchnera Americana,** L.

Anthers 2-celled. Flowers pretty............338

338. Filaments hairy. The upper 2 corolla-lobes smaller, and more united. Pod many-seeded. Leaves linear. Flowers axillary. Peduncles 1-flowered. **Gerardia.**

Filaments not hairy. Anthers sagittate. Corolla almost regularly 5-cleft. Pod 8- to 12-seeded, stalked. Leaves ovate or elliptical, nearly entire. Flowers solitary, few or clustered in the axils, with a pair of (wing-like) leafy bracts. **Dipteracanthus.**

339. (331.) Herbs. Stamens included, anthers of the longer pair rarely tipped with a glandular appendage, or upper pair sometimes antherless. Corolla salver-form, 5-lobed, often with its tube curved; lobes slightly unequal. Fruit splitting into 4 nutlets. Flowers sessile, in simple or often panicled spikes, bracted. **Verbena.**

Shrub. Stamens exserted; anthers opening at the apex. Corolla tubular-bell-shaped, 5- (or 4-) lobed, lobes nearly equal. Fruit a small, violet-colored drupe, with 4 nutlets. **Callicarpa Americana,** L.

340. (330.) Lobes of the corolla all turned forward. Filaments capillary, curved, very long, exserted much beyond the corolla. Flowers bluish. Peduncles usually 1-flowered. Leaves linear or lance-oblong. **Trichostema.**

Lobes almost equally spreading. Stamens nearly included. Peduncles 1- to 3-flowered. Flowers pale blue. Leaves lance-oblong, 3-nerved. **Isanthus cæruleus,** Mx.

341. (322.) Stamens 4..........................342
Stamens 5 or more......................355

342. Ovary entire...............................343

Ovary deeply 4-lobed, forming 4 achenia in fruit. Corolla short-bell-shaped, 4-cleft; upper lobe broadest. Odorous herbs. Flowers small, pale purple or whitish, in axillary capitate whorls. **Mentha.**

343. Leafless herbs..............................344
Herbs with leaves......................345

344. Twining around other plants. Calyx 4- or 5-cleft. Corolla globular-urn-shaped, or campanulate, or funnel-form. Stamens 4, or 5. Flowers cymose-clustered. **Cuscuta** (392).

Not twining. Root-parasites, slender, much-branched, purplish, or yellowish-brown, 6–12 inches high. 329

345. I. Ovary 1-celled. Herbaceous plants.........347
II. Ovary 2-celled........................354
III. Ovary 4-celled. Fruit a drupe, with 4 nutlets. Shrubs..................................346

346. Style 1. Calyx 5- (or 4-) toothed............339
Styles 4...............................130

347. Apparent calyx consisting of 3 bractlets under the ovary. Apparent corolla-tube (the real calyx) coherent with the ovary. Flowers in cylindrical, terminal spikes. Leaves alternate, pinnate...................................700

Ovary free in the corolla. Leaves entire......348

348. Leaves opposite, or rarely whorled. Ovary with 2 parietal placentæ.......................................349

Leaves alternate. Ovary with a central placenta.

Corolla usually withering on the summit of the pod, which opens by a lid. Stems 2-5 inches high. Leaves ovate, oblong, or spatulate-oblong. Flowers nearly sessile, small, white, axillary. **Centunculus minimus**, L. (365).

349. Style distinct and slender, with a capitate, or 2-lipped stigma, deciduous. Anthers spiral. Corolla funnelform, or salver-shaped, 4- (or 5-) cleft, in withering twisted on the pod. Low herbs, 2-3 inches high. Flowers rose-purple or reddish. **Erythræa** (375).

Style (if any) and stigmas persistent. Anthers straight...350

350. Corolla with a glandular spot or a little, hollow spur to each lobe..351

Corolla without glands or spurs.............352

351. Corolla 4-parted, wheel-shaped, with a glandular and fringed pit on the upper side of each segment. Filaments awl-shaped, usually monadelphous at the base. Stigma 2-lobed. Leaves whorled. Flowers greenish-yellow, numerous, pedicelled, in open cymes.
Frasera Caroliniensis, Walt.

Corolla 4- (or 5-) cleft, bell-shaped, and with a little spur at the base of each lobe. Stem leafy, 9-18 inches high, simple or branched above. Leaves 3- to 5-nerved, lance-oblong or oblong-spatulate. Flowers yellowish, in terminal fascicles. **Halenia deflexa**, Griseb. (376).

352. Calyx of 2, leaf-like sepals. Corolla tubular-bell-shaped, 4-lobed. **Obolaria Virginica**, L.

Calyx of 4 or 5 segments...................353

353. Calyx and corolla 4-parted, the latter with no plaits at the sinuses. **Bartonia**.

Calyx 4- (or 5-) cleft. Lobes of the corolla sometimes fringed or toothed; its sinuses often with a crown of plaited folds; its bottom sometimes bearing glands between the bases of the filaments; anthers either separate or cohering with each other, and either fixed by the middle or by the deep sagittate base. Seeds winged or wingless. Flowers solitary or cymous. **Gentiana** (376).

354. (345.) Herbs with scapes. Stamens much exserted.

Corolla membranaceous, withering. Flowers whitish, small, in a bracted spike, or head, on a naked scape.
Plantago (878).

Herb with a stem, diffusely much branched, forming patches. Stamens included. Corolla broadly bell-shaped. Flowers small, white, sessile, cymose. Leaves linear, subulate, opposite.
Polypremum procumbens, L.

355. (341.) I. Ovary 1, entire.............. 356
  II. Ovary 1, deeply 4-lobed, with the style rising from between the lobes, forming 4 achenia in fruit......377
  III. Ovaries 2................................410

356. Ovary 1-celled..............................357
  Ovary 2- to 6-celled.......................385

357. Leaves submersed or floating. Aquatics......358
  Leaves not submersed......................359

358. Leaves round-heart-shaped, on very long petioles, which bear an umbel of polygamous flowers, and often a cluster of roots near the summit. Corolla wheel-shaped. Two parietal placentæ.
Limnanthemum lacunosum, Griseb.

Leaves pectinately dissected into filiform segments, scattered on the floating and rooting stems, and crowded at the base of clustered peduncles, the latter much inflated between the joints. Calyx with linear lobes. Corolla salver-shaped. Stamens included. A central placenta.
Hottonia inflata, Ell.

359. Stemless plants. Flowers on scapes. Leaves all radical...360
  Plants with leafy stems...................363

360. Leaves trifoliolate, at the top of long petioles, which are stipulate (membranaceous and sheathing) at the base, springing from a thickish, creeping root-stock. Throat and inner surface of the lobes of the short funnel-form corolla white-bearded. Flowers racemed on the naked scape.
Menyanthes trifoliata, L.

  Leaves simple and entire...................361

361. Stamens exserted; filaments short, united at the base. Corolla reflexed, 5-parted. Leaves oblong-spatulate. A simple, naked scape, bearing an ample umbel of purple, rose-colored, showy, usually nodding flowers.
Dodecatheon Meadia, L.

Stamens included. Corolla salver-shaped, or funnel-form..362

362. Corolla open at the throat. The simple scapes bearing an umbel of lilac or flesh-colored flowers.
Primula.

Corolla contracted at the throat, short. Flowers small, white. Scapes 2–4 inches high, many-flowered.
Androsace occidentalis, Pursh.

363. Leaves all alternate, entire, or partly alternate and palmately or pinnately parted......................364

Leaves, opposite, or whorled and entire........370

364. Leaves entire, and all alternate. Placenta free, central..365

Leaves lobed or parted. Two parietal placentæ. Ovary hairy, few- (rarely several-) seeded. Herbs pubescent.
366

365. Ovary half superior; calyx adherent to its base. Corolla somewhat bell-shaped, with 5 sterile filaments in the sinuses of the limb (besides the fertile). Pods opening by 5 valves. Flowers small, white, in simple or panicled racemes. Leaves obovate. Bractlets on the middle of the slender pedicels. **Samolus Valerandi**, L. (523)

Ovary perfectly free. Corolla wheel-shaped, 5- (or 4-) cleft. Pod opening by a circumscissile line, the top falling away like a lid. (348) **Centunculus minimus**, L.

366. Corolla-lobes convolute in the bud. Calyx constantly, or often with a tooth or other appendage in each sinus. Flowers white, or blue......................367

Corolla-lobes imbricate (quincuncial) in the bud. Sinuses of the calyx naked. Flowers white, blue, or purple.
368

367. Stamens exserted. Flowers in forked scorpoid racemes. Calyx naked, or with minute appendages at the sinuses. Leaves large, either pinnately parted into 5-13 divisions, or palmately 5- to 7-lobed. **Hydrophyllum.**

Stamens included. Flowers solitary, opposite the leaves. Leaves parted into 3 to 5 wedge-obovate, sparingly toothed segments.
Nemophila microcalyx, Fisch & Meyer.

368. Flowers in 1-sided racemes. Leaves pinnately parted into 3-9 divisions. Filaments exserted. Flowers blue, purple, or white..............................369

Flowers solitary. Calyx triangular, and in fruit much enlarged. Stamens included. Leaves pinnately parted into 7-13 lanceolate or linear-oblong, cut-toothed divisions. Peduncles 1-flowered, in the forks, or opposite the leaves. Corolla whitish. **Ellisia Nyctelea,** L.

369. Lobes of the corolla fringed. **Cosmanthus,** Nolte.
(*Phacelia,* Juss.)

Lobes of the corolla entire. **Phacelia.**

370. A central free placenta......................371
Two parietal placentæ........................374

371. Pod opening by valves or teeth..............372

Pod opening by a circumscissile line, the top falling off, like a lid. Corolla longer than the calyx, 5-parted, with almost no tube, scarlet, purple, blue, or white. Petals fringed with minute teeth. Filaments bearded. Leaves ovate, sessile, shorter than the axillary solitary peduncles. A low, procumbent annual. **Anagallis arvensis,** L.

372. Corolla mostly 7- (5-12-) parted, almost polypetalous. Filaments united in a ring at the base; anthers revolute after flowering. Stem with a whorl of lanceolate leaves at the summit. Alternate scales on the stem. Peduncles one or more, very slender, bearing a star-like, white flower.
(137) **Trientalis Americana,** Pursh.

Corolla 5 or 6-parted........................373

373. A little tooth between each of the 5 or 6 linear-lanceolate segments of the corolla. Pod few-seeded. Filaments exserted, distinct. Leaves lanceolate, dotted with purple glands. Flowers yellow, small, in axillary, peduncled spikes or racemes. Marsh herbs.
(143) **Naumburgia thyrsiflora**, Reichenb.

Teeth none in the sinuses of the corolla-limb. Pod few- to many-seeded, 5- to 10-valved. Filaments often united in a ring, mostly with the rudiments of a sterile set between the fertile filaments, at the base. Leaves opposite, or in whorls of 4, 5 (rarely 3 or 6), with or without dots. Flowers axillary, or racemed, mostly yellow. **Lysimachia.**

374. Style distinct and slender, deciduous. Anthers erect, curved or spiral. Flowers rose-color or pink......375

Style, if any, and stigmas persistent. Anthers straight. Flowers blue, purple, yellowish, or greenish-white. 376

375. Corolla wheel-shaped, 5- to 12-parted. Stamens 5-12. Anthers soon recurved. Style 2-parted. **Sabbatia.**

Corolla funnel-form or salver-shaped, 5- (or 4-) lobed, twisted and withering on the pod. Anthers spiral. Style not parted; stigma capitate or bilabiate.
(349) **Erythræa.**

376. Corolla with a small spur at the base of each lobe.
(351) **Halenia deflexa**, Griseb.

Corolla without spurs, mostly with plaited folds at the sinuses of its border. (353) **Gentiana.**

377. (355.) Corolla-limb somewhat unequally and obliquely 5-lobed; lobes rounded; tube bell-shaped; throat expanded, open and naked. Stamens mostly exserted. Style filiform. Nutlets rough or wrinkled. Leaves lanceolate, sessile. Flowers reddish-purple, or blue, in spikes or panicled racemes. **Echium vulgare**, L.

Corolla-limb equally 5-lobed..................378

378. Nutlets prickly. Style lateral. Corolla-throat closed by very conspicuous scales.....................379

Nutlets not prickly, fixed by their base, separate from the style. Coarse herbs. Corolla-throat closed by convex scales or folds, or rarely naked..................380

379. Nutlets erect, tuberoled at the back, their margins beset with barbed prickles. Corolla salver-formed. Flowers small, blue, in bracted racemes.
Echinospermum Lappula, L.

Nutlets oblique or depressed, roughened all over with short, barbed or hooked prickles. Corolla funnel-form. Herbs with a strong, unpleasant scent. Flowers blue, purple, or white, in bracted or bractless racemes.
Cynoglossum.

380. Nutlets hollowed out at the base. Throat of the corolla closed with 5 scales; its tube enlarged, and often much incurved. 381

Nutlets not hollowed out at the base. Throat of the corolla naked, or with folds rather than scales; its tube straight..382

381. Nutlets rough-wrinkled. Scales in the throat of the corolla blunt, oval, hairy; the tube funnel-form, much incurved. **Lycopsis arvensis, L.**

Nutlets smooth. Scales in the throat of the tubular-bell-shaped and nearly straight corolla awl-shaped.
Symphytum officinale, L.

382. Nutlets fleshy, when fresh, at length wrinkled. Corolla purple or blue, rarely whitish; tube cylindric; lobes rounded; throat open, naked, or with small glandular folds. **Mertensia.**

Nutlets not fleshy. Corolla rarely blue.......383

383. Anthers sagittate or linear-oblong, mucronate, sessile in the naked throat of the corolla. Style much exserted. Calyx-lobes linear. Corolla-limb ventricose; its lobes converging. Nutlets bony. **Onosmodium.**

Anthers not sagittate. Corolla with the throat closed by folds, and the tube funnel-form, or salver-shaped.
384

384. Calyx 5-parted. Nutlets bony. Racemes bracted. Flowers white, or yellow. Corolla-lobes imbricated in the bud. **Lithospermum.**

Calyx 5-cleft. Nutlets not bony. Racemes bractless. Flowers mostly blue, rarely white. Corolla-lobes convolute in the bud. **Myosotis.**

385. (356.) Fruit separating into 4 or 2 nutlets. Flowers in scorpoid spikes, blue or purple. Herbs pubescent. 386
Fruit not splitting into nutlets............... 387

386. Fruit separating into 4 nutlets. Tube of the corolla cylindrical; throat open. Flowers white, or purple. Leaves oval, long-petioled. **Heliotropium Europæum,** L.

Fruit separating into two nutlets, each 2-seeded (that is, 4 nutlets in pairs). Leaves petioled, ovate, or somewhat heart-shaped. Fruit 2-cleft, mitre-shaped before splitting. Flowers purple, or blue.
Heliophytum Indicum, D.C.

387. Fruit a drupe or a berry.................... 388
Fruit a pod (rarely twin).................. 392

388. Trees or shrubs. Fruit a drupe, sometimes berry-like.. 389
Herbs. Fruit a berry....................... 390

389. Corolla 5-cleft, with a row of 10 narrow appendages on the edges of the lobes. Stamens 5, with sagittate anthers, opposite the corolla-lobes, alternate with 5 petaloid sterile stamens. Ovary 5-celled. Drupe ellipsoid, 1-seeded. Spiny trees and shrubs. **Bumelia.**

Corolla of 6 segments. Stamens 6. Drupe berry-like, with 6 nutlets. Flowers perfect, or, by abortion, imperfect. Shrubs. **Prinos** (862).

390. Corolla wheel-shaped, 5-parted, or cleft. Anthers connivent, and opening by pores or chinks at the apex. Berry usually 2-celled. Leaves sometimes prickly along the midrib. **Solanum.**

Corolla bell-shaped or bell-funnel-form, somewhat 5-lobed, or entire, plaited in the bud. Anthers separate and opening lengthwise. Calyx enlarged and bladdery in fruit, enclosing the berry. Leaves sharply toothed, or sinuate-angled... 391

391. Calyx 5-cleft, reticulated in fruit. Berry juicy, 2-celled. The plaited border of the greenish-yellow corolla somewhat 5-lobed or toothed. Anthers blue, or yellow. The 1-flowered peduncles extra-axillary, nodding. **Physalis.**

Calyx 5-parted, the segments arrow-shaped. Berry

dry, 3- to 5-celled. Corolla-limb nearly entire. Flowers pale blue, solitary, on axillary and terminal peduncles.
Nicandra physaloides, Gærtn.

392. Leafless, parasitic herbs, yellowish or reddish, with thread-like stems, beset with minute scales, twining around other plants. Corolla globular-urn-shaped, bell-shaped, or somewhat tubular, the spreading limb 5- (rarely 4-) cleft. Flowers small, commonly white, in clustered, cymose glomerules. (344) **Cuscuta.**

Plants with green leaves..................393

393. Leaves alternate (sometimes crowded).........394
Leaves opposite, at least most of them (sometimes crowded)...405

394. Anther-cells opening by terminal pores, or by a transverse line...395
Anther-cells opening lengthwise..............396

395. Prostrate, creeping herbs, with crowded, evergreen leaves. Anthers opening by a transverse line, awn-pointed at the base. Flowers solitary and sessile, very numerous, white, or rose-color. **Pyxidanthera barbulata,** Mx.

Erect shrubs. Flowers showy, rose-color, pink, or whitish, often glandular-viscid, often fragrant. Stamens long, exserted; anthers opening by 2 terminal pores. Corolla funnel-form, 5-lobed, slightly irregular; the lobes spreading. Calyx often minute. Pod 5-celled, 5-valved, many-seeded. Shrubs. **Azalea.**

396. Leaves pinnate; leaflets 7–11. Ovary 3-celled. Style 3-lobed. Flowers few, in corymbs, blue, nodding. Filaments hairy-appendiculate at the base.
Polemonium reptans, L.

Leaves simple..........................397

397. All the 5 filaments, or the 3 upper ones, woolly. Corolla wheel-shaped, 5-lobed; the lobes rounded. Pod globular, 2-celled, the placentæ in the axis. Flowers in large, terminal racemes, ephemeral. Leaves sessile, or decurrent.
Verbascum.

Filaments not woolly. Corolla often plaited...398

398. Pod opening by a lid, and enclosed in the urn-shaped calyx. Corolla with an oblique and short tube, the border somewhat unequal and plaited, dull yellowish, strongly reticulated, with purple veins. Clammy-pubescent, fetid, narcotic. Leaves sinuate-toothed, or angled. Flowers sessile in 1-sided spikes. **Hyoscyamus niger,** L.

Pod opening lengthwise....................399

399. Calyx prismatic, 5-toothed, separating transversely above the base in fruit, the upper part falling away. Corolla funnel-form, with a large, 5- to 10-toothed, plaited, spreading border. Stigma 2-lipped. Pod globular, prickly, 2- (or rather imperfectly 4-) celled, 4-valved. Rank, narcotic, poisonous weeds. Leaves ovate, angular-toothed. Flowers large, showy, white, in the forks of the stem.

Datura Stramonium, L.

Calyx not prismatic.......................400

400. Pod 2- to 6-seeded, globular. Herbs twining, trailing, creeping, or prostrate............................401

Pod many-seeded. Plants erect, or rarely procumbent. Calyx tubular-bell-shaped, 5-cleft. Corolla funnel-form, or salver-shaped, yellowish-green, the plaited border 5-lobed. Pod 2-celled. **Nicotiana rustica,** L.

401. I. Styles 2, or rarely 3. Stigma depressed-capitate. Leaves linear or linear-lanceolate. Peduncles longer than the leaves, 1- to 3-flowered. Corolla white, slightly downy outside. **Stylisma evolvuloides,** Choisy.

II. Styles united into 1 to far above the middle. Leaves narrowly-linear. Bracts resembling the leaves, equalling the flower. **Stylisma Pickeringii,** Gray.

III. Styles united. Leaves cordate, sagittate, hastate, fiddle-shaped (or sometimes pinnatifid)........402

402. Calyx enclosed by 2 broad, leafy, mostly cordate bracts. Style with 2 linear-oblong stigmas. Pod half 2-celled. Leaves heart-shaped, or sagittate. Peduncles axillary, 1-flowered. Corolla white, sometimes tinged with **Calystegia.**

Calyx not enclosed by bracts...............403

403. Stamens exserted. Corolla cylindrical-tubular, with a spreading border. Stigma capitate-2-lobed. Pod 4-celled;

cells 1-seeded. Peduncles 1-, or about 5-flowered. Corolla red or crimson. Leaves cordate, or pinnatifid.
Quamoclit.

Stamens included..........................404

404. Stigmas 2, linear, often revolute. Corolla bell-shaped. Pod 2-celled, the cells 2-seeded. Stems twining, or pcorumbent. Leaves ovate-oblong, arrow-shaped. Peduncles usually 1-flowered. Corolla white, or tinged with red.
Convolvulus arvensis, L.

Stigma capitate, often 2- to 3-lobed. Corolla bell-shaped, white or purple. Pod 2- to 3-celled, cells 2-seeded. Leaves heart-shaped, fiddle-shaped, sometimes 3-lobed. Peduncles 1- to 5-flowered.
Ipomœa (including **Pharbitis**, Choisy).

405. (393.) Leaves connected by stipules...........406
Leaves not connected by stipules............407

406. Corolla narrowly funnel-form, 4 times longer than the calyx, crimson outside, yellowish within. Style 1. Pod twin, the 2 cells few-seeded. Flowers large, spiked in 1-sided cymes. Leaves sessile, ovate, or lanceolate, acute.
Spigelia Marilandica, L.

Corolla tubular, short, hairy in the throat. Ovary (and pod) mitre-shaped, or 2-beaked, the 2 short styles separate below, but at first united at the summit. Seeds many. Flowers small, white, spiked along one side of a terminal, stalked cyme. Leaves petioled, thin, oblong-lanceolate.
Mitreola petiolata, Torr. & Gr.

407. Calyx of 5 sepals. Anther-cells opening by an obliquely transverse line. Flowers white, peduncled, 3-bracted under the calyx. Low evergreens, occurring in convex tufts. **Diapensia Lapponica, L.**

Calyx 5-cleft, or parted. Anther-cells opening lengthwise 408

408. Style 3-lobed. Calyx narrow, somewhat prismatic, plaited or angled. Corolla salver-form, with a long tube. Flowers cymose, terminal, or in the upper axils, pink-purple, or whitish. **Phlox.**

Style not 3-lobed..........................409

409. A dwarfy evergreen, much branched and tufted. Pod 3- (or 4-) celled. Flowers small, white or rose-color, 2–5 in a cluster, from a terminal, scaly bud. Leaves coriaceous, elliptical. **Loiseleuria procumbens,** Desv.

A twining, shrubby plant. Anthers sagittate. Pod 2-celled. Leaves ovate, or lanceolate, shining, short-petioled. Flowers large, showy, yellow, fragrant, 1–5 together in the axils. Corolla open, funnel-form, somewhat oblique. Style short, with 2 stigmas, each 2-parted.
Gelsemium sempervirens, Ait.

410. (355.) Styles and stigmas of the 2 ovaries separate. Pods utricular, 1- to 2-seeded, distinct. Stigmas thick. Creeping herbs. Leaves alternate, round-kidney-shaped, pubescent. Peduncles axillary, 1-flowered. Corolla small, greenish-white. **Dichondra repens,** Forst.

Stigmas and sometimes styles united into one. Leaves entire. Pods elongated 411

411. Pollen in ordinary grains. Filaments distinct. Anthers sometimes somewhat coherent with the stigma.
412

Pollen in masses. Anthers always united with the stigma. Filaments commonly monadelphous, rarely distinct. Pods often with warty projections. Seeds with a silky tuft. Leaves opposite, whorled, rarely scattered.
414

412. Seeds comose. Leaves opposite.............413

Seeds naked. Leaves alternate. Corolla-tube bearded inside, and at the summit also outside. Anthers longer than the filaments. Leaves short-petioled, ovate-lanceolate. Flowers pale blue, in terminal panicled cymes.
Amsonia Tabernæmontana, Walt.

413. Corolla bell-shaped, with 5 triangular appendages in the throat. Filaments short, broad, flat. Style none; stigma large, ovoid, slightly 2-lobed. Leaves ovate or oblong. **Apocynum.**

Corolla funnel-form, not appendiculate. Filaments slender. Calyx with 3–5 glands at its base inside, its lobes taper-pointed. Anthers arrow-shaped, with an inflexed tip, adherent to the stigma. Twining, slightly woody plants. Leaves oval-lanceolate. Flowers pale yellow, in cymes.
Forsteronia difformis, A. D. C.

414. Filaments monadelphous. Pollen-masses 10, fixed to the stigma in pairs.................................415

Filaments distinct, or nearly so. Pollen-masses 5, granular, separately attached to the stigma. Corolla wheel-shaped, with 5-awned scales in the throat. Anthers bearded on the back. Twining, shrubby plants. Leaves ovate, or ovate-lanceolate. Flowers brownish-purple, in panicled cymes. Corolla-lobes linear-oblong, very hairy above.
Periploca Græca, L.

415. Pollen-masses pendulous and vertical..........416

Pollen-masses horizontal. Corolla wheel-shaped, sometimes reflexed-spreading. Crown a wavy-lobed, fleshy ring. Anthers horizontal, partly hidden under the flat stigma. Pods ribbed, or with soft prickles. Twining, herbaceous, or shrubby plants. Leaves heart-shaped, usually hairy. Flowers dark purple, in extra-axillary umbels.
Gonolobus.

416. Calyx- and corolla-lobes reflexed, or spreading. Crown of 5 hooded-fleshy bodies. Flowers in simple, many-flowered umbels. Corolla greenish, white, yellowish, purple, or pink...417

Calyx and corolla erect. Crown of 5 membranaceous bodies, each terminated by a 2-cleft tail. A twining herb. Leaves ovate-cordate, long-petioled. Flowers small, whitish, in racemose-umbellate clusters on axillary peduncles. **Enslenia albida**, Nutt.

417. Hooded bodies with an incurved horn in the cavity of each, from its base curved toward the stigma. Pods ventricose, sometimes with warty or spinous projections, one of the 2 often abortive. Herbs upright. Leaves rarely scattered. Umbels on terminal, or mostly lateral peduncles. **Asclepias.**

Hooded bodies without a horn. Leaves obovate, lanceolate, or linear. Umbels peduncled or nearly sessile.
Acerates.

418. (322.) Leaves twice-pinnate. Ovary 1-celled. Calyx 5-toothed, minute. Stamens 10–12, distinct or slightly united at the base. Pod long, narrow, prickly. Flowers rose-color, small, in axillary, peduncled, rounded heads. Herbs, with the procumbent stems and petioles prickly.
Schrankia.

Leaves simple, often entire, sometimes none. Ovary 3- to 8-celled. Corolla 5- (rarely 4-) lobed.......419

419. Fruit a drupe, or a berry, or drupe- or berry-like. 420

Fruit a pod............................423

420. Fruit a drupe, or a berry. Filaments smooth, distinct. Anthers sometimes awned. Flowers perfect, or polygamous 421

Fruit drupe- or berry-like. Filaments either united into a tube at the base, or distinct and hirsute..........422

421. Fruit a drupe, with 5 firmly united nutlets, red, or black. Corolla ovate, or urn-shaped, with a short, revolute, 5-toothed limb. Style 1. Stamens 10, included. Anthers opening by terminal pores, and bearing 2 reflexed awns on the back. Flowers white, all alike, scaly-bracted, in terminal racemes or clusters. Trailing and depressed evergreens. Leaves obovate or spatulate. **Arctostaphylos.**

Fruit a berry, 8-celled, 8-seeded, fleshy, plum-like, yellow, when ripe. Calyx and leathery corolla 4-lobed. Styles 4, each 2-lobed. Flowers polygamous, greenish-yellow, the fertile usually solitary, and with 8 stamens, the sterile ones mostly clustered and 16-androus. Anthers extrorse. A small tree. Leaves ovate-oblong.
Diospyros Virginiana, L. (860).

422. Fruit coriaceous, drupe-like, 1-celled, 1-seeded. Calyx truncate, or slightly 5-toothed, coherent with the base of the 3-celled, many-seeded ovary. Corolla 5-parted. Filaments united at the base in a short tube. Flowers showy, white, drooping in axillary racemes. Shrubs.
Styrax (511).

Fruit an apparent berry, but really a 5-celled, many-seeded pod, enclosed by the fleshy, scarlet calyx. Calyx 5-cleft, with 2 bractlets at the base. Corolla ovoid-tubular, with 5 revolute lobes, white. Filaments distinct, hirsute. Anthers opening by terminal pores. Stem prostrate, mostly hidden under ground. Branches ascending, 3 inches high. Leaves obovate, mucronate, denticulate, crowded at the top of the stem-like branches. Flowers terminal, few, drooping. **Gaultheria procumbens**, L.

423. A tree, shrubs, or shrublets..................424
Leafless scaly-bracted herbs..................433

424. Corolla saucer-form, holding the anthers in 10 pouches. Leaves alternate, opposite, or in threes, oblong to lanceolate. Shrubs, 1–8 feet high. Flowers scattered, or in simple, or clustered umbel-like corymbs, rose-color, crimson, or lilac purple, often very showy. **Kalmia.**
Corolla without pouches..................425

425. Corolla salver-form, hairy inside. Calyx with 3 bractlets at the base. Anthers opening lengthwise. Trailing shrublet, bristly with rusty hairs. Leaves rounded, or heart-shaped, reticulated, alternate, long-petioled. Flowers rose-color, in small, axillary clusters, fragrant.
Epigæa repens, L.
Corolla of another shape. Anthers opening by terminal pores or chinks..................426

426. Corolla bell-shaped, slightly unequal, or regular; its lobes large and spreading. Calyx minute. Stamens exserted, commonly declined. Pod 5-celled, 5-valved. Flowers large, showy, rose-color, or lilac-purple, in dense, terminal, umbel-like corymbs, from scaly buds. Low trees, shrubs, or dwarfish shrublets. (Pods opening septicidally.)
Rhododendron.
Flowers smaller. Corolla urn-shaped (ovoid, cylindric, or globular); lobes small, equal..................427

427. Pod opening loculicidally—that is, into the cells, so that the valves, after dehiscence, bear the partition on their middle..................428
Pod opening septicidally—that is, between the cells, so that the valves bear half the thickness of the partition on their margins. Flowers purple..................432

428. Leaves needle-shaped, imbricated. A moss-like, procumbent shrublet. Stems filiform, 1–4 inches long. Corolla globular. Anthers with 2 recurved awns. Flowers white or rose-color, solitary, on slender erect peduncles.
Cassiope hypnoides, Don.
Leaves ample..................429

429. A fine tree, 40–50 feet high. Leaves deciduous,

petioled, oblong-lanceolate, serrulate. Flowers white, 3 lines long, in terminal panicles of slender racemes. Pod oblong pyramidal. **Oxydendron arboreum**, D. C.

Shrubs. Anther-cells often awned...........430

430. Corolla cylindrical, white....................431

Corolla urn-shaped, globular, or ovoid-oblong. Anthers mostly awned, fixed near the middle. Flowers commonly white, umbelled, clustered, or in panicled racemes. Shrubs ½–10 feet high. **Andromeda.**

431. Calyx 5-parted, with 2 bractlets at the base. Anthers awnless. Pericarp of 2 layers—the outer 5-, the inner 10-valved. Leaves coriaceous, obscurely serrulate, ferruginous underneath, oval-oblong. Flowers in about 25-flowered, almost 1-sided, leafy terminal racemes. Shrub 2–4 feet high. **Cassandra calyculata**, Don.
(*Andromeda calyculata*, L.)

Calyx of 5 sepals, naked. Pericarp simple; valves 5. Anthers awned, or naked. Leaves petioled, serrulate, the teeth sometimes ciliate or spinulose. Flowers scaly-bracted, in dense, axillary, or terminal spikes, or racemes.
Leucothoe, Don.
(*Andromeda,* L.)

432. Flowers 4-parted, 8-androus, in terminal panicles, slender-pedicelled, nodding. Leaves deciduous, tipped with a gland, alternate. Shrub, 4 feet high.
Menziesia ferruginea, Smith.

Flowers 5-parted, 10-androus, at the end of the branches, pedicelled, nodding. Leaves evergreen, linear, obtuse, rough-margined. Procumbent, 6 to 10 inches long.
Phyllodoce taxifolia, Salisb.
(*Menziesia,* Robbins.)

433. (423.) Anthers 2-horned on the back, opening lengthwise. Calyx 5-parted. Corolla ovate, with 5 reflexed teeth, withering-persistent. Flowers in a loose raceme.
Pterospora Andromedea, Nutt.

Anthers opening at the top. Calyx of 5 erect scaly sepals. Corolla broadly bell-shaped, 5-lobed, slightly 5-gibbous at the base. Flowers in a short spike, exhaling the fragrance of violets. **Schweinitzia odorata**, Ell.

434. (321.) I. Stamens 10, distinct. Upper lip of the corolla 3-lobed, or cleft, lower 2-parted or of 2 distinct petals. Style slender, declined. Flowers rose-purple, 3–5, sessile in terminal clusters. A low shrub, 2–3 feet high. Leaves obovate-oblong, downy-canescent beneath, rather later than the flowers. **Rhodora Canadensis**, L.

    II. Stamens 2, 4, or 5, distinct..............437

    III. Stamens 6, 8, or 10, monadelphous, or united in 2 bundles..435

435. Corolla 1- or 2-spurred, or saccate at the base. Stamens 6, their filaments united into 2 bundles (apparently 2 broad filaments, each bearing 3, stalked anthers).....221

    Corolla not spurred, nor saccate at the base....436

436. Calyx of 5 sepals, the 3 outer ones smaller. Petals 3, placed between the 2 inner, larger sepals, connected with each other and the split tube of the 6–8 stamens, the middle one boat-shaped and often crested..............228

    Calyx tubular, 5-toothed. Corolla papilionaceous, the claws of the petals partly united with one another and with the filaments. Leaves trifoliolate. **Trifolium** (256).

437. I. Stamens, fertile ones, 2..................438

    II. Stamens, fertile ones, 3, and a fourth sterile, or even 2 sterile ones, see 328 and 450.

    III. Stamens 4 (fertile)......................452

    IV. Stamens 5 (all perfect in *Echium*, see 377, or one of them rudimentary in *Martynia*, see 450).

438. Lower corolla-lip spurred. Pod 1-celled......439

    Lower corolla-lip not spurred................440

439. Calyx 5-cleft, somewhat 2-lipped. Corolla 2-lipped; upper lip 2-, lower 3-lobed. Leaves simple, radical. Scape 1-flowered. In wet places, greasy to the touch.
                            Pinguicula vulgaris, L.

    Calyx of 2 sepals. Corolla personate, 2-lipped; upper lip bifid, lower longer. Aquatics. Leaves submersed, dissected into capillary divisions, bearing little bladders.
                                   Utricularia.

440. Ovary deeply 4-lobed, or 4 nutlets at the bottom of the calyx. Stems usually 4-angled...................441

    Ovary entire................................446

441. Calyx 2-lipped.............................442

Calyx regularly or nearly equally 5-lobed or toothed 445

442. Calyx hairy in the throat, ovoid, or tubular, gibbous near the base. Upper corolla-lip notched at the top. Fertile stamens 2, the upper pair reduced to sterile filaments, or wanting. **Hedeoma.**

Calyx not hairy in the throat................443

443. Calyx 13-nerved ; its upper lip with 3 awned teeth, the lower with 2 nearly awnless teeth. Flowers small, pale bluish-purple, crowded in axillary and terminal, globose, capitate whorls. **Blephilia** (482).

Calyx not 13-nerved.......................444

444. Upper lip of the yellow corolla consisting of 4, nearly equal, small lobes; lower lip undivided, long and fringed. Stamens much exserted, divergent. (Corolla exhaling the odor of lemons.) **Collinsonia Canadensis,** L.

Upper lip of the ringent corolla straight, scythe-shaped; lower spreading or pendent, 3-lobed. Connectile transversely jointed to the filament, bearing at each end a cell of the halved anther, one end ascending, the other descending. Flowers blue, or violet-purple. **Salvia.**

445. Flowers large, bluish, pale purple, or bright red; upper lip of the ringent corolla linear; lower lip reflexed and 3-lobed, middle lobe largest. **Monarda.**

Flowers small, pale red. Corolla pubescent; upper lip flattish, usually notched; lower spreading, 3-cleft. Herb delightfully fragrant. **Cunila Mariana,** L.

446. Leaves opposite..........................447

Leaves radical, roundish, petioled. Scapes with foliaceous, partly clasping bracts. Calyx 4-parted. Corolla slightly bell-shaped, either with the upper lip entire and a little longer and narrower than the lower, 3-toothed lip; or the upper lip notched and the lower 3-parted. Very rarely 2 additional lower stamens.

(326) **Synthyris Houghtoniana,** Benth.

447. A tree. Calyx deeply 2-parted. Corolla bell-shaped, 4- or 5-cleft, slightly 2-lipped, large. Pod long and slender. Seeds winged. Leaves ovate, cordate, ample..328

Herbs. Calyx 5-lobed, or 4-toothed..........448

448. Calyx 5-parted, or cleft.....................449

Calyx 4-toothed; segments shorter, than the very unequal corolla-lips. Upper lip of the corolla very short, entire; the lower 3-lobed, the middle lobe largest and spreading. A scale at the base of each filament. Herb a few inches long, branched. Leaves roundish-ovate, obscurely 3-nerved, sessile, crowded. Flowers minute, sessile, axillary, white. **Micranthemum micrantha,** Rich.
(*Hemianthus micranthemoides*, Nutt.)

449. The cells of each anther placed one lower down than the other. Pod 4-seeded. Leaves ensiform. Spikes oblong, dense, long-peduncled. Flowers purplish. Herbs erect, 2-3 feet high, with an angular, smooth stem. Borders of streams and ponds. **Dianthera Americana,** L.
(*Rhytiglossa pedunculosa*, Nees.)

Anther-cells parallel, or divergent. Pod many-seeded...450

450. Flowers very large. Corolla gibbous, bell-shaped, 5-lobed, somewhat 2-lipped, pale dull yellow, spotted with brownish purple. Fertile stamens 2, or 4. Pod fleshy, terminated by a long, incurved beak, 4-celled. Low, branching herbs, clammy-pubescent, with an unpleasant scent. Leaves cordate, entire or wavy. Seeds wingless.
(323) **Martynia proboscidea,** Mx.

Flowers smaller. Pods not beaked..........451

451. Flowers white or yellow. Sterile filaments short, or none. Upper lip of the corolla entire or slightly bifid, the lower 3-cleft. Leaves not clasping. The 1-flowered pedicels usually with 2 bractlets, near the calyx.
Gratiola.

Flowers purple. Sterile filaments exserted, forked, one of the forks glandular, or rarely with half an anther. Upper lip of the corolla short, erect, 2-lobed, the lower longer and spreading, 3-cleft. Upper leaves partly clasping.
Ilysanthes gratioloides, Benth.

452. (437.) Ovary entire......................453

Ovary deeply 4-lobed, forming in fruit 4 achenia.
481

453. Leafless root-parasites, thick and fleshy, bearing scales in place of leaves, lurid yellowish or brownish throughout. Flowers solitary, or spiked, or panicled..........454

Leafy plants..............................456

454. Flowers solitary, without bractlets. Corolla slightly 2-lipped. Upper lip 2-, lower 3-cleft. Herb glandular-pubescent.
          Aphyllon.

Flowers spiked, or panicled. Calyx with 2 bractlets. Corolla ringent; upper lip notched, lower 3-parted. Anthers sagittate....................................455

455. Stamens exserted. Calyx deeply cleft on the lower side. Corolla swollen at the base. Herb smooth.
    Conopholis Americana, Wallr.

Stamens included. Calyx nearly regularly 5-cleft. Corolla not swollen at the base. Herb glandular-pubescent.
    Phelipæa Ludoviciana, Don.
      (*Conopholis,* Wood.)

456. Ovules or seeds solitary in the cell, or cells of the ovary, or fruit.................................457

Ovules or seeds several or many, rarely few in the 2, rarely 4 cells of the ovary, or pod................458

457. Fruit reflexed and closely appressed to the rhachis, 1-celled, 1-seeded, dry, oblong, striate. Calyx 2-lipped, cylindrical; the upper, longer lip of 3 subulate, bristle-pointed teeth, the lower 2-toothed. Corolla with the upper lip notched, the lower 3-lobed, much larger. Leaves large, ovate-oblong, coarsely toothed, 3–6 inches long, short-petioled. Flowers spiked.
    Phryma leptostachya, L.

Fruit in axillary, peduncled heads, a dry, 2-celled, 2-seeded drupe. Calyx membranaceous, enclosing the fruit. Leaves oblanceolate, serrate above. Corolla bluish-white; upper-lip notched, lower much longer and 3-lobed. Procumbent and creeping. Flowers small, in heads, raised on slender axillary peduncles. **Lippia lanceolata,** Mx.

458. Pod 2-celled............................459

Pod 4-celled, with a long beak, which is longer than the pod itself, and becomes 2-horned by dehiscence. Calyx 5-cleft. Corolla gibbous, bell-shaped, 5-lobed, slightly 2-lipped. Stamens 5, one sterile. Low, branching, glandular-hairy, strong-scented annuals, with large flowers. Leaves cordate.450

459. Placentæ parietal. Corolla scarcely 2-lipped, 2-3 inches long. Pods elongated. Seeds winged. Leaves compound. Woody vines..............................460

Placentæ central........................461

460. Leaves pinnately 5- to 11-foliolate. Corolla 2-3 inches long. Plant climbing by rootlets.
(333) **Tecoma radicans**, Juss.

Leaves of 2 ovate, or oblong leaflets, with a branched tendril between. Corolla bell-shaped, 5-lobed, and slightly 2-lipped, orange-colored, 2 inches long.
Bignonia capreolata, L.

461. Corolla with a spur or sac at the base, the throat with a palate. Pod opening by chinks or holes. Flowers in simple racemes, or axillary. Lower leaves usually opposite, or whorled..462

Corolla not spurred, nor saccate..............463

462. Corolla spurred at the base; the palate rarely closing the throat. Flowers yellow and large, or blue, purplish, and smaller; either racemed or scattered. Leaves either all or only the upper ones alternate, linear, lanceolate, ovate, or halberd-shaped. **Linaria.**

Corolla merely saccate at the base, the palate closing the throat. Flowers purplish, in loosely flowered spikes. Sepals longer than the corolla. Leaves lance-linear.
Antirrhinum Orontium, L.

463. Calyx conspicuously 2-lobed or lipped, the lobes or lips entire, or incised; or calyx cleft in front..........464

Calyx 4- or 5-toothed, cleft or parted........467

464. Calyx-tube 10-ribbed, oblique, the posterior of its 5 teeth much smallest, the two anterior united much higher, than the rest. Stem 2 feet high, pubescent, simple. Leaves sessile, lanceolate-ovate, or oblong, 3-nerved, with ciliate edges. Flowers dull purple, or brownish-yellow, in elongated spikes. **Schwalbea Americana**, L.

Calyx not 10-ribbed..........................465

465. Upper lip of the corolla linear, very long, arched and keeled, lower short, 3-lobed; tube included in the calyx. Anther-cells unequal, the outer fixed by the middle, the in-

ner pendent. Flowers dull yellow, nearly sessile in the axils of the scarlet, or bright yellow bracts, which are crowded near the summit of the stem. Leaves entire, or incised.
Castileja.

Upper corolla-lip not linear. Anther-cells equal.
466

466. Upper lip of the corolla vaulted, short-beaked at, or 2-toothed below the apex; lower erect, and nearly closing the throat. Pod ovate or lanceolate, mostly oblique, several-seeded. Leaves alternate, or somewhat opposite. Flowers large, pale yellow, in a spike. **Pedicularis.**

Upper lip of the corolla not vaulted; notched, or 2-cleft, the lower 3-lobed. Leaves entire, many-nerved, or nerveless, obovate, or roundish, commonly clasping. Flowers solitary, axillary, blue. **Herpestis** (479).

467. Calyx 4-toothed or cleft. Lower lip of the corolla covering the upper in the bud........................468

Calyx 5-toothed, cleft or parted..............472

468. Calyx 4-toothed, inflated, flat, ovate, broader than the ringent corolla. Upper lip of the corolla arched, minutely 2-toothed, lower lip 3-lobed, the middle lobe conduplicate. Seeds broadly winged, rattling, when ripe, in the large inflated calyx. Leaves oblong or lanceolate. Flowers yellow, solitary, axillary, sessile.
Rhinanthus Crista-galli, L.

Calyx 4-cleft, not inflated...................469

469. Calyx 10-ribbed...........................464

Calyx not 10-ribbed.........................470

470. Upper corolla-lip linear......................465

Upper corolla-lip not linear..................471

471. Seeds ribbed. Pod oblong, many-seeded. Flowers in spikes. Calyx tubular, or bell-shaped. Upper lip of the corolla scarcely arched, 2-lobed; lobes broad and spreading; lower lip 3-cleft, spreading, the lobes obtuse or notched. Leaves ovate or oblong, the cauline crenate, the floral cut-serrate, with cuspidate teeth. Herb 2 to 6 inches high. Flowers small, greenish-white. **Euphrasia officinalis,** L.

Seeds smooth. Pod flattened, oblique, 1- to 4-seeded. Flowers solitary, in the upper axils, small, yellowish-green.

Calyx tubular. Upper lip of the corolla arched, triangular-compressed, the margin folded back, the lower erect-spreading and biconvex, 3-lobed at the apex. Floral leaves nearly triangular, with 3 or 4 setaceous teeth at each edge, near the base; the rest of the leaves lanceolate or linear. Herb about 1 foot high. **Melampyrum Americanum**, Mx.
(*M. pratense*, Wood.)

472. Corolla deeply 2-lipped. Upper lip of the corolla covering the lower in the bud (with occasional exceptions in Mimulus)...473
Corolla scarcely 2-lipped..................480

473. The rudiment of a 5th stamen present, or 2 of the 4 stamens with globular bodies imitating anthers.......474
No rudiment of a 5th stamen; the 4 stamens always all fertile...................................477

474. The outer pair of stamens sterile and exserted, their forked filaments bearing either a glandular body, or half an anther on one of the branches................451
A rudiment of the 5th stamen present........475

475. The rudiment of the 5th stamen a filament....476
The rudiment of the 5th stamen a scale, at the summit of the tube of the corolla, the limb of which consists of 4 upper lobes and a lower lip. Stem 4-sided. Leaves oblong or lanceolate, coarsely serrate, rounded or cordate at the base. Flowers greenish-yellow, in a cymose panicle.
Scrophularia nodosa, L.

476. Seeds winged. Sterile filament shorter than the rest. Calyx with 3 bractlets. Stem upright, branching. Leaves lance-oblong, or lanceolate, short-petioled. Flowers white, tinged with rose-color, in spikes, closely imbricated with large, round-ovate, concave bracts and bractlets. Filaments woolly. **Chelone glabra**, L.
Seeds angular. Sterile filament equalling the regular stamens, and either naked or bearded. Stem branched. Leaves lanceolate, ovate, or rounded, clasping. Flowers purple, bluish, or white, in simple or compound racemes. **Pentstemon.**

477. Calyx prismatic, 5-angled. Corolla-tube long. Leaves lanceolate, oblong, or kidney-shaped. Flowers violet, purple, or yellow, on solitary, axillary peduncles.
Mimulus.

Calyx not prismatic, 5-parted..............478
478. Corolla with the middle lobe of the lower, 3-cleft lip saccate, the sac enclosing the declined stamens and style. Flowers in umbel-like clusters, in the axils of the upper leaves. **Collinsia.**

Corolla with the lower lip not saccate. Flowers not clustered in the axils.........................479
479. Calyx equally 5-parted. Leaves many-parted into linear-wedge-shaped divisions. The small, solitary, greenish-white flowers on axillary, 2-bracteated peduncles.
Conobea multifida, Benth.

Calyx unequally 5-parted, or 2-lipped, the upper lip broadest. Leaves entire, many-nerved, or nerveless, obovate, or roundish, usually clasping. Flowers solitary, axillary, blue. (466) **Herpestis.**
480. Corolla purple. Leaves linear or lanceolate, entire.
338

Corolla yellow. Leaves pinnatifid...........336

---

481. (452.) Stamens rarely 4; the upper pair shorter, or rudimentary..482

Stamens always 4, perfect..................483
482. Corolla pale blue; upper lip entire, lower 3-lobed. Flowers in dense, axillary, or nearly terminal verticils.
(443) **Blephilia.**

Corolla yellow; upper lip of 4, very short, almost equal lobes; lower lip long, fringed. Flowers in a long, terminal, leafless raceme, or panicle.
(444) **Collinsonia Canadensis**, L.
483. Upper or inner pair of stamens longer, than the lower, outer. None of the filaments appendiculate at the base..484

Upper or inner pair of stamens shorter, than the lower (or in *Phlomis*—see 506—often longer, with their filaments appendiculate at the base)....................488

484. Stamens diverging, exserted, the upper pair curved downward, the lower ascending. Anther-cells nearly parallel. Tall perennials. Leaves petioled, serrate. Flowers numerous, purple or greenish-yellow, in interrupted terminal spikes. Middle lobe of the lower corolla-lip crenate.
Lophanthus.

Stamens all ascending..................485

485. Anther-cells divergent; anthers approximate in pairs..................................486

Anther-cells parallel. Flowers large. Corolla hairy inside, about 1 inch long; upper lip flattened, or concave, 2-lobed, the lower trifid, middle lobe largest. Herbs hairy, about 1 foot high. Leaves cordate, obtusely crenate. Flowers purplish, in approximate whorls at the summit of the stems. **Cedronella cordata,** Nutt.

486. Upper lip of the corolla rather flat, notched. Anther-cells much divergent.........................487

Upper lip of the corolla vaulted, notched. Upper tooth of the 5-dentate, 13- to 15-nerved calyx usually much largest. Anther-cells little divergent. Leaves petioled, lanceolate, serrate. Bracts leafy, awn-toothed or fringed. Flowers small, bluish, verticillate, forming a terminal head or spike. **Dracocephalum parvifolium,** Nutt.

487. Anthers approximate, forming a cross, the cells diverging at a right angle. Corolla bluish, thrice the length of the calyx; its lower lip flat. Herbs creeping and trailing. Leaves round-kidney-shaped, crenate, petioled. Flowers few, in loose axillary verticils. **Glechoma,** L.
(*Nepeta glechoma,* Benth.)

Anthers not forming a cross. Filaments after flowering bent aside. Corolla whitish, dotted with purple, its lower lip very concave. Leaves cordate, oblong, crenate, whitish downy underneath. Verticils dense, forming interrupted spiked racemes. **Nepeta cataria,** L.

488. Stamens divergent (usually exserted)..........489

Stamens parallel.........................493

489. Calyx 15-nerved. Stamens exserted. Upper lip of the corolla erect, flat, obscurely notched, the lower 3-cleft, middle, obcordate lobe largest. Herbs growing in tufts.

Leaves lanceolate, entire, sessile. Flowers blue, in racemose, secund verticils. **Hyssopus officinalis**, L.

Calyx 10- to 13-nerved.....................490

490. Calyx-throat hairy.........................491

Calyx-throat naked..........................492

491. Calyx erect in fruit, 5-toothed. Bracts large, roundish, tinged with purple. Stamens exserted. Leaves round-ovate, petioled. Flowers purplish, in dense, oblong, or cylindrical spikes. **Origanum vulgare**, L.

Calyx nodding in fruit, 2-lipped; upper lip 3-toothed, spreading, lower bifid, with the subulate segments ciliate. Bracts narrow, minute. Herbs prostrate. Leaves oval or ovate, short-petioled. Flowers purplish, or whitish, capitate, or verticillate. **Thymus Serpyllum**, L.

492. Anther-cells fixed by a triangular connective. Calyx 10-nerved. Stamens connivent, arching, scarcely exserted. Herbs much branched. Leaves lance-linear, acute. Corolla lilac, or white, the throat dotted with purple. Flowers about 5 together, in axillary cymes. **Satureja hortensis**, L.

Anther-cells parallel. Stamens exserted, or included; upper pair scarcely shorter than the lower. Calyx about 13-nerved. Corolla whitish or purplish, the lips dotted with purple. Floral leaves often whitened. Verticils dense, with crowded bracts, usually forming terminal heads or close cymes. **Pycnanthemum**.

493. Stamens exserted through a cleft in the upper side of the tube of the corolla. Upper corolla-lip consisting of 4, small, nearly equal lobes, which are turned forward; lower lip truncate, larger and roundish. Leaves short-petioled, ovate-lanceolate, serrate. Flowers pale purple, in a long wand-like spike. **Teucrium Canadense**, L.

Stamens ascending in pairs beneath the upper lip. 494

494. Calyx 13-nerved, 5-toothed, and somewhat 2-lipped...495

Calyx 5- to 10-nerved, or irregularly netted, sometimes 2-lipped..497

495. Tube of the white or cream-colored corolla curved upward. Calyx 2-lipped; upper lip 3-toothed; tube curved. Leaves broadly ovate, coarsely crenate, exhaling, when bruised, the odor of lemons. Clusters few-flowered, loose, secund, with a few leaf-like, ovate bracts.
Melissa officinalis, L.

Tube of the corolla straight, or nearly so; throat inflated 496

496. Calyx gibbous below. The sessile globular clusters of pale purple flowers crowded with awl-shaped bracts, which are as long as the calyx. Bracts and calyx hispid with whitish hairs. Leaves ovate, petioled, somewhat crenate. **Clinopodium**, L.
(*Calamintha Clinopodium*, Benth).

Calyx scarcely gibbous. Bracts less conspicuous. Flowers purplish or whitish. Leaves various; broadly ovate, oblong, oblong-linear, or linear, crenate, sparingly cut-toothed, or entire, petioled. **Calamintha**.

497. Calyx strongly 2-lipped; upper lip truncate, closed in fruit..498

Calyx not (or scarcely) 2-lipped..............499

498. Calyx-lips toothed; upper 3 teeth minute, lower 2 large. Stems simple, or somewhat branched, ascending. Leaves oblong-ovate, dentate, petioled. Flowers blue, in clusters of 3, forming closely spiked verticils. Bracts imbricated, reniform, 2 under each verticil.
Brunella vulgaris, L.

Calyx-lips entire, the upper with a helmet-like appendage on the back. **Scutellaria**.

499. Calyx 4-cleft. Corolla 1½ inch long, yellowish-white. The upper pair of anthers connate; each anther with one cell fertile, and the other smaller one sterile. Filaments hairy. Herbs, 1 foot high. Leaves broadly ovate, cordate, crenate, lower ones long-petioled, upper sessile and passing into bracts, each with a solitary flower.
Synandra grandiflora, Nutt.

Calyx 5- or 10-toothed......................500

500. Calyx almost equally 5-toothed..............501

Calyx unequally 10-toothed, woolly, the teeth recurved, surrounded with many bristle-pointed bracts. Corolla white, pubescent. Leaves round-ovate, rugose, crenate-toothed, hoary-pubescent. Flowers in dense, axillary, sessile, hairy whorls. **Marrubium vulgare**, L.

501. Calyx-teeth spiny-tipped....................502
Calyx-teeth not spiny-tipped.................504

502. Anther-cells opening transversely by 2 valves; the inner valve ciliate-fringed. Leaves notched. **Galeopsis.**
Anther-cells opening lengthwise..............503

503. Achenia rounded at the top. Whorls 2- to many-flowered, approximate in a terminal, spike-like raceme.
Stachys.

Achenia truncate, 3-angled at the top. Calyx-teeth, when old, very hard. Upright herbs. Leaves cut-lobed. Flowers pale purple, or whitish, in close axillary whorls. **Leonurus.**

504. Corolla inflated..........................505
Corolla not inflated, short. Calyx salver-form, 10-ribbed, 5-toothed, the teeth longer than the corolla. Anthers exserted. Herbs erect, hairy. Leaves ovate, dentate. Flowers purplish, in dense whorls. **Ballota nigra**, L.

505. Anther-cells divergent. Corolla slightly inflated; lips large...506
Anther-cells parallel. Corolla much inflated in the throat; lips very small; upper lip erect, lower 3-parted. Herbs smooth. Leaves oblong-ovate to narrowly lanceolate, remotely toothed, or sometimes nearly entire. Flowers large, flesh-color, tinged with purple, in terminal leafless, simple, or panicled dense spikes.
Physostegia Virginiana, Benth.

506. Upper corolla-lip helmeted, keeled, broad, entire or notched, incurved, bearded with white hairs inside; lower lip 3-cleft. Filaments of the upper pair of stamens with an awl-shaped appendage at the base. Herb 3–5 feet high. Leaves ovate, cordate, crenate, petioled. Flowers purple, in dense, axillary, bracted whorls. **Phlomis tuberosa**, L.

Upper corolla-lip ovate, arched, bearded, or naked; lower lip 3-lobed, middle lobe obcordate and contracted at the base, as if stalked; lateral lobes acute. Herbs decumbent at the base. Leaves rounded or cordate, crenate-toothed, upper ones sometimes clasping. Flowers purple, in axillary, whorled clusters. Nutlets 3-sided, with the apex truncate. **Lamium.**

---

507. (320.) Plants without tendrils..............508

 A tendril-bearing, slender vine. Flowers polygamous (or monœcious), the fertile with the calyx-tube constricted above the ovary. Calyx funnel-bell-shaped, with 5 subulate segments. Corolla bell-shaped, deeply 5-parted. Stamens 5, triadelphous; anthers 3 or 5, contorted, united. Stigmas 3. Flowers small, yellow, axillary. Ovary with 3 parietal placentæ. Fruit a small, ovoid, many-seeded berry. Leaves roundish, cordate, 5-lobed, or angled. Tendrils simple. **Melothria pendula,** L. (807).

508. Corolla 2-lipped...........................536
 Corolla regular, or nearly so................509
509. I. Stamens 8–12. Leaves simple............510
 II. Stamens 5...............................515
 III. Stamens 4..............................525
 IV. Stamens 3 (rarely 2)....................534

510. Filaments united into a ring or short tube at the base. Anthers linear-oblong, their cells opening lengthwise. Fruit dry or drupe-like, 1- or 4-celled. Shrubs or small trees. Leaves alternate......................................511

 Filaments not united. Fruit a berry or berried drupe, crowned with the calyx-teeth. Shrubs or shrubby plants...512

511. Corolla 5-parted. Stamens 10, jointed to the base of the corolla; filaments united into a short tube at the base. Fruit coriaceous, a sort of dry drupe, 1-celled, mostly 1-seeded. Flowers showy, white, drooping, racemed.
(422) **Styrax.**

 Corolla 4-parted. Stamens 8–12, connate into a tube at the base. Style thread-form, pubescent. Fruit dry, 4-winged, the wings equal, or alternately smaller, 1- to 3-seeded. **Halesia tetraptera,** L.

512. Anthers opening at the apex 513

Anthers opening lengthwise down to the middle, awnless. Filaments very broad and short. Calyx and corolla 4-cleft, the latter bell-shaped, white. Berry globular, 4-celled, many-seeded, $\frac{1}{4}$ inch broad, bright white. A trailing evergreen. Leaves short-petioled, ovate, pointed, with revolute margins, small, their lower surface, like the branches, clothed with rusty bristles. Flowers very small, short-peduncled, nodding, bicracteolate, solitary, axillary.
Chiogenes hispidula, Torr. & Gr.

513. Ovary 10-celled, 10-ovuled. Flowers 5-merous, white or reddish, small, in lateral, bracted racemes. Stamens 10; anthers awnless, the cells prolonged upward into tubular beaks, opening at the apex. Drupe berry-like, globular, 10-celled, 10-seeded, black or dark-blue, sweet. Leaves often resinous-dotted beneath. **Gaylussacia**, H. B. K.

Ovary 4- or 5- or spuriously 8- to 10-celled, many-ovuled. Flowers 4- or 5-merous 514

514. Stems usually trailing. Flowers 4-merous. Corolla with narrow, reflexed segments. Stamens 8, convergent. Anthers tubular, 2-parted, opening by oblique pores. Berry globose, 4-celled, many-seeded, red, or purple. Leaves alternate. **Oxycoccus**, Pers.
(*Vaccinium*, L.)

Stems always erect. Flowers 5- or 4-merous. Corolla urn-shaped, campanulate, or cylindric, with the lobes reflexed. Stamens 10 or 8. Anthers awned or awnless, the 2 cells produced into a tube at the apex. Berry 4- or 5- (or by a false partition 8- to 10-) celled, black, bluish, greenish, rarely red. Leaves scattered. **Vaccinium**, L. (657).

515. Leaves not stipulate...................... 516

Leaves stipulate. Flowers 5- (or 4-) merous. Stamens 5 (or 4). Style short, or none; stigmas 2. Fruit a 2-celled, many-seeded pod. Herbs erect or prostrate. Stipules with 2 to 4 subulate points each side. Flowers small, axillary, white. **Oldenlandia** (533).

516. Shrubs, or shrubby plants. Leaves opposite...517

Herbaceous plants.......................... 521

517. Corolla tubular, funnel-form, or bell-shaped. Style filiform.. 518

Corolla wheel- or urn-shaped. Stigmas sessile. Flowers white, in flat, compound cymes............520

518. Calyx-teeth slender, awl-shaped. Corolla twice as long as the calyx, greenish-yellow. Ovary slender, 2-celled (apparently, from the projecting placentæ, 4-celled). Fruit a pod. Shrubs about 2 feet high. Leaves opposite, ovate, acuminate, serrate, deciduous. Peduncles axillary and terminal, mostly 3-flowered. **Diervilla trifida**, Mœnch.

Calyx-teeth short. Fruit a berry.............519

519. Ovary 4-celled, only two of the cells with a fertile ovule. Calyx-teeth and corolla-lobes 5 (or 4). Flowers small, white, tinged with rose-color, in dense short spikes or clusters. Berry globose, white, rarely dark red. Low, erect, branching shrubs. **Symphoricarpus** (529).

Ovary 2- to 3-celled. Calyx-teeth and corolla-lobes 5, the latter often unequal. Corolla-tube often gibbous at the base. Stamens exserted. Flowers usually showy and fragrant, often in pairs. Berry few-seeded, red, purple, blue, or orange, 2 berries sometimes united into one. Stems climbing, or erect. Leaves opposite, entire, often crenate.
Lonicera.

520. Shrubs with pinnate leaves. Stigmas 3. Ovary 3-celled. Berry 3-seeded, black-purple, or bright red (rarely white). **Sambucus.**

Shrubs with simple leaves. Stigmas 1–3. Fruit a 1-celled, 1-seeded drupe, blue, black, glaucous, or red. Cymes rarely with marginal neutral flowers, furnished with large, showy corollas. **Viburnum.**

521. Stamens distinct..........................522

Stamens united into a tube, commonly by their filaments, and always by their anthers. Corolla split down to the base on one side. Flowers axillary, or chiefly in bracted racemes, red and showy, or blue, variegated with white, commonly smaller. Leaves alternate. **Lobelia.**

522. Leaves alternate...........................523

Leaves opposite, large, connate, oval or lanceolate. Calyx-lobes linear-lanceolate, leaf-like. Ovary mostly 3-celled. Fruit a dry drupe, with 3 bony nutlets, crowned with the calyx, orange-colored. Herbs coarse, hairy. Flowers axillary. **Triosteum.**

523. Stamens alternate with the corolla-lobes. Style 1, beset with collecting hairs above; stigmas 3. Ovary 3-celled. Pod opening on the sides by holes or valves. Corolla wheel-shaped. 524

Stamens opposite the corolla-lobes. Style and stigma 1. Ovary 1-celled, with a free central placenta. Corolla somewhat bell-shaped, usually with 5 sterile filaments in the sinuses. (365) **Samolus Valerandi**, L.

524. Pod prismatic, elongated. Flowers axillary and terminal, sessile, erect, 2–3 together; the lower ones fruiting, without expanding. Filaments hairy, membranaceous. Leaves small, roundish-ovate, toothed, clasping by the cordate base. **Specularia perfoliata**, A. D.C.

Pod short, not prismatic. Flowers panicled, or crowded in a long, leafy spike. Filaments naked, broad and membranaceous at the base. **Campanula.**

525. (509.) Stamens as many as the corolla-lobes....526

Stamens fewer than the corolla-lobes, 2 longer than the other. The 5-lobed corolla purple and whitish, hairy inside. Berry 3-celled, indehiscent, 1-seeded (2 cells abortive). A trailing, evergreen herb. Stem creeping. Leaves small, roundish, opposite, petioled, obtusely lobed or toothed. Peduncles 2-flowered, about 3 inches high, thread-form, slightly pubescent (the only erect part of the plant).
Linnæa borealis, Gron.

526. Numerous flowers in a naked or involucrate head. Fruit dry..527

Flowers not in heads.....................528

527. Heads not involucrate, globular, about 1 inch in diameter, or larger. Style much exserted. Shrubs, with stipules. **Cephalanthus occidentalis**, L.

Heads involucrate, oblong, about as large as a hen's egg. Stipules none. Involucre many-leaved, longer than the chaffy, leafy-tipped and pointed bracts among the flowers. Flowers bluish, the middle zone of the head expanding first, each with a 4-sided involucel, closely investing the calyx in fruit. Fruit 1-seeded, crowned with the calyx. Corolla 4-cleft. Style slender. Stout and coarse, prickly perennial herbs. Leaves lance-oblong, opposite, connate.
Dipsacus sylvestris, Mill.

528. Leaves without stipules.....................529
    Leaves with (sometimes minute) stipules, or a connecting stipular membrane (which are often furnished with bristles)..530

529. Leaves whorled, in fours, sixes, or eights (rarely in fives). Corolla wheel-shaped, 4-cleft. Styles 2. Ovary 2-celled. Fruit dry, or fleshy, globular, twin, separating, at maturity, into two 1-seeded, indehiscent nutlets. Stems 4-angled. Herbs. (276) **Galium** (645).
    Leaves opposite. Calyx-tube globose, limb 4- (or 5-) toothed. Corolla funnel or bell-shaped, the border 4- (or 5-) lobed. Stigma capitate. Berry globose, 2-seeded, snow-white, or rarely dark red. (519) **Symphoricarpus.**

530. Flowers one to each ovary. Fruit dry........531
    Flowers 2 on each double ovary, fragrant. Calyx 4-parted. Corolla funnel-form, white, tinged with purple within. Stigmas 4. Fruit red, a berry-like double drupe, crowned with the calyx-teeth of the 2 flowers, each containing 4 nutlets. Evergreen herbs, smooth, trailing.
Mitchella repens, L.

531. Ovules, or seeds, solitary in each cell. Flowers small, white. Stipules fringed with bristles............532
    Ovules, or seeds, many or several in each cell of the 2-celled ovary, or pod. Stipules small, united to the petioles..533

532. Flowers in dense axillary sessile clusters. Corolla scarcely exceeding the calyx; throat bearded. One of the 2 carpels dehiscent. Glabrous, spreading herbs.
Spermacoce glabra, Mx.
    Flowers 1-3 in each axil. The 2 carpels of the fruit indehiscent. Stem hairy, or smooth. **Diodia.**

533. Seeds angular, minute, very numerous in each cell. Pod completely enclosed in the calyx-tube. Corolla wheel-shaped, white, much shorter than the calyx. Anthers short. Stems branched, spreading, pubescent. Leaves oblong. Flowers in sessile, axillary clusters.
(515) **Oldenlandia glomerata,** Mx.
    Seeds saucer-shaped, with a ridge down the middle of the concave inner face, rather few in each cell (4-20). Pod partly included. Corolla salver- or funnel-form, purplish,

bluish, or white, much exceeding the calyx, often hairy inside. Anthers linear. **Houstonia**, Wood.
(*Oldenlandia*, Gray.)

534. (509.) Leaves in whorls of 4 to 6, linear, or oblanceolate, obtuse; the margins and the midrib rough. Corolla-lobes and stamens mostly 3.
(529) **Galium trifidum**, L.
Leaves opposite..........................535

535. Calyx-limb at first small and involute, at length evolving a pappus of plumose bristles. Corolla funnel-form, 5-cleft, commonly evidently gibbous near the base. Fruit 1-celled, 1-seeded. Flowers in umbel-like, close cymes, pink, rose-color or white, sometimes imperfectly diœcious, or dimorphous. Leaves mostly pinnately divided.
Valeriana (878).
Calyx 1- to 5-toothed, or cup-shaped, 6-cleft or 6-toothed at the summit. Stamens 3 (rarely 2). Fruit 3-celled. Flowers in dense, terminal cymelets, white or bluish. Leaves entire or toothed, oblong or linear, sessile.
Fedia, Gærtn.
(*Valerianella*, D. C.)

536. (508.) Corolla-tube split down to the base on one side. Anthers, and usually also the filaments, connate..521
Corolla not split. Stamens not united........519

C. INCOMPLETE FLOWERS.
(537–773.)

537. (77.) Three to five, twelve, or many flowers in a calycine or corolline involucre......................538
Involucre to several flowers none, at least not a calycine or corolline..............................539

538. Flowers with a perianth, 3–5, all of the same sort, in the same 5-lobed, open, funnel-form, pubescent involucre, which is thin and reticulated in fruit. Tube of the perianth bell-shaped, rose-color or purple; its limb deciduous. Stamens (in each flower) usually 3. Style filiform. Stem repeatedly forked. Leaves oblong-ovate, triangular, or cordate, opposite. **Oxybaphus nyctagineus**, Nees.
Flowers greenish, with an obsolete, or minutely 4-toothed calyx, 4 petals, 4 stamens, and a 1-styled ovary;

these flowers numerous, in a close head, which is surrounded by a 4-leaved, petaloid, white involucre. A low herb, 3–8 inches high. Upper leaves crowded, apparently whorled.
Cornus Canadensis, L.

Flowers without a perianth, of 2 sorts: the central one fertile—that is, an ovary with 3 bifid styles, and the rest, 12 or so, sterile—that is, stamens of unequal length, each jointed to a pedicel, rising from the axil of a little bract; all these naked flowers in a colored, 4- to 5-lobed involucre, which has a gland in each sinus between the lobes. Herbs with a milky juice. Leaves whorled or alternate.
Euphorbia.

539. Perianth either none, or merely an obscure margin, bristles, a scale, a cup, etc....................540

An evident perianth present, being either a (commonly petaloid) conspicuous calyx, or consisting of 2 or several scales, or bracts.............................564

540. Grass- or rush-like plants, with terminal (sometimes capitate) spikes, formed of imbricated, 1-flowered scales (glumes). Flower usually of 3 stamens and a 1-styled ovary, commonly surrounded with bristles (very rarely with scales). Style 2- or 3-cleft. Culms solid. Leaves grass- or rush-like. Sheaths not split. (*Cyperaceæ*, Sedges)................551

Plants not grass- nor rush-like................541

541. Stem- and scapeless-plants in the form of roundish or oblong, small leaves (frond), ½–4 lines in length, grouped or single, floating upon the water, furnished below with loosely pendent roots. Flowers at the margin of the frond, sessile. In place of a perianth, or involucre, an unequally 2-lobed spathe. Stamens 2 (sometimes 1) below the ovary. Style simple. **Lemna.**

Plants with stems or scapes, striking root into the ground, or fixing themselves to stones or pebbles.......542

542. Maritime, or salt-water plants, submersed......543

Terrestrial, or fresh-water plants..............544

543. Sessile anthers and pistils arranged alternately in 2 rows on the inner side of a flat-linear spadix, enclosed by a spathe. Stems trailing, with tufts of fibrous roots at the joints, and alternate, linear, entire, sheathing leaves. On sandy banks in the sea. **Zostera marina, L.** (809).

Spike (spadix) axillary, 2-flowered, in an inflated spathe. Perianth none. Anthers 4, sessile, surrounded by 4, at first sessile, at length stalked pistils. Stigmas depressed. Leaves capillary, 1 to 2 feet long, with a dilated, sheathing base, alternate. Flowers rising to the surface of the water, at the time of expansion. **Ruppia maritima,** L.

544. Leafless, succulent, fleshy, low plants, with jointed stems, growing in salt marshes. Flowers forming club-shaped spikes on the thickened upper joints. Scales 1-flowered, appressed, 3 together, and forming a triangle. Stamens 1 or 2, protruded beyond the margin of the scale. Styles 2, partly united, short. **Salicornia,** Tourn.

Plants leafy.................................545

545. Leaves whorled, linear (8 or 12 in a whorl). Flowers solitary, axillary. Calyx-limb short, on the top of the ovary. Stamen 1, on the margin of the calyx; anther large. Style thread-form, received into the groove between the anther-cells. **Hippuris vulgaris,** L. (816).

Leaves not whorled. Stamen below the pistil..546

546. Flowers on a spadix, surrounded by a spathe. The radical leaves sometimes sheathing, by the petiole, the scape, so as to be, at first sight, mistaken for cauline ones. Berries red..550

Flowers spiked, axillary, or lateral. Herbs with a stem...547

547. Aquatics. Flowers perfect or monœcious......549

Non-aquatics............................548

548. Marsh herb. Flowers naked, each consisting of 3 or 4 pistils, and 6 to 8 hypogynous stamens, white. Spike terminal, wand-like, lengthened, drooping at the summit. Leaves ovate-oblong, cordate, petioled, alternate. Stem weak, furrowed, 1½ to 2 feet high. **Saururus cernuus,** Willd.

Shrub. Flowers in threes, axillary...........660

549. Stamen 1, being a naked, sterile flower, on the outside of a fertile flower, which consists of a bell-shaped perianth with 3 or 4 short-styled pistils. They are sessile in an amplexicaul stipule. A whorl of 3 or 4, long linear leaves at each joint, subtended by a 2-flowered stipule of the above sort. Stem terete, branching, floating.

Zanichellia palustris, L. (818).

Stamens 2, borne on one side of the stalk of the ovary, with their long filaments united below (and 2 short, sterile filaments sometimes on each side). Stigmas 2, sessile, recurved. Flower naked, from a tubular sac-like involucre. Leaves 2-ranked, dilated into a sheathing base, rigid and dichotomously dissected into filiform divisions. Herbs small, submersed, adhering to stones and pebbles.
Podostemon ceratophyllum, Mx.

550. The whole spadix covered with stamens and pistils, intermixed. Ovary roundish; stigma sessile. Lower flowers always perfect, the upper often of stamens only. Scape thick. Spathe broad, open, green without, white within. Leaves cordate. In cold bogs. **Calla palustris,** L.

Spadix club-shaped, the upper third naked. Flowers, by abortion, diœcious; or monœcious; when numerous, the pistillate in the lower, the staminate in the middle third of the spadix. Leaves compound, of 3, or of 7–11 leaflets. Scape sheathed by the petioles of the veiny leaves, and therefore stem-like. **Arisæma**, Martius (812).
(*Arum*, L.)

551. (540.) I. Flowers perfect, or sometimes polygamous 552

II. Flowers strictly monœcious.
Scleria and **Carex** (see both in 821).

III. Flowers strictly diœcious. Glumes imbricated or alternate. Perianth of united scales (perigynium), sac-like, enclosing the achenium. **Carex** (821).

Gr. T. VI.

552. Glumes of the spikes 2-rowed. Flowers perfect.
553

Glumes imbricated in several rows. Flowers perfect, or polygamous............................555

553. Inflorescence terminal. Perianth none.........554

Inflorescence axillary. Perianth of 6–10, downwardly barbed bristles. Peduncles solitary, axillary, from the sheaths of the leaves. **Dulichium spathaceum**, Pers.

Gray, T. I.

554. Spikes 2- to many-flowered, collected in an involucrate, or compound, terminal umbel or head. **Cyperus.**

Gr. T. I.

Spikes 1-flowered, glomerate in a sessile head.
Stamens 1-3. **Kyllingia pumila**, Mx.
Gr. T. I.

555. Each glume flower-bearing, or sometimes the lowest empty. Inflorescence exclusively either terminal or axillary. Flowers all perfect...................................556

Many of the lower glumes empty. Inflorescence both terminal and axillary (with a single exception). Flowers perfect or polygamous..........................560

556. I. Perianth none..........................559

II. Perianth of 3 large scales and mostly as many alternate bristles. **(Fuirena)**, 646.

III. Perianth of 3 to many bristles............557

557. Achenium crowned with a tubercle. Bristles mostly 6. Spike solitary, terminal. Stems leafless. **Eleocharis.**
Gr. T. III.

Achenium not tuberculate. Stems mostly leafy, rarely leafless 558

558. Bristles 3-6, short, or else tawny. Style 2- to 3-cleft. Spikes 1 to many, terminal (appearing lateral sometimes). **Scirpus.**
Gr. T. III.

Bristles many (rarely only 6), long, white, or woolly. Stem generally leafy. Spikelets mostly in umbels, at length clothed with long, silky hairs. **Eriophorum.**
Gr. T. III.

559. Involucre 1-leaved (usually with another minute leaf). Spikes one or few in an apparent lateral cluster, which is in fact terminal. Stamen 1.
Hemicarpha subsquarrosa, Nees.
Gr. T. II.

Involucre 2- to 3-leaved. Style 2- or 3-cleft (3-cleft in § *Trichelostylis*). Stamens 1-3. Spikes in a terminal head or umbel. **Fimbristylis**, Gray
(including **Trichelostylis**, Listib.)
Gr. T. III.

560. Inflorescence strictly terminal; the few-flowered spikes crowded in a leafy-involucrate head. Perianth none. Style 2-cleft. Bracts colored. Achenium truncate at the summit, transversely wrinkled.
Dichromena leucocephala, Mx.
Gr. T. IV.
Inflorescence both terminal and axillary.......561

561. Achenium beaked with the dilated, persistent style, or its bulbous base. Perianth of bristles, or none. Stems leafy..562
Achenium not beaked, corky, or only pointed at the top. Perianth none. Style 3-cleft. Stamens 2. Spikes clustered in heads, 3–8 together on 2–4 peduncles. Culm obscurely 3-cornered. **Cladium mariscoides**, Torr.
Gr. T. V.

562. Perianth of 5–12 bristles, rarely wanting......563
Perianth none. Stamens 2. Spikes terete, ovoid, cymose. **Psilocarya scirpoides**, Torr.
Gr. T. IV.

563. Style 2-cleft. Achenium tubercled with the base of the style. Bristles 6–12. Spikelets few-flowered, ovate, the scales loosely imbricated. Stem 3-sided.
Gr. T. IV. **Rhynchospora**.
Style simple, entire crowning the achenium as an awned beak. Spikelets 2- to 5-flowered, one flower perfect, lower ones staminate. Bristles 5 or 6.
Gr. T. IV. **Ceratoschœnus**.

---

564. (539.) True grasses. Perianth a pair of small *bracts* (called *paleæ*, *pales*, or *perianth*, R. Br.), enclosing each particular flower, which is furnished also with 2 or 3 minute, hypogynous scales (squamulæ). Another pair of bracts (glumes or calyx, L.) include 1, 2, several, or many flowers, forming thus 1-, 2-, several-, or many-flowered spikelets, which are collected into spikes or panicles. Stamens 3 (rarely 1, 2, or 6). Styles 2 (rarely 1). Stem usually hollow and closed at the joints (culm), with alternate, 2-ranked leaves, their sheaths split on the side opposite the blade, or rarely not split, extended above the base of the blade into a scarious appendage (ligule). All the bracts 2-ranked..........565

No grasses, but plants with a perianth of 3 or more sepals, or segments..................................639

Aquatic, small annuals. Flowers monœciously polygamous. Flowers naked, between two bracts (an apparent calyx). Stamen solitary in the sterile flower. Ovary of the fertile flower 2-styled. A fertile and a sterile flower sometimes together in an axil. **Callitriche** (817).

GRASSES.
(565–638.)

565. I. Spikelets of a solitary perfect (rarely staminate or pistillate) flower..................................566

II. Spikelets, some of them 1-, some 2-flowered, either pistillate, or staminate, at the joints; the pistillate 1-flowered, consisting of 2 glumes, 3 pales and 1 pistil, occupying alternately about the lower third of the rhachis, each separate, and deeply sunk into a boat-shaped recess of it; the staminate, making up the 2 upper thirds of the spike, alternately disposed in pairs, every pair slightly imbedded by the base only, each spikelet consisting of 2 glumes, 2 pales (sometimes with a few additional minute ones), and 3 stamens with red anthers. Culm 5–8 feet high. Leaves broadly linear. Spikes (6–8 inches long) commonly 2 or 3 at the top of the culm (sometimes a solitary spike).

Tripsacum dactyloides, L.
Gr. VIII. 62.
W. V. 63.

III. Spikelets containing 2 or more flowers (when several, rarely only one of them perfect)..............603

566. Spikelets of 2 pales only, and without glumes (or the latter rarely present, rudimentary, and forming a little cup)..567

Spikelets consisting of 2 glumes, or at least 2 (or 1) glume-like scales, and 1, 2, 3, or 4 pales...............568

567. Flowers perfect, flattened laterally, awnless. Glumes none. Stamens 2 or 3. **Leersia**.
Gr. I. 1.
W. I. 1.

Flowers monœcious, convex on the back. Lower pale long-awned in the fertile spikelets, awnless in the sterile. Stamens 6. **Zizania** (837).
Gr. I. 2.
W. I. 3.

568. I. Spikelets with 1-pale....................569
II. Spikelets with 2 pales..................570
III. Spikelets with 3 or 4 pales.............598

569. Panicle. Style short; stigmas plumose, emerging laterally from the bottom of the flower. The solitary pale sometimes awned on the back. **Agrostis** (585).
Gr. I. 7.
W. I. 4.

Spike cylindric, with crowded spikelets. Stigmas protruded at the apex of the flower. The solitary pale thin, with the margins united, bag-like, awned below the middle.
Gr. I. 3. **Alopecurus.**
W. I. 10.

570. Spikelets solitary at each joint, and each partly imbedded into a cavity of the rhachis. Glumes 2, transverse, enclosing the thin pales. Stamens 3. Spikes triangular, subulate, 1-3 inches long, 6-10 in number, disposed in a raceme. **Lepturus paniculatus**, Nutt. (610).
Gr. V. 41.
W. IV. 47.

Spikelets not imbedded......................571

571. Spikes, 2 or several, placed digitately upon the culm, or peduncle...572
Solitary spikes, racemed spikes, or panicles.....574

572. Spikelets of 2 sorts, 2 at each joint, one sessile and fertile, the other pedicelled and sterile, or abortive. Flowers 3-paled..602
Spikelets of one sort. Flowers 2-paled. Spikes 1-sided...573

573. Spikes 4-7 at the summit of the culm. Glumes narrow, keeled; upper pale linear, excavated by a furrow, lower one laterally compressed, ovate. Stigmas 2, on an elongated style. A rudiment of a second flower, in the form of a pedicel, close to the upper pale.

Cynodon Dactylon, Pers.

Gr. III. 20.
W. V. 56.

A pair of spikes at the summit of the peduncle. Glumes not keeled, membranaceous, rounded (one of them, by theory, a pale !). Styles 2. No pedicel-like rudiment.

Paspalum (in two spec.).

Gr. VII. 58.
W. II. 15.

574. Flower surrounded at the base by a tuft of copious, white, bristly hairs, their length at least one third of their own. Spikelets lanceolate, acute. Glumes keeled or boat-shaped. Pales naked ; lower one either awnless, or with a dorsal or almost terminal awn. Flower often with a pedicel at the base (a rudimentary flower). (*Calamagrostis.*)...575

Flower without, or with scarcely perceptible hairs at the base. Pales naked or hairy. Glumes sometimes awned, or bristle-pointed..............................576

575. Lower glume shorter than the upper, and shorter than the pales. Lower pale 1-nerved and entirely awnless, the upper strongly 2-keeled. Panicle open, pyramidal. Rudiment none. § **Calamovilfa**, Gray.
(*Calamagrostis brevipilis*, Torr., and *Calamagrostis longifolia*, Hook.)

Lower glume longer than, or about as long as the upper; both glumes commonly longer than the pales. Panicle open, or spiked-contracted. A rudimentary 2d flower, in the form of a plumose pedicel, sometimes present.

§ **Calamagrostis**.
(§ *Calamagrostis Proper*, Gray, and
§ *Ammophila*, Host.)

Gr. II. 12.
W. I. 9.

576. Glumes both, rarely only one of them, with an awn or an awn-like bristle................................577

Glumes awnless, or merely sharp-pointed......580

577. Glumes awn-pointed, or altogether bristle-form, 6 in number, placed side by side, in front of the 3 spikelets

upon each joint of the rhachis, 2 glumes belonging to each particular spikelet, and altogether forming a sort of involucre, only the middle spikelet fertile ; lower pale of its lower long-awned from the apex. Spike dense.
Hordeum.

Gr. V. 44.
W. IV. 48.

Glumes not forming an involucre............578

578. Apparent spike oblong-linear, ovate-oblong, or cylindric—that is, a much-contracted panicle..........579

One-sided spikes racemed, sometimes 2 only. Spikelets spiked in two rows on the outer side of a triangular rhachis. Glumes coriaceous, laterally compressed, keeled, bristle-pointed, upper one often awned. Styles long, commonly united below. **Spartina** (597).

G. 16.
W. 60.

579. Pales unequal, mucronate ; and the nearly equal, linear glumes with their bristle-like awns twice their length. Panicle spicate, densely conglomerated, interrupted, 2–3 inches long. **Muhlenbergia glomerata,** Trin. (587).

Lower pale with a few, minute teeth at the apex, and awned ; the upper one 2-toothed. Glumes hairy all over. Contracted, spike-like panicle interrupted.
Polypogon Monspeliensis, Desf.

Gr. II. 8.
W. I. 8.

Lower pale truncate and awnless. Glumes tipped with either short bristles about $\frac{1}{4}$ of their length or awn-like ones, as long as themselves. Spike oblong-ovate, or long, cylindric. **Phleum.**

Gr. I. 4.
W. I. 11.

580. Spikelets in a loose, more or less open, or contracted panicle (or raceme).................................581

Spikelets arranged in a solitary spike, in a spike-bearing raceme, or in a spike-like, ovoid panicle........594

581. Lower pale awned...........................582
Lower pale awnless..........................588

582. Awn deeply 3-parted (or 3 awns). Panicle contracted or racemose. **Aristida.**
  Gr. II. 15.
  W. I. 12.
  Awn not 3-parted..........................583

583. Awn twisted, very long, from the apex, stout below. Lower pale coriaceous, cylindrical-involute, closely embracing the smaller, upper one, and the cylindric grain. Flower with a conspicuous, obconical, bearded stalk (callus). Panicle loose, or somewhat contracted. **Stipa.**
  Gr. II. 14.
  W. I. 13.
  Awn not twisted..........................584

584. Lower pale awned at the apex...............586
  Lower pale awned at the back..............585

585. Flower stalked in the keeled glumes. Pales lanceolate, pointed. Stamen only 1. Panicle large, compound, terminal. **Cinna arundinacea, L.**
  Gr. II. 9.
  W. I. 6.
  Flowers sessile in the glumes. Pales pointless; upper one minute, or obsolete. (569) **Agrostis** (591).
  Gr. I. 7.
  W. I. 4.

586. Conspicuous squamulæ 2 or 3. Flower with a very short callus. Lower pale coriaceous, at length involute; its awn deciduous. Spikelets nearly terete. Styles short, or united below. Panicle various, simple and raceme-like, sparingly branched, or finally contracted, with the branches usually in pairs. **Oryzopsis.**
  Gr. II. 13.
  W. II. 14.
  Conspicuous squamulæ none. Panicle mostly contracted.................................587

587. Glumes minute, the lower one scarcely perceptible. Pales rough, with scattered, short bristles; lower one 5-nerved and contracted at the apex into a stout, long, straight, smooth

wn, the upper 2-pointed. An awn-like rudiment lodged in groove on its back. **Brachyelytrum aristatum**, Beauv.
(*Muhlenbergia aristata*, Pers.)
Gr. II. 11.

Glumes various, sometimes pretty large, the lower naller, or sometimes both small or minute, the lower always iorter. Flower usually hairy-bearded at the base, the wer pale 3-nerved. Rudiment none.
(579) **Muhlenbergia** (592).
Gr. II. 10.
W. I. 7.

588. (581.) Flowers all alike....................589

Flowers of 2 sorts; one sort under ground. The rminal flowers in a strict, slender, contracted panicle, the idical ones solitary on a slender peduncle.
Amphicarpum Purshii, Kunth.
Gr. VII. 57.
W. II. 17.

589. Two hairy scales at the base of the flower, one on ach side. Stigmas protruded at the apex of the flower. 'anicle branched, clustered, at length somewhat spreading.
Phalaris arundinacea, L.
Gr. VII. 55.
W. II. 22.

No scales at the base of the flower. Stigmas merging from the side...........................590

590. Glumes ventricose, nearly as long as, or scarcely nger, than the shining and at length cartilaginous pales (in ieory, the lower glume is wanting, and its place occupied y a pale!). Grain coated by the cartilaginous pales. 'anicle spreading. **Milium effusum**, L.
Gr. VII. 56.
W. II. 16.

Glumes convex-compressed and frequently (at ast one of them) longer than the pales, sometimes shorter· ian they.................................591

591. Lower pale obtuse, upper one small, often minute, r wanting. (585) **Agrostis**.

Lower pale acute..........................592

592. Lower pale 3-nerved, mucronate at the apex, usually hairy-bearded at the base. (587) **Muhlenbergia.**

Gr. 10.
W. I. 7.

Lower pale 1-nerved, or obscurely 3-nerved, not bearded at the base, usually longer than the glumes.....593

593. Seed adherent to the linear-cylindric pericarp (caryopsis). Panicle contracted, or spiked. Leaves involute, usually bearded at the throat. **Vilfa**, Beauv.
Gr. I. 5. (*Sporolobus*, Wood.)

Seed loose in the ovoid or globular pericarp (utricle). Panicle open, pyramidal, or racemose-elongated, rarely contracted and spike-like. **Sporolobus**, R. Br.

Gr. I. 6.
W. I. 5.

594. (580.) Flower with a smooth scale on each side, at the base, half as long as the pales. Glumes wing-keeled. Panicle spiked, close, ovoid (spike-like).

Gr. 55. (589) **Phalaris Canariensis**, L.
W. 22.

Flower without 2 opposite basilar scales.......595

595. Lower pale awned, or 3-toothed. At the base of the flower 1–3 sterile flowers, in the form of thin, linear scales, or awns, and, when 3 in number, sometimes raised on a pedicel, so as to resemble a 3-tined fork..............596

Lower pale awnless. Rudiments none. Spikelets in 2 or 4 rows, on one side of a flat, or triangular rhachis. Spikes racemed, or sometimes in pairs toward the summit of the naked peduncle 597

596. Lower pale 3-cleft and pointed, or 3-awned. Glumes keeled, the upper one larger. Spikes either 20–40, very short, in one long, 1-sided raceme, or 1–5, longer and racemed on both sides toward the summit of the axis. Anthers red.

Gr. III. 18. **Bouteloua.**
W. V. 61.

Lower pale and the single rudiment, that is, a mere

pedicel, awned. Stigmas pencil-form, purple. Spikes filiform, in a crowded, naked raceme. **Gymnopogon.**
    Gr. III. 19.
    W. V. 54.

597. Glumes laterally compressed, keeled, coriaceous, pointed. Styles long, exserted, commonly united below. Rhachis of each spike triangular, sometimes continued into a bract-like appendage. The 1-sided spikes racemed, sometimes only 4, or 2. (578) **Spartina.**
    Gr. III. 16.
    W. V. 60.

Glumes orbicular, or ovate-orbicular, not keeled. Rhachis of each spike flattened, or filiform, bearing the plano-convex spikelets in 2 or 4 rows. Spikes racemed, or rarely, toward the summit of the naked peduncle, in pairs, and therefore looking sometimes like digitate.
    (573) **Paspalum.**

598. (568.) Only one, or none of the 3 or 4 pales awned.
    599

Two of the four pales awned—that is, the two outer, one at the base, the other near the apex, both emarginate and hairy. Glumes 2, the upper one nearly double the length of the lower. Stamens and styles 2, much exserted. Panicle spiked, apparently a cylindric spike.
    Anthoxanthum odoratum, L.
    Gr. VII. 54.
    W. II. 23.

599. Spikelets all perfect, arranged in pairs or threes, or scattered..600
    Spikelets, 2 or three at each joint—only one of them, the sessile, perfect 602

600. One of the 3 pales awned. Spikelets in pairs upon each joint of the slender rhachis—one of them sessile, the other stalked. Spikes crowded in a panicle and clothed with long silky hairs, especially in a tuft around each spikelet.
    Erianthus.
    Gr. VIII. 63.
    W. V. 67.

Flowers not awned 601

601. Spikelets each subtended by an involucre of several or few awned pedicels. Spiked panicle dense (an apparent cylindric spike). **Setaria.**
Gr. VII. 60.
W. II. 20.

No involucre of awned pedicels. Lower glume usually shorter, sometimes minute, rarely wanting. Pales 3 or 4. Stigmas usually purple. Spikelets either 2 or 3 together, or imbricated, or scattered. Spikes often 1-sided, panicled, racemed, or sometimes almost digitate.
Panicum (618).
Gr. VII. 59.
W. II. 18.

602. Spikes lateral or terminal, commonly clustered or digitate, hairy, mostly silky. Spikelets in pairs, not in threes, the peduncled sterile one usually neutral, often a mere pedicel, and only in one species, *A. furcatus*, Muhl, staminate.
Andropogon.
Gr. VIII. 64.
W. V. 69.

Panicle narrowly oblong, clothed with fawn-colored hairs, at length drooping. Spikelets in pairs or threes, the one or two, stalked sterile reduced to hairy pedicels.
Sorghum nutans, Gray.
(*Andropogon nutans*, L.)
Gr. VIII. 65.
W. 69.

603. (565.) Spikelets (3- to 7-flowered) disposed in 2–4 (rarely more) 1-sided, digitate spikes. Lower pale keeled.
604.

Spikelets disposed in solitary spikes, racemes, or panicles..605

604. Upper glume awn-pointed. Lower pale boat-shaped, pointed. Spikelets 3-flowered, the upper flower sterile. Spikes usually 4; rhachis mucronate at the naked tip.
Dactyloctenium Ægyptiacum, Willd.
Gr. III. 21.
W. 59.

Upper glume larger than the lower, both obtuse.

Lower pale ovate, pointless. Spikelets 5- to 7-flowered, with
a terminal rudiment of a flower. Spikes usually 2-4.
Eleusine Indica, Gærtn.
Gr. III. 22.
W. V. 58.

605. Spikelets (awnless, 2-4 together) in a globular, bur-like involucre, beset with numerous, retrorsely hispid spines. Glumes 2, 2-flowered, the outer one smaller. Lower flower sterile. Involucres sessile in a terminal spike, composed of 3-20 spherical heads. **Cenchrus tribuloides**, L.
Gr. VIII. 61.
W. II. 21.
No prickly involucre........................606

606. An arborescent, shrubby grass. Flowers polygamous (perfect and staminate). Glumes very small, membranaceous, the upper one largest. Lower pale convex, mucronate, or bristle-pointed. Panicle simple, with few spikelets.
Arundinaria macrosperma, Mx.
Gr. V. 40.
W. IV. 46.
Grasses not arborescent........ 607

607. Glumes awnless, or sometimes both bristle-awned.
608
Upper glume with a stout, recurved, horn-like awn from the middle of the back, and warty-glandular outside; lower one much smaller. Spikelets 4- to 6-flowered, all flowers, but one, neutral, the one or two lower consisting of empty, awned pales, the one or two uppermost of empty, awnless pales. Spike bearing the imbricate spikelets on one side in 2 rows, solitary and arcuate-curved.
Ctenium Americanum, Spreng.
Gr. III. 17.
W. V. 62.

608. Spikelets sessile, single, in pairs, threes or fours, upon the joints of the rhachis, forming a 2- or 4-rowed spike.
609
Spikelets (single, or 2 to 3 together) raised, at least the terminal spikelet, on a distinct, sometimes very short pedicel. Panicle, or raceme........................612

609. Sessile spikelets single........................610

Sessile spikelets 2–4 together (1- to 8-flowered). Glumes nearly side by side, in front of the spikelets, 2 for each of them, forming an involucre. The lower pale awned, or awnless, rigid, or soft. Both glumes often bristle-awned.
Elymus.
Gr. V. 45.
W. IV. 49.

610. Spikelets (2- or 1-flowered) sessile in excavations of the rhachis, forming, before flowering, a cylindric spike. Glumes 2, or 1, placed in front of the spikelet.
(570) **Lepturus paniculatus,** Nutt.
Gr. V. 41.
W. IV. 47.

Spikelets sessile upon the joints, not in cavities. Glumes, at least those of the terminal spikelet, opposite..611

611. Spikelets (5- to 9-flowered) placed edgewise on the rhachis, the terminal one with 2 glumes, the rest with 1 only (the second glume being supplied by the rhachis). Flowers awned, or awnless.
Lolium.
Gr. V. 42.
W. IV. 50.

Spikelets (4- to 9-flowered) placed with the side against the rhachis, all 2-glumed, their flowers alternate on the axis of the spikelet. Glumes ovate, or ovate-lanceolate. Lower pale pointed, or awned from the tip.
Triticum.
Gr. V. 43.
W. V. 51.

612. (608.) Spikelets (2- to 4-flowered) raised at each joint in pairs or threes (rarely single) on a very short, thick, callous pedicel (callus). Glumes none, or small, awn-like rudiments. Spike upright, loose.
Gymnostichum Hystrix, Schreb.
Gr. V. 46.
W. 49.

Stalked spikelets single.....................613

613. I. Spikelets composed of staminate flowers on one specimen of the species, and of pistillate ones on another. Ovary stalked. Sterile flowers slightly rounded on the back,

and smaller than the pistillate, the latter more rigid, with lengthened, plumose stigmas.
Brizopyrum spicatum, Hook.
Gr. IV. 32.
W. IV. 41.

II. Spikelets composed of perfect and staminate flowers. Panicles.................................614

III. Spikelets with the flowers always perfect (or with one or a few additional neutral or rudimentary ones). 619

614. Spikelets (5- to 10-flowered) all either staminate or pistillate, except one or a few perfect ones. Styles much protruded at the apex of the flower. Lower pale flattened-boat-shaped, obscurely many-nerved, acute, somewhat coriaceous. Spike oblong, flattened, 1 inch long.........613, I.

Spikelets (2- to 5-flowered) with only 1 or 2 staminate flowers..615

615. Flowers surrounded by a tuft of long, silky hairs from the axis of the spikelet, equalling them in length. Spikelets narrowly lanceolate, with 3 to 5, rarely more flowers—the lowest one staminate, the rest perfect. Glumes very unequal. Lower pale narrowly-awl-shaped, thrice the length of the upper. Panicle loose, nodding.
Phragmites communis, Trin.
Gr. V. 39.
W. IV. 45.

Flowers not surrounded by a tuft of long hairs. Spikelets 2- to 3-flowered............................616

616. Spikelets 3-flowered, the 2 lower flowers staminate, awnless, or awned, the middle one perfect, with 2 stamens, awnless. Glumes as long as the flowers. **Hierochloa.**
Gr. VII. 53.
W. II. 24.

Spikelets 2-flowered—one flower staminate, the other perfect.................................617

617. Lower flower staminate. Glumes unequal.....618

Lower flower perfect, awnless; upper one staminate, its lower pale awned from the back, a little below the tip.
Holcus lanatus, L.
Gr. VI. 52.
W. II. 25.

618. Lower pale in the perfect flower bristle-pointed from near the tip, in the staminate furnished with a long, kneed awn from below its middle.
Arrhenatherum avenaceum, Beauv.
(*Avena elatior*, L.)
Gr. VI. 51.
W. 28, § 1.

Flowers not at all awned. (601) **Panicum**.

619. (613.) Flowers awned on the back, or from the base. 620

Flowers awnless, or awned at, or from a little below the tip.........623

620. Awn of the lower pale from about the middle of its back. Spikelet 2- to 3-flowered..........621

Awn from near the base of the membranaceous pale, which is 3- to 5-nerved on the back and eroded or denticulate at the truncate apex. Awn either barely equalling the pale, or twice its length. § **Deschampsia**, Beauv.
(*Aira flexuosa et A. cæspitosa*, L.)
Gr. VI. 47.
W. III. 26.

621. Pale with the apex truncate and entire, or nearly so. Awn borne at or above the middle, stout, twice the length of the pale. **Aira atropurpurea**, Wahl.
Gr. 47.
W. 26.

Pale with the apex 2-toothed. Awn bent.....622

622. Pale rounded on the back and obscurely 3- to 5-nerved. Awn from the middle, or a little below it.
§ **Airopsis**.
(*Avena præcox*, Beauv.)
Gr. 50.
W. III. 28.

Pale compressed-keeled. Awn from a little above the middle. Panicle long, narrow, loose, or oblong to linear, much contracted, dense. **Trisetum**.
Gr. VI. 49.
W. III. 29.

623. Upper pale with the margins pectinately ciliate by stiff, distant hairs (and usually 2-toothed at the apex)...624

Upper pale more or less evidently ciliate on the nerves, or apparently naked..........................626

624. Lower pale awned just from below the 2-cleft, or sharply 2-toothed apex..............................625

Lower pale awned at the very tip of the 2-toothed, or entire apex, 3-nerved. Stamens 2 or 3. The 1-sided spikes racemed. **Leptochloa.**

Gr. III. 23.
W. V. 53.

625. Lower pale either convex on the back, or compressed-keeled, 5- to 9-nerved. Style or stigmas proceeding from below the tip of the ovary, on its anterior face. Stamens 3. Panicle simple or compound. **Bromus.**

Gr. IV. 37.
W. III. 30.

Lower pale rounded on the back, about 7-nerved, much longer than the upper one. Styles from the tip of the ovary. A short-bearded tuft at the base of each flower.
Avena striata, Mx.
(*Trisetum purpurascens*, Torr.)

Gr. VI. 50.
W. 29.

626. Flowers rounded on the back................627

Flowers compressed-keeled on the back.......634

627. Flowers all equally developed, or nearly so... .628

Spikelets oblong, or ovate-oblong, awnless, with 3 to 5 flowers, and only the 2 lower of them developed and fertile, the rest imperfect, dissimilar, convolute around each other. Lower glume almost equalling the spikelet. Lower pale minutely scabrous on the nerves, blunt at the summit. Panicle simple and sparingly branched.
Melica mutica, Walt.

Gr. IV. 30.
W. III. 38.

628. Glumes both, or one of them shorter than the lowest flowers, or, if longer, at least shorter than the spikelet.
629

Glumes large, as long as, or a little longer than the flowers, which are enclosed by them. Lower pale 2-toothed, oblong, or ovate, rounded-cylindraceous, 7- to 9-nerved, bearing between the sharp-pointed teeth an awn, composed of the 3 middle nerves; awn flattish and spirally twisted at the base. Spikelets 7-flowered, appressed. Panicle simple, raceme-like, 2 inches long. **Danthonia spicata**, Beauv.
Gr. VI. 48.
W. III. 27.

629. Spikelets ovate, sublanceolate, or linear........630

Spikelets roundish, heart-shaped at the base. Glumes convex. Flowers 2-rowed, imbricated, awnless. Panicle loose, erect; branches spreading. Spikelets on delicate, mostly drooping pedicels. **Briza media**, L.
Gr. IV. 35.
W. IV. 43.

630. Axis of the spikelet bearded below each flower. Lower pale 3-nerved, pointed, or short-awned...........631

Axis of the spikelet not bearded below the flowers, or the hairs belonging to the lower pale..............632

631. Upper pale apparently smooth; lower one bearded toward the base on the nerves, and with 3 cusps and 2 intermediate short, obtuse, membranaceous teeth at the summit. Spikelets purple. **Tricuspis seslerioides**, Torr.
Gr. IV. 24.
W. III. 31.

Both pales conspicuously fringed; the lower furnished with 1 short awn and 2 cuspidate teeth.
Uralepis purpurea, Nutt.
(*Tricuspis purpurea*, Gray.)
Gr. 24.
W. III. 32.

632. Spikelets always acute at the apex............633

Spikelets obtuse at the apex, truncate (rarely slightly acute), oblong, awnless, several- to many-flowered. Pales membranaceous, lower one 5- to 7-nerved; the nerves parallel and separate. Stamens 3 or 2. Stigmas plumose, mostly compound. Panicle racemose. Flowers mostly early deciduous. **Glyceria**.
Gr. IV. 31.
W. IV. 42.

633. Lower pale rigidly coriaceous; its 3 nerves terminating into a strong and abrupt awl-shaped tip. Stamens 2. Spikelets shining, 2-flowered. Panicle very simple and slender; its branches erect.

Diarrhena Americana, Beauv.
(*Diarrhena diandra*, Raf.)

Gr. IV. 26.
W. III. 35.

Lower pale papery, not rostrate, 3- to 5-nerved, acute, pointed, or often bristly awned, rarely bluntish. Upper pale finely ciliate. Stamens mostly 3. Flowers rather dry and harsh. **Festuca.**

Gr. IV. 36.
W. III. 36.

634. (626.) Glumes unequal-sided, one of them broader, rounder, the other narrower, more flattened, or somewhat concave..635

Glumes equal-sided. Tip of the flower straight. 636

635. Pales awn-pointed, the points bent inward; lower one keeled and 5-nerved, with the nerves converging into the tip; keel rough-ciliate. Glumes keeled, awn-pointed, with the tips bent inward, lower one broader and more convex than the other, which is flattish or slightly concave. Panicle 1-sided, with glomerulate (3- or 4-flowered) spikelets. Sheaths compressed. **Dactylis glomerata,** L.

Gr. IV. 27.
W. III. 33.

Pales obtusish, or acute; lower one oblong, compressed-boat-shaped, papery, upper narrower, flattish, hyaline. Glumes very dissimilar; lower one narrowly linear, keeled, 1-nerved, upper broadly obovate, convex, 3-nerved, scarious-margined. Spikelets commonly 2-flowered, with an abortive rudiment or pedicel. Panicle loose, or contracted.
Eatonia.

Gr. IV. 29.
W. III. 37.

636. Spikelets compressed-2-edged (7- to 20-flowered). Lower flowers, 1–3, neutral and 1-paled. Lower pale flattened and wing-keeled, upper one double-wing-keeled.

Glumes keeled. Stamen 1 (in *U. paniculata*, L., however, 3 stamens). Panicle spiked, open. **Uniola.**
Gr. V. 38.
W. IV. 44.

Spikelets not 2-edged, lanceolate, or ovate.....637

637. Flowers lanceolate. Lower pale acute, mucronate, or bristle-pointed. Glumes keeled, as are the pales, almost as long as the spikelet and 3-nerved. Spikelets 2- to 4-flowered, silvery and shining. Panicle narrowly spiked, spike-like, ovoid, or oblong-ovate, interrupted, or lobed at the base. Lower leaves sparingly hairy, or ciliate.
Kœleria cristata, Pers.
Gr. IV. 28.
W. III. 34.

Flowers ovate or lanceolate, not mucronate, nor bristle-pointed. Spikelets 2- to 70-flowered. Panicles spreading 638

638. Ligule and throat of the sheaths bearded with long villous hairs. Upper pale and axis of the spikelet persistent after flowering. Lower pale keeled, 3-nerved, membranaceous, like the glumes (not webby-hairy at the base).
Eragrostis.
Gr. IV. 34.
W. IV. 39.

Ligule and throat of the sheaths naked. Flowers deciduous, together with the joints of the axes. Lower pale keeled, 5-nerved, usually cobwebbed at the base. **Poa.**
Gr. IV. 33.
W. IV. 40.

---

639. (564.) Stamens 5–60, inserted on the 5-parted calyx. Succulent, prostrate, maritime herbs, with opposite leaves. Pod 3- to 5-celled, opening transversely. Styles 3–5. Flowers axillary, short-peduncled. (50) **Sesuvium**

Stamens numerous. Calyx colored. In place of a corolla, sometimes staminodia. (*Ranunculaceous plants*).
30

Stamens not over 12 (or, if more, rudimentary). Ovaries or styles 1–20.......................................640

640. Perianth regular, the segments either all, or the alternate ones equal............................641

Perianth decidedly irregular; segments, or sepals, of very unequal size and shape; or the perianth-tube curiously bent..755

641. Perianth of 3 to 10 divisions or sepals........642

Perianth (apparently) of 2 sepals. Stamen 1. Styles 2, filiform. Fruit indehiscent, nut-like, 4-lobed, 4-celled. Flowers polygamous. The solitary stamen in the sterile flowers between a pair of bracts (the apparent perianth), in the fertile between the pistil and stem. Other flowers, with 1 stamen and 1 pistil, between 2 bracts. Aquatics. Flowers axillary........................564

642. I. Perianth of 3 segments (lobes or teeth), or sepals...643

II. Perianth of 4 or 5 segments, or sepals......650

III. Perianth of 6 divisions...................701

IV. Perianth of 7–10 divisions................753

643. Ovary superior.............................646

Ovary inferior..............................644

644. Stamens 12, the tips of the filaments usually produced beyond the anther into a point. Perianth 3-cleft or parted, greenish- or brownish-purple. Fruit fleshy, 6-celled, crowned with the calyx, globular. Stemless herbs; the branched rhizome bearing 2 or 3 long-petioled, kidney-shaped leaves, and a short-peduncled flower, close to the ground. **Asarum.**

Stamens 3................................645

645. Leaves whorled. Corolla 3-cleft. Calyx-limb obscure. Style 1; stigmas 2. (534) **Galium.**

Leaves alternate. Perianth-tube 3-sided; limb 3-parted. Stigmas 3. Stems creeping at the base. Leaves lanceolate and serrate, or pectinately dissected. Flowers small, axillary, solitary, or 3–4 together. **Proserpinaca.**

646. Perianth of 3 lance-ovate, awn-pointed, petaloid scales. Stamens 3. Style 3-cleft. Culm leafy. Leaves (parallel-veined) and sheaths hairy. Spikes oblong, clustered in heads. (556) **Fuirena.**

Perianth not consisting of 3 scales............647

647. Perianth (3-cleft) becoming juicy, or berry-like and red in fruit, enclosing the 1-seeded fruit (utricle). Stamen (usually) 1. Styles 2. Flowers in glomerules.
Blitum capitatum, L.
Perianth not becoming berry-like............648

648. Flowers imbricated with (commonly 3) dry, scarious, persistent bracts. Stamens and stigmas 3 (or 2). Flowers sometimes perfect, but usually monœcious. Leaves entire...649

Flowers destitute of scarious bracts. Stamen 1. Ovary 2-styled. Perianth rather fleshy, but not becoming berry-like; its lobes 3 (2–4). Seed shining, with the margin acute. Leaves triangular-lanceolate, sparingly and coarsely toothed. Flowers all perfect.
Blitum maritimum, Nutt.

649. Utricle opening by a circumscissile line, 3- (or 2-) beaked. Flowers greenish, in small, close, axillary glomerules. Sepals mucronate. Bracts pungently pointed. Stems spreading. Leaves spatulate-oblong, very obtuse, long-petioled. Herb pale green. **Amarantus albus,** L. (827).

Utricle indehiscent or tearing open. Herb either smooth, livid-purplish, erect, or puberulent, ashy green, ascending. **Euxolus** (827).

650. Styles and ovaries 5–15. A shrubby plant with clustered stems, 1–2 feet high, and pinnate leaves. Sepals deciduous. Flowers polygamous, dull-purple, in compound racemes. (270) **Xanthorrhiza apiifolia,** L'Her.

Styles, or ovaries, 4–10. Herbs..............651
Styles, or ovaries, 1–3........................654

651. Stamens 8–10. Perianth of 5 sepals..........652
Stamens 4. Perianth of 4 segments, or sepals.653
Styles 4. Ovary 1. Sepals lanceolate, acute...692

652. Styles 5. Ovaries 5, united below, forming a 5-angled, 5-horned and 5-celled pod, which opens by the falling off of the beaks, many-seeded. Stamens 10. Flowers yellowish-green, in loose spikes. Leaves scattered.
(271) **Penthorum sedoides,** L.
Styles 10. Ovaries 10, green, united in a ring, in

fruit becoming a depressed, globose, 10-celled, 10-seeded, purple berry. Flowers white, racemed. Leaves alternate.
Phytolacca decandra, L.

653. Ovaries 4. Perianth 4-parted. Flowers in spikes or little heads. Leaves stipulate, mostly alternate or almost opposite. Aquatics. **Potamogetom**, L.
Ovary 1. Styles 4. Perianth of 4 sepals. (Petals wanting.) Leaves opposite, linear-subulate, or spatulate.
Sagina apetala, L.

654. Trees or shrubs.............................655
Herbaceous plants..........................664
655. Ovary superior.............................659
Ovary inferior..............................656
656. Stamens 8–10 (occasionally fewer). Flowers perfect, or polygamous...............................657
Stamens 4, or 5. Flowers never polygamous. Fruit a drupe, or berry............................658

657. Trees. Flowers diœciously-polygamous. Ovary 1-celled. Drupe 1-seeded, ovoid, blackish-blue. *Pistillate flowers* in peduncled clusters of 3 (2–5), much larger than the staminate ones (rarely single); calyx truncate, or a mere rim; petals very small, or early deciduous; stamens 5–10. *Staminate flowers* in peduncled clusters of 5–9; calyx truncate; petals, 5; stamens 10 (5–12) on a disk; pistil none.
- **Nyssa** (859).

Shrubs. Flowers all perfect. Ovary 4- to 5-celled. Berry many-seeded. Stamens 8–10. Calyx-teeth minute, or a mere rim. Anthers opening by terminal pores. Corolla bell- or urn-shaped, 4- to 5-toothed. (514) **Vaccinium.**

658. Stamens 4. Style 1. Calyx very small, minutely 4-toothed. Fruit a drupe, with a 2-seeded stone. Leaves opposite, entire. (277) **Cornus.**
Stamens 5. Styles 2. Petals small or minute, on the top of the 5-lobed calyx-tube, alternate with its at length reflexed segments. Fruit a berry. Low shrubs, often prickly. Leaves alternate and lobed. (286) **Ribes.**

659. Styles or stigmas 2. Flowers polygamous. Trees.
662

Style 1. Flowers all perfect. Leaves alternate.
Shrubs..660

660. Ovary 4- to 5-celled. Fruit a drupe, or a pod.
Stamens 4 or 5..661

Ovary 1-celled. Fruit a reddish drupe. Stamens 8. Perianth funnel-shaped, 4- (or 5-) toothed, wavy, or truncate. Flowers light yellow, 3 together, from a bud of 3 dark hairy scales, preceding the alternate leaves.
Dirca palustris, L.

661. Leaves evergreen, elliptical, rusty beneath, the margins replicate. Flowers all perfect. Petals 5, spreading. Calyx very small, obscurely 5-toothed, or almost obsolete. Stamens 5. Fruit a pod. (153) **Ledum latifolium**, Ait.

Leaves deciduous, not rusty beneath, nor revolute. Flowers polygamous. Fruit a berry-like drupe, black. Calyx 4- or 5-cleft. Stamens 4 or 5. (103) **Rhamnus**.

662. Ovary 1-celled. Fruit dry, nut-like, or a globular drupe. Stamens 4 or 5. Flowers monœciously-polygamous.
663

Ovary 2-celled. Fruit a double samara with opposite wings. Stamens 3–8. Calyx 5-cleft. Leaves deeply 5-lobed, silvery white or whitish. (167) **Acer**.

663. Anthers introrse. Fruit a drupe. Calyx of the perfect flowers 5-, that of the staminate 6-parted. Stamens as many as the calyx-lobes. Flowers greenish, axillary, the upper fertile (solitary or in pairs) peduncled, the lower staminate, clustered along the base of the young branches. Leaves cordate, ovate, or very long and taper-pointed, entire.
Celtis (785).

Anthers extrorse. Fruit nut-like, stalked in the calyx, beset with irregular, rough projections. Flowers in small, axillary clusters. Leaves small.
Planera aquatica, Gmel. (785).

664. (654.) Ovary superior.........................665
Ovary inferior..................................693

665. Perfect, staminate and pistillate (or only staminate and pistillate) flowers on the same plant..............666

Flowers all perfect, and the perianth 5- or 4-parted, cleft, or sepalled.....................................668

666. Flowers all with a perianth, often imbricated with bracts..667

The pistillate flowers consisting of a naked, 2-styled ovary, placed between two leafy bracts, which might be mistaken for a perianth. The staminate flowers furnished with a 5-sepalled perianth, also with a rudimentary pistil. Fruit an utricle. Herbs scurfy. Leaves triangular, or halberd-shaped. **Atriplex hastata,** L. (826).

667. Perianth 4-lobed. Stamens 4. Stigma sessile, tufted. Leaves chiefly alternate. Many of the (generally monœcious) flowers perfect. Achenium polished. **Parietaria Pennsylvanica,** Muhl. (831).

Perianth of 5 sepals, silvery-white, as are the bracts. Stamens 5, with 1-celled anthers. Stigmas 2 or 3. Leaves opposite, petioled. Utricle roundish-ovate, enclosed in the perianth. **Iresine celosioides,** L. (896).

668. Perianth 4-toothed, with the sinuses appendiculate, or 4-sepalled, with the two outer sepals smaller. Stamens 4 or 6. Low herbs..669

Perianth of 4 or 5 equal, or not appendiculate segments, or sepals...670

669. Calyx globular, or bell-shaped, 4-angled, 4-toothed, with a little horn-like appendage at each sinus (corolla wanting). Style 1. Stamens 4. Fruit a pod. Leaves narrow, opposite. Flowers small, greenish, axillary, solitary, or 3 together. Stem branched. (104) **Ammannia.**

Perianth 4-sepalled, the 2 inner sepals broader, erect, the 2 outer smaller and spreading, or reflexed. Achenium lens-shaped, girt with a broad wing. Stigmas 2, sessile, tufted. Stamens 6. Low acaulescent herbs. Leaves kidney-form, long-petioled. Flowers racemed, or somewhat panicled. (85) **Oxyria digyna,** Campd. (*O. reniformis,* Hook.)

670. Stamens 1–5, all fertile.....................671

Stamens 6–12, the alternate ones often sterile..684

671. Perianth 4-parted or sepalled.................672

Perianth 5- (rarely 4-) toothed, cleft, parted or sepalled..675

672. Flowers on a spadix........................673
Flowers not on a spadix....................674

673. Spadix (subglobose) enveloped in a shell-form spathe. Perianth deeply 4-parted; segments hood-form or wedge-shaped, with the apex bent inward, truncate, fleshy, becoming spongy. Style 4-cornered. Stamens 4. Berries globose. Leaves cordate, oval, acute.
Symplocarpus fœtidus, Nutt.

Spadix not in a spathe. Perianth 4- to 6-sepalled. Stamens 4–6. Stigma sessile. Fruit a green utricle. Flowers yellow, at the summit of the scape, which thickens upward into the yellow spadix. Leaves lanceolate.
Orontium aquaticum, L.

674. Perianth 4-parted, rotate (white). Stamens 4. Ovary globular. Stigma simple. Berry. Flowers in terminal racemes. Stem with 2 (sometimes 3) cordate leaves.
Majanthemum bifolium, D.C.
(*Smilacina bifolia*, Ker.)

Calyx of 4 sepals (petals wanting). Stamens 2. Silicles compressed. Cauline leaves incised, those of the branches linear. (115) **Lepidium ruderale**, L.

675. Flowers subtended by 2 or 3 bracts...........676
Flowers not subtended by 2 or 3 bracts.......677

676. Leaves prickly-pointed, awl-shaped, fleshy. Divisions of the perianth at length horizontally winged. Flowers solitary in the axils, with 2 leaf-like bracts. Embryo spiral.
Salsola Kali, L.

Leaves not prickly, nor fleshy. Lobes of the perianth not winged. Filaments united into a tube, bearing 5 1-celled anthers and 5 strap-shaped appendages. Stem not leafy above. Flowers in cottony spikelets, crowded into an interrupted spike. Embryo annular.
Frœlichia Floridiana, Moq.

677. Styles or stigmas 4, 3, 2, or solitary and bifid...678

Style 1; stigma capitate. Perianth bell-shaped, deeply 5-cleft (white, tinged with red); lobes oblong. Stamens 5. Flowers solitary, axillary. Leaves fleshy.
Glaux maritima, L.

678. Leaves opposite, or whorled..................690
Leaves alternate............................679

679. Sheathing stipules (ochreæ) present.
Polygonum (687.)
Stipules none. Fruit an utricle.............680

680. Styles constantly 3. Perianth urn-shaped, 5-cleft, or toothed. Flowers glomerate in racemes or panicles..681

Styles or stigmas usually 2, rarely 3. Perianth not (or hardly) urn-shaped, usually 5- (rarely 3-) lobed, remaining dry, or becoming juicy and red after flowering......682

681. Perianth with a scarious, horizontal wing, after flowering. Utricle depressed. Seed horizontal. Stem furrowed. Leaves sinuate-toothed.
Cycloloma platyphyllum, Moq.

Perianth not winged after flowering. Utricle podlike, glandular-dotted. Seed vertical. Leaves once- or twice-pinnatifid. **Roubieva multifida,** Moq.

682. Seed vertical. Perianth 5- (or 3-) parted, becoming juicy or berry-like, or remaining dry, after flowering. Styles or stigmas 2. Embryo in a complete ring. Flowers glomerate, forming axillary or terminal spikes.
(647 and 648.) **Blitum.**

Seed horizontal. Perianth 5- (or 3-) cleft or parted, remaining dry after flowering. Embryo spiral or coiled.
683

683. Embryo spiral. Flowers glomerulate (2 or 3), axillary. Stigmas 2 or 3, sessile. Leaves linear, fleshy.
Chenopodina maritima, Moq.

Embryo coiled. Flowers glomerate in panicled spikes. Styles 2. Leaves broad, often rhombic. Herbs often glaucous, or glandular. **Chenopodium.**

684. (670.) Perianth campanulate, or top-shaped, its limb apparently 8- to 10-lobed (that is, parted into 4 or 5 petaloid segments, alternate with as many glands). Stamens about 12, each jointed on a pedicel, which has a little bract at its base. Ovary stalked, 3-lobed. Styles 3, bifid.
538

Perianth 4- or 5-parted, without interposed glands.

Stamens not over 10, not jointed on a pedicel. Ovary not stalked. Herbs without milky juice..................685

685. Stipules of the alternate leaves sheathing the stem above the tumid joints. Stamens 8 (rarely 6 or 5). Styles 3; or 2. Achenium 3-cornered.......................686

Stipules not sheathing, or altogether wanting..688

686. Pedicels usually (not always) fascicled 687

Pedicels solitary, with imbricated, truncate bracts. Sepals erect-spreading, 3 of them at length closed, withering-persistent on the achenium. Leaves linear, caducous from the top of the tubular, truncate sheath.

Polygonella articulata, Meisn.
(*Polygonum articulatum*, L.)

687. Sepals all at length closed on the achenium. Embryo semilunarly curved around half of the albumen.

Polygonum.

Sepals all open. Stamens with nectariferous glands between. Embryo straight in the centre of the albumen; cotyledons twisted-plaited. Leaves triangular, or halberd-form. **Fagopyrum esculentum**, Mœnch.

688. Perianth 4-cleft or sepalled...................689

Perianth 5-parted or sepalled.................690

689. Perianth 4-cleft, greenish-yellow, with purple lines. Ovary half superior, 2-styled. Stamens 8, very short; anthers orange-colored. Flowers in leafly cymes, the terminal one 10-androus. Leaves orbicular, crenate, chiefly opposite. Aquatic herb. **Chrysosplenium Americanum**, Schw.

Perianth 4-sepalled. Ovary entirely superior, 1-styled. Stamens 6, 4 of them antherless. Silicle kidney-form, or bitubercular, 2-seeded. **Senebiera didyma**, Pers.

690. (678.) Styles or stigmas 2, or style 1, bifid.....691

Styles or stigmas 4 or 3....................692

691. Perianth bell-shaped, 5-parted. Stamens 5 (or 10, only 5 of them fertile). Styles 2. Joints of the stem swollen. Stipules none. **Seleranthus annuus**, L.

Perianth 5-sepalled. Sometimes bristles in place of the corolla. Stamens 5, 3, or 2. Style bifid, or sessile. Stigmas 2. Flowers small, enveloped in dry bracts. Stipules scarious, silvery. **Paronychia & Anychia.**

692. Leaves broadly lanceolate, opposite. Styles usually 4. Stamens 10 (sometimes 5 only fertile). Pod twice the length of the calyx. (Petals sometimes present.) Stems flaccid, many times forked, finally resolved into a leafy cyme. Stipules none. **Stellaria borealis**, Big.

Leaves spatulate, whorled at the joints. Perianth of 5 sepals, white inside. Stigmas usually 3, alternate with the 3 ovary-cells (or sometimes 5, and alternate with the sepals). Herbs prostrate, in patches. Pedicels 1-flowered, forming sessile umbels at the joints. Stipules obsolete.
Mollugo verticillata, L.

693. (664.) Submersed aquatics. Flowers monœcious, polygamous, or sometimes mostly perfect. Stigmas 4, pubescent, sessile......................................828

    Non-aquatics 694

694. Styles 2, or more.........................695

    Style 1..................................698

695. Stamens 8–10............................697

    Stamens 5. Calyx-limb nearly entire. Flowers in simple or compound umbels, or in umbellate corymbs, or panicles...696

696. Styles 2, distinct or united at the base. Umbels compound. Fruit dry, splitting in two, at maturity......290

    Styles 2–5, separate, or united into one. Fruit fleshy, or berry-like. Umbels corymbed, or panicled, or simple. **(Aralia & Panax)** 287.

697. Leaves roundish, somewhat cordate, obscurely crenate-lobed. Leaves chiefly opposite.................689

Leaves linear, opposite. The 5 sepals united below in an indurated cup, enclosing the 1-seeded utricle..691

698. Leaves alternate, rarely opposite. Involucre to many flowers none..699

    Leaves opposite. Calyx-limb obsolete. Petals and stamens 4. Drupe small, with a 2-seeded stone. Cyme (or head) of flowers subtended by a 4-leaved, white, rather showy involucre. Upper leaves crowded into an apparent whorl, in 6's or 4's. Stem 4 to 8 inches high...........538

699. Stamens (4 or 5) connected to the lobes of the 4- to 5-parted perianth by a tuft of hairs and inserted on a disk. Fruit drupaceous. Stems 8–10 inches high, very leafly. **Comandra.**

Stamens not connected by hairs to the perianth. 700

700. Perianth 3-bracted at the base. Ovary 4-angled. Stigma tufted. Stamens 4. Flowers in terminal, cylindric (in fruit much elongated) spikes. Leaves pinnate, alternate. **Sanguisorba Canadensis,** L.

Calyx not bracteolate at the base. (Corolla often present.) Ovary, or pod, short-cylindrical, or cubic. Leaves simple and entire, alternate, or opposite. Flowers axillary. (277) **Ludwigia.**

701. (642.) Number of stamens the same in all flowers of the species (or in some of them none)..............704

Number of stamens not the same in all flowers. Stamens 9 in the sterile flowers, in 3 rows, the inner row gland-bearing at the base (this sort of flowers nearly always exclusively borne by one specimen of the species), and either 6, or 15–18 in the fertile flowers (growing, mostly exclusively, on another individual). Shrubs, or trees.....702

702. Leaves uniform. Fertile flowers with 15–18 rudiments of stamens in 2 forms, and a globular ovary. Drupe red. Shrubs..703

Leaves of 2 forms : one sort ovate, entire, the other 3-lobed. Fertile flowers with 6 short rudiments of stamens, and an ovoid ovary. Anthers 4-celled, 4-valved. Drupe blue, on a red pedicel. Tree 15–50 feet high. **Sassafras officinale,** Necs. (871).

703. Anthers 2-celled, 2-valved. Involucre of 4 scales. The inner row of the 9 stamens of the sterile plant 1- to 2-lobed. **Benzoin** (872).

Anthers 4-celled, 4-valved. Branchlets zigzag. involucre of 2 to 4 scales.
Tetranthera geniculata, Nees (872).

704. I. Stamens 3, rarely 9.....................705
II. Stamens 6............................712
III. Stamens 12, 3 of them sterile. Perianth 6-

parted. The anthers of 9 stamens 4-celled. Trees. Leaves oblong, entire, alternate. Flowers small, panicled. Berries dark blue on a red pedicel. **Persea Caroliniensis**, Nees.

705. The 3 stigmas petaloid, covering the stamens. Style none, or scarcely any. Perianth funnel-form, adherent to the ovary; segments in 2 sets (colored alike), somewhat unequal; the 3 outer ones reflexed, the 3 inner erect. Flowers large, showy. Herbs from horizontal rhizomes. Leaves sword-shaped, equitant, 2-ranked. **Iris.**

Stigma, or stigmas never petaloid. Perianth segments greenish, or of another color 706

706. Ovary superior 710
Ovary inferior 707

707. Submersed, diœciously-polygamous aquatics. Stems elongated, branching, crowded with lanceolate, oblong or linear, serrulate, verticillate leaves. Perianth of the fertile flowers slender, hair-like, 6-parted. Flowers dingy white, solitary and sessile, from a sessile and tubular axillary spathe.
Anacharis Canadensis, Planchon (885).

Herbs not submersed 708

708. Leafless herbs, beset with minute scales. Stems capillary, 2 to 3 inches high. Tube of the bell-shaped perianth strongly 3-winged. Flowers 1 or 2, sometimes more, small, light blue, at the top of the stem. Capsule prismatic.
Burmannia biflora, L.

Plants with parallel-veined, grass-like leaves ... 709

709. Anthers linear, fixed by the middle; filaments distinct. Perianth smooth and yellow within, scurfy or woolly outside; the 3 exterior sepals linear, the 3 inner lance-oblong. Leaves sword-shaped, equitant, mostly radical. Marsh herbs.
Lachnanthes tinctoria, Ell.

Anthers extrorse; filaments monadelphous. Perianth blue, or purple, glabrous, with the segments mucronate. Flowers on filiform pedicels, small. Scape simple, strongly winged, so as to resemble a leaf. Leaves linear, as long as the scape. Spathe 2-leaved.
Sisyrinchium Bermudianum, L.

710. Perianth petaloid; tube slender. Low, or grass-like, creeping, floating or submersed herbs, with a 1- or few-flowered spathe. Leaves parallel-veined..............711

Perianth herbaceous. Rushes. Flowers cymose or panicled. **Juncus** (741).

711. Anthers of 2 forms; the upper 2 filaments thickened in the middle, bearing ovate, yellow anthers; the lower, longer one an arrow-shaped, green-colored anther. Spathe 3- to 5-flowered. Leaves round-reniform. Flowers white. Muddy margins of streams. **Heteranthera.**

Anthers alike, oblong-arrow-shaped; filaments awl-shaped. Herb growing wholly under water. Leaves linear, thin, sessile. Spathe terminal, 1-flowered, rising to the surface of the water; flower small, pale yellow.

Schollera graminea, Willd.

712. (704.) Ovary or ovaries superior..............713
Ovary (entirely or only half) inferior..........748

713. Tree..663
Herbs...714

714. Stamens separately opposite the perianth-segments. 715

Stamens in pairs, opposite the 3 inner sepals. Flowers perfect, or diœcious, in simple or compound racemes. Perianth of 6 nearly distinct sepals, the 3 inner (valves) larger, petaloid, convergent over the achenium, one of them usually tubercled on the back, the 3 others herbaceous and reflexed in fruit. Styles 3, short; stigmas tufted.

(94) **Rumex** (884).

715. Flowers crowded on a lateral cylindric spadix: Perianth of 6 scales. Stigma minute, blunt, sessile. Ensiform leaves and leaf-like scape radical, aromatic.

Acorus calamus, L.

Flowers solitary, or in a terminal spike, umbel, simple or compound raceme, etc......................716

716. Flowers terminal............................720
Flowers axillary............................717

717. Leaves thread-like, or bristle-form. Perianth bell-shaped, on a jointed pedicel, the 3 inner segments somewhat broader. Ovary ovoid. Stigmas 3. Berry red.

Asparagus officinalis, L. (890).

Leaves ovate-oblong, elliptic, or lanceolate, often clasping. Fruit a berry or a pod.....................718

718. Flowers axillary throughout the flowering period.
719

Flowers primarily terminal, at length axillary, nodding..740

719. Flowers solitary, axillary, on distorted and abruptly bent peduncles, drooping. Peduncles ascending first and then, forming a sling or knee in the middle, deflexed. Perianth bell-form ; sepals, each with a nectariferous pore at the base,. recurved-spreading from a little above the bell-shaped base. Stamens inserted at the base ; anthers sagittate. Berry red. Roots fibrous, matted. **Streptopus.**

Flowers on undistorted peduncles, which are 1- to 4- (usually 2-)flowered, axillary, drooping. Stamens inserted near the middle of the erect-lobed perianth-tube. Berry blue, or black. Rhizome creeping, thick, knotty.
Polygonatum multiflorum, Desf.

720. Flowers in terminal umbels, or heads. Scape naked, or with sheathing leaves near the base ; rarely a stem with 2 whorls of leaves. Perianth always corolline.....721

Flowers in racemes, panicles, corymbs, or solitary (rarely in umbels, and then with the 6 sepals glumaceous).
723

721. Two whorls of leaves. Stem simple, clothed with flocculent, deciduous wool, bearing 2 verticils of sessile obovate-lanceolate leaves, one of 5-9 near the middle, and another of 3 (smaller ones) at the top. Flowers small, recurved, pale greenish-yellow, about 4, in a terminal sessile umbel. Style none ; stigmas 3. Berry dark purple. Rootstock having the taste of the cucumber.
Medeola Virginica, L.

Leaves not whorled........................722

722. Perianth of 6 distinct sepals, bell-shaped, lily-like, slightly downy outside, deciduous. Filaments thread-like. Flowers rather large, greenish-yellow, or white. Berry ovoid, blue, few- to many-seeded. Leaves 2-4, large, oblong or oval, ciliate, sheathing at the base of the scape.
Clintonia.

Perianth deeply parted; segments 1-nerved, persistent, at length dry and scarious. Filaments awl-shaped (dilated below). Pod lobed, 3-valved. Seeds 1 or a few, oval-reniform. Flowers rather small, some of them changed to bulblets. Scape naked, or sheathed by leaves at or near the base. **Allium.**

723. Perianth funnel-form, or bell-shaped; limb 6-cleft down to the middle (or farther).....................724
Perianth of 6 sepals, barely united at the base.726

724. Ovary half superior..........................748
Ovary entirely superior. Scapes.............725

725. Perianth funnel-form, lily-like, tubular at the base; limb campanulately dilated, 6-parted, very wavy. Stamens and curved style long, filiform, ascending, the style longer than the stamens, which are inserted on the throat of the perianth. Ovary oblong. Pod 3-cornered, fleshy, many-seeded. Flowers large, about 3 inches long, orange-color. Leaves long, linear, keeled, 2-ranked. **Hemerocallis fulva**, L.

Perianth ventricose-campanulate; limb cleft down to the middle. Stamens inserted on the base, included. Ovary ovoid, tapering into a short style. Flowers racemed, snow-white. Berry few-seeded, red. Leaves 2, oblong, with long, sheathing petioles. **Convallaria majalis**, L.

726. Style none; stigmas sessile. Flowers in terminal racemes. Root creeping.........................727
Styles 1 or 3, conspicuous...................729

727. Leaves coriaceous, lanceolate, with filaments on their margins, covering the short trunk, which is from a running rhizome and 1 foot high. Flowers large, showy, white, panicled. The 3 inner sepals broader. Pod oblong, 6-sided. **Yucca filamentosa**, L.
Leaves not coriaceous, nor ciliate. Flowers small. 728

728. The 3 outer of the (greenish) sepals concave, tubercled at the base, the 3 inner more straightened. Anthers oval. Ovary 1; stigmas 3–6, downy. Pod linear, acute at the base, dehiscent from below by 3 teeth. Scape with a terminal raceme. Leaves rush-like, fleshy. **Triglochin.**

Sepals (greenish-yellow) perfectly equal. Ovaries 3-6, connate at the base; stigmas flat, blunt. Pods 3-6, obliquely ovoid, diverging. Flowers peduncled, solitary, from the axils of long sheathing bracts, in a raceme. Stem zigzag, partly sheathed by the bases of the conduplicate, grass-like leaves. **Scheuchzeria palustris**, L.

729. Styles 3, or style 1, and 3-cleft or parted......730
 Style 1; stigma simple, blunt..............742

730. Styles 3....................................731
 Style simple, 3-cleft or parted. Flowers drooping or nodding..738

731. Anthers heart-shaped or kidney-form, confluently 1-celled (shield-shaped after flowering). Styles awl-shaped. Pod 3-horned, septicidal. Stem simple. Leaves lanceolate, or linear. Flowers racemed or panicled..............732
 Anthers 2-celled. Pod loculicidal. Flowers racemed or spiked...735

732. Sepals bearing 1 or 2 glands on the inside, next the base. Flowers panicled. Seeds winged..............733
 Sepals not glandular, nor clawed. Perianth widely spreading..734

733. Flowers polygamous. Sepals cordate or halberd-form, perfectly free from the ovary; their long claws coherent with and shorter than the filaments. Glands confluent. Flowers dull yellowish-green, or cream-colored.
Melanthium Virginicum, L.

Flowers perfect. Sepals oblong, or ovate, scarcely clawed, either free, or coherent with the base of the ovary. Stamens free from the sepals, and about their length. Flowers whitish-green. **Zygadenus.**

734. Flowers polygamous, panicled, or racemed. Sepals longer than the stamens, greenish or brownish, contracted at the base, but free from the ovary, or nearly so. Filaments recurved. Leaves 3-ranked, broadly oval, strongly plaited and pointed, or more lanceolate, scarcely plaited, and sometimes conduplicate-keeled. Stem very leafy.
Veratrum (including *Stenanthium*, Gray.)

Flowers perfect, racemed. Perianth free; sepals oval or obovate, shorter than the stamens. Stem scape-like,

few-leaved. Raceme simple. Flowers white. Leaves grass-like, broadly linear, elongated.
Amianthium muscætoxicum, Gray.

735. Styles stigmatic down the inner side. Anthers extrorse. Pod loculicidal, ovoid, or globular............736

Styles subulate, with terminal stigmas. Anthers innate, slightly introrse. Pod septicidal, 3-angled. Leaves equitant, linear. Stem scape-like, glutinous, or roughened with dark glands. Flowers usually involucrate with 3 bracts. Sepals concave, sessile. (263) **Tofielda.**

736. Flowers diœcious. Fertile flowers with rudimentary stamens. Styles linear, club-shaped. Seeds linear, winged. Stem wand-like, leafy. Leaves flat, lanceolate, the lowest spatulate, tapering into a petiole.
Chamælirium luteum, Gray (889).

Flowers perfect. Styles filiform. Pod globose, 3-lobed..737

737. Perianth white; sepals as long as the filaments. Stem simple, from a bulbous base. Leaves rush-like, the radical 1 foot long, 1 inch wide, in a dense tuft, those of the stem needle-shaped, passing upward gradually into bristle-form bracts. Ovary-cells few-ovuled. Seeds not appendiculate. **Xerophyllum asphodeloides**, Nutt.

Perianth purplish-green; sepals shorter than the filaments. Anthers blue. Scape naked, hollow, from a tuberous rhizome, and sheathed with broad bracts at the base. Leaves oblong-spatulate, or oblanceolate. Ovary-cells many-ovuled. Seeds linear, with a tapering appendage at each end.
Helonias bullata, L.

738. (730.) Leaves grass- or rush-like. Root fibrous. Stigmas 3, filiform, hairy. Pod 3-celled, 3-valved.......741

Leaves ample, not grass- nor rush-like. Root a bulb, or fibrous.......................................739

739. Stem forking, from a root-stock, or a fibrous root. Style trifid. Anthers extrorse. Leaves ovate-oblong, or oblong..740

Scape 1-flowered, from a solid, scaly bulb. Style elongated. Anthers introrse. Leaves 2, elliptical-lanceolate, spotted, not dotted, from a little above the base of the scape, their petioles sheathing its base. Perianth white, or bluish-white. Filaments subulate. Pod obovate. Seeds numerous. **Erythronium albidum**, Nutt.

740. Perianth about 1 inch long. Filaments shorter than the long linear anthers. Pod 6- to many-seeded. Style deeply 3-cleft. Flowers pale yellow, commonly solitary (rarely in pairs). Peduncles terminal, at length lateral, by the growth of the branches. Leaves sessile, or clasping-perfoliate. **Uvularia.**

Perianth half an inch long. Filaments filiform, much longer than the linear-oblong, blunt anthers. Berry 3- to 6-seeded. The 3 stigmas of the style short. Leaves rounded or cordate at the base, ovate-oblong, taper-pointed, closely sessile. Flowers solitary, or mostly in pairs, greenish-yellow. Herb downy. **Prosartes lanuginosa**, Don.

741. Cells of the pod 1-seeded. Leaves mostly hairy; grass-like, flat. **Luzula.**

Cells of the pod many-seeded. Leaves never hairy, channelled, or semiterete, or wanting, and in their place sheaths only. (710) **Juncus.**

742. (729.) Filaments woolly; anthers linear. Sepals linear, lanceolate, yellow. Pod cylindrical-oblong, pointed with the style, which is terminated by a simple stigma, many-seeded. Seeds furnished with a long bristle-form tail at each end. Rhizome creeping, with linear, equitant leaves, and a scape, 6–18 inches high, bearing a single raceme. **Narthecium Americanum**, Ker.

Filaments not woolly. Leaves not equitant ... 743

743. The 3 inner sepals of the lily-like perianth each with a tooth on each side, at the base. Leaves 2, elliptical-lanceolate, pale green, spotted with purplish, and dotted, a little above the ground. Style club-shaped; stigmas united. Flower pale yellow, at least 1 inch long, drooping. Scape 6–9 inches long.

(739) **Erythronium Americanum**, Smith.

Sepals without teeth 744

744. Flowers very large, 2–4 inches long, orange-color, or scarlet, spotted with purple. Perianth funnel-form, or bell-shaped; sepals each with a nectariferous, linear furrow at the base, spreading, or recurved above, and at the base sometimes distant and long-clawed. Anthers versatile. Style elongated, somewhat club-shaped; stigma 3-lobed. Seeds in 2 rows in each of the 3 cells of the pod. Leaves sessile, numerous, alternate-scattered, or whorled. Stems from scaly bulbs. **Lilium.**

Flowers much smaller. Nectariferous furrows none 745

745. Sepals white, marked with a vertical green stripe outside, on the middle. Style 3-sided; stigma 3-cornered. Scape and channelled leaves linear, from a coated bulb. Flowers 5–8, on long, spreading pedicels; corymbed.
Ornithogalum umbellatum, L.

Sepals not striped on the outside, deciduous. Style not 3-sided.. 746

746. Stem leafy. Perianth-segments spreading, white. Berries pale red, speckled with purple, or red and not speckled, or blackish. Leaves alternate. Raceme simple, or compound. **Smilacina.**

Scape terete, or angled. Raceme bracted...... 747

747. Flowers white, nodding, sweet-scented, in a secund raceme. Perianth bell-shaped. Style stout; stigma 3-angled. Leaves 2, oblong, with long, sheathing petioles 725

Flowers blue, or purple. Sepals spreading. Style filiform; stigma minutely 3-cleft. Pod 3-angled; seeds several in each cell, black. Leaves long, linear, keeled.
Scilla Fraseri, Torr.
(*S. esculenta*, Ker.).

748. (712.) Perianth adherent to the whole surface of the ovary. Stemless herbs; their scapes usually from a bulb, rarely from a fibrous-rooted crown................... 750

Perianth adherent only to the lower half of the ovary. Herbs with a creeping root-stock, or a fibrous root.
749

749. Perianth 6-cleft, yellow within, and woolly without. Stamens inserted near its base. Style conical, tripartible. Root creeping. Stem flexuous, white-woolly above. Leaves equitant. Flowers in a loose, woolly corymb.
Lophiola aurea, Ker.
(*L. Americana*, Wood.)

Perianth 6-cleft, white or yellowish, minutely warty outside, looking scurfy, or mealy. Stamens inserted in its throat. Style awl-shaped, or 3-sided, 3-cleft at the apex. Root fibrous. Scape bearing the numerous flowers in a wand-like spiked raceme. Leaves lanceolate, clustered at the base.
Aletris.

750. The throat of the slender perianth-tube bearing a top-shaped, ample, delicate, 12-toothed membrane (crown), which connects the bases of the exserted stamens. Anthers linear, versatile. Pod thin. Scape and long, strap-shaped leaves from a coated bulb. Flowers large, showy, about 3 inches long, white, 2 to 6 (usually but 2).
Pancratium rotatum, L.

Crown none..................................751

751 Anthers versatile........................752

Anthers erect. Perianth parted nearly down to the ovary, the segments yellow within, hairy and greenish outside. Scapes several, slender, hairy, bearing 1–4, peduncled flowers at the top. Leaves linear, grass-like.
Hypoxis erecta, L.

752. Flower solitary, large (over 3 inches long), white and pink, very showy, peduncled, from a 1-leaved, 2-cleft spathe. Stamens inserted into the throat of a funnel-form perianth. Scape and long, linear leaves from a coated bulb.
Zephyranthus Atamasco, Herb.
(*Amaryllis Atamasco*, L.)

Flowers several, 1 inch long, greenish-yellow, very fragrant, sessile, scattered along the scale-bearing, elongated scape. Leaves linear, lanceolate, very thick and fleshy, the margin with cartilaginous teeth, clustered at the base. Root tuberous. **Agave Virginica, L.**

753. (642.) Trees, with simple deciduous leaves, and fasciculate, rarely racemose flowers. Perianth 7- to 9-cleft. Styles or stigmas 2. Stamens 7–9. **Ulmus.**

Herbaceous plants. Perianth apparently 8- to 10-cleft 754

754. Stamens 1 or 4. Perianth inferior, tubular, apparently 8-, but really 4-parted, with as many alternate

bractlets. Leaves 3-parted, the wedge-form lobes 2- to 3-cleft, pubescent. Flowers greenish, axillary.
Alchemilla arvensis, L.

Stamens 8-10. Leafless plants, with a simple scale-bearing stem. (Sepals 4 or 5, concave and erect, like the petals.) Stigma funnel-form. Pod 4- to 5-valved......126

755. (640.) Ovary within the perianth. Stamens 6, either distinct or diadelphous........................756

Ovary under the perianth............757

756. Perianth-leaves more or less distinct, or readily separable; 2 sepals and 4 petals—the two outer, or one of the latter spurred or sac-like at the base; the two inner ones callous and coherent at the top, including 6 diadelphous stamens, the middle one in each set of 3 anthers largest and 2-celled, the lateral ones 1-celled. Leaves compound.....221

Perianth 2-lipped, tubular at the base, consisting of 6 sepals, the 3 upper of them united into the upper lip, the 3 lower spreading, scarcely coherent and separable down to the base of the tube. Stamens 6—the 3 lower exserted, the 3 upper very short and often imperfect. Anthers blue. Ovary 3-celled, 2 cells empty, the third one 1-seeded. Fruit an utricle. Scape 1-leaved, terminating into a spike of violet-blue flowers. Leaf cordate, oblong, obtuse.
Pontederia cordata, L.

757. Perianth tubular, variously bent or curved and inflated above the ovary (sometimes S-form), its short limb obtusely or obscurely 3-lobed. Stamens 6; the sessile anthers adnate to the short, 3- to 6-lobed or angled stigma. Flowers axillary, greenish or lurid purple. Stem either erect and flexuous, or climbing and twining. **Aristolochia.**

A ringent perianth of 6 very irregular sepals, adherent below to the 1-celled ovary, with 3 parietal placentæ. The solitary stamen (or 2 stamens in *Cypripedium*) completely coherent with the style, and forming with it the column. The 3 outer and 2 of the 3 inner sepals consimilar in texture and shape; the middle (upper, or posterior) of the 3 inner ones becoming (one case excepted) the lower and anterior, from the twisting of the ovary, and differing from the rest in shape and direction (*labellum*, or lip). Lip frequently lobed, and often spurred. Pollen cohering in waxy or mealy masses (*pollinia*). (Class: *Gynandria*, L.) 758

758. Anther 1, either terminal or dorsal on the stigma. 759

Anthers 2, both fertile, fixed below the lateral lobes of the column; moreover a petaloid sterile stamen. Lip large, inflated, forming a somewhat slipper-shaped sac.
Cypripedium.

759. Lip anterior. 760

Lip posterior, on a narrowed stalk, dilated and bearded above with white, yellow, and purple hairs. Pollen-masses 2, bipartible. Column petal-like, incurved, winged at the apex. Scape 3–5 inches high, 2- to 5-flowered, sheathed below by the base of the grass-like leaf.
Calopogon pulchellus, R. Br.

760. Lip spurred, the spur perfectly free from the ovary. Anther terminal. 761

Lip spurless, or spur adnate to the ovary. Anther terminal, or dorsal. 764

761. Anther erect. Pollen-masses 2, granular, borne on a stalk, its base attached to the 2 glands of the stigma. . 762

Anther lid-like, over the stigma, at length deciduous. Pollen-masses 4 (or rather 2, and 2-parted), connected by a linear stalk with the transverse small gland. Column narrow. Lip short and flat; its spur long and thread-like. Raceme many-flowered, on a naked scape. Flowers small, greenish, tinged with purple. Leaf solitary, ovate, plaited and tinged with purple beneath.
Tipularia discolor, Nutt.

762. Glands naked. Lip entire, toothed, lobed, dissected, or fringed. 763

Glands confined in a little pouch or hooded fold. Lip entire. Flowers showy, large, white with pink, in a spike. Scape 4-angled, 4–7 inches high. Leaves 2, oblong-obovate. Bracts leaf-like. **Orchis spectabilis**, L.

763. Anther-cells parallel, contiguous. Glands close together. Lip entire or toothed. Flowers small, pale-yellowish green, or orange-yellow, spiked. Stem either only with 2 or 3, or with several leaves. **Gymnadenia.**

Anther-cells diverging below. Glands widely separated. Lip entire, or lobed, cleft, dissected, ciliate-fringed. **Platanthera**, Rich.

764. Leafless, tawny, or brownish-purple herbs, with scaly scapes, rarely with a root-leaf. Flowers spiked or racemed..765

Herbs green, and with leaves.................767

765. Lip concave, not hooded. Pollen-masses 4.....766

Lip hooded, jointed, crested along the upper face, often 3-lobed. Column half-cylindric. Pollen masses 8, in pairs, with a stalk to each pair. **Bletia aphylla**, Nutt.

766. Lip sessile, with 2 projecting ridges on the face below, slightly adherent at the base to the column, and often with a short spur, which is adnate to the summit of the ovary. Rhizome much branched and toothed, coral-like. **Corallorhiza**, Haller.

Lip short-clawed, 3-lobed, free, the palate 3-ridged. Not any vestige of a spur. Radical leaf, appearing late in the season, petioled, elliptic or ovate, large, 3–5 inches long. Root of 2 or more tubers, connected by thread-like fibres. **Aplectrum hyemale**, Nutt.

767. Lip sac-like, inflated, 3-lobed at the apex, middle lobe bearded above and 2-pointed underneath, near the end. Column broadly winged, petal-like. Scape 1-leafed, 1-flowered, from a corm. Flower large, whitish, variegated with purple and yellow. **Calypso borealis**, Salisb.

Lip not sac-like..................................768

768. Anther dorsal (attached to the back of the column). 769

Anther terminal (fixed to the apex)..........771

769. Lip 2-cleft, usually drooping. Stigma with a rounded beak. Stem with a pair of opposite, ovate leaves in the middle. Flowers small, greenish, or brownish-purple, spiked or racemed. **Listera.**

Lip entire.......................................770

770. Lip free from the column, strap-pointed. Pollen-masses elastic. Root-stock bearing a tuft of ovate petioled leaves next the ground. Scape, spike, and small, greenish-white flowers glandular-downy. **Goodyera**, R. Br.

Lip embracing the column, pointless, channelled. Lateral outer sepals decurrent on the ovary, covering the base of the lip. Column arching. The stigma usually with a 2-cleft beak. Spike usually twisted. Flowers, small white. **Spiranthes**, Rich.

771. Lip bearded inside, its base adherent to the linear column, which is petal-like below. Pollen-masses 4 (2 in each anther-cell). Scape, from a globose, solid bulb, bearing a single, large, rose-purple, fragrant flower. Leaf solitary, linear. **Arethusa bulbosa, L.**

Lip not bearded. Anther lid-like............772

772. Lip auricled, cordate or sagittate at the base. Column minute, round. Scape from a solid bulb, with 1 or 2 leaves, and a raceme of minute greenish flowers.
Microstylis, Nutt.

Lip not auricled at the base...............773

773. Lip crested or 3-lobed. Column club-shaped. Anther stalked. Stem 1- to 5-leaved. Flowers purple, either a few axillary, or a solitary one. **Pogonia.**

Lip entire, flat, often with 2 tubercles above the base. Inner set of sepals thread-like. Column long, incurved, margined at the apex. Scape low. Leaves 2. Flowers purplish or greenish, racemed. **Liparis.**

D. DICLINOUS FLOWERS.
(774–903.)

774. (4.) Sterile (staminate) and fertile (pistillate) flowers separate, but on the same plant, monœcious........775

Sterile and fertile flowers on two individuals of the species, the one or other of these sorts borne exclusively (or nearly so) by each, diœcious....................840

775. Trees or shrubs........................776
Herbaceous plants........................797

776. Pistil (fertile flower) an open scale; the ovules at its base fertilized by the direct application of the pollen; perianth none. Sterile flowers also naked. Both fertile and sterile flowers in aments; the fertile aments called cones, or strobiles. Cotyledons often more, than 2. Leaves needle-form, awl-shaped, filiform, or scale-like, flat, squamous. *Conifers*..777

Pistil closed; ovules, or seeds, enclosed. Sterile flowers either in aments, or spikes, or rarely solitary; sometimes clustered in the axils..........................782

777. Fertile scales many, each subtended by a bract, with 2 inverted ovules. Seeds winged................778

Fertile scales few, bractless, each with 2 to 8 erect ovules 780

778. Leaves evergreen 779

Leaves deciduous, filiform, very slender, many in the fascicles, on short, lateral branchlets. The sterile flowers with 2-celled anthers, their cells opening lengthwise. Cones erect, ovoid; scales colored, persistent.
Larix Americana, Mx.

779. Leaves fascicled, 2–5 together. Sterile aments in terminal clusters; stamens many; anthers 2-celled, with a scale-like connectile; pollen of 3 united grains. Fertile aments (cones) conical, or cylindric, the carpellary scales bracted, imbricate, often thickened or awn-pointed at the tip, at length hardened. Seed nut-like. **Pinus.**

Leaves separate, scattered. Sterile aments axillary, clustered toward the ends of the branches. Scales of the fertile aments (cones) flat, not thickened, nor awn-pointed.
Abies.

780. Leaves evergreen, scale-like, or awl-shaped. Flowers monœcious on different branches. Filaments scale-like, with 2–4 anther-cells. Branchlets 2-edged, compressed..781

Leaves deciduous, linear, 2-ranked. Flowers monœcious on the same branches. Scales of the fertile, closed, clustered aments with 2 ovules at the base. Sterile aments spiked-panicled, of few stamens; filaments scale-like, peltate, bearing 2–5 anther-cells.
Taxodium distichum, Rich.

781. Scales of the globular, closed cones in 4 ranks, shield-shaped, angular, bearing 4–8 bottle-shaped, erect ovules at the base inside. Cones opening at maturity. Leaves flat, squamous, tuberculate at the base, imbricated.
Cupressus thyoides, L.

Scales of the ovoid cones few, imbricated, oblong,

loose, flattish, 2-ovuled, spreading at maturity. Leaves scale-like, imbricate and appressed.
Thuja occidentalis, L.

782. Leaves pinnate. Sterile flowers in aments, with an irregular perianth; fertile ones solitary or clustered. Calyx of the fertile flowers 3- to 5-lobed; its tube adherent to the 2- to 4-celled ovary. Fruit a tryma.....................783

Leaves simple, sometimes lobed, often later than the flowers...784

783. The fertile flowers with a 4-cleft calyx and no corolla. Style none; stigma divided, 2-lobed, the lobes 2-cleft. Shell (epicarp) 4-valved; nut (nucleus) nearly quadrangular, even. Sterile flowers in mostly 3-parted aments; calyx-scale 3 parted; stamens 4–6; anthers hairy. **Carya.**

The fertile flowers with a 4-cleft calyx and a 4-parted corolla; stigmas 2. Epicarp indehiscent; nut rugose and irregularly furrowed. The sterile flowers in simple aments; calyx-scale 5- or 6-parted, somewhat bracteate at the base; stamens about 20. **Juglans.**

784. Flowers in terminal racemes, or else umbellate-corymbed, or clustered, monœcious by abortion. Stamens mostly 8, or 3–6. The 2-styled ovary 2-lobed. Fruit a double samara. Leaves palmately lobed. (167) **Acer.**

Flowers all axillary, monœciously-polygamous. Fruit nut-like, or a drupe............................785

Flowers, both sorts, or only one kind, in aments, ament-like heads, or spikes..........................786

785. Calyx bell-shaped, 4- to 5-cleft; stamens 4 or 5; stigmas 2, oblong. Fruit nut-like, indehiscent, roughened with scale-like points. Flowers in clusters of 2–5.
(663) **Planera aquatica**, Gmel.

Calyx of the sterile flowers 6-parted; stamens 6; that of the fertile or perfect ones 5-parted, mostly with (5) stamens; stigmas 2, subulate, recurved. Drupe globular, 1-seeded, as large as a bird-cherry. Flowers solitary or clustered. (663) **Celtis.**

786. Only one sort of flowers in aments, or ament-like heads. Fertile flowers solitary, or few together, bud-like.
787

Both sorts, sterile and fertile flowers, in aments, heads, or spikes..................................790

787. Involucre 2- to 3-flowered, prickly. Stamens 8–15.
788

Involucre 1- to 2-flowered, smooth. Stamens 6–12.
789

788. Sterile flowers in long, cylindric aments, interruptedly clustered in them; calyx 5- to 6-parted; stamens 8–15. Fertile flowers in pairs, or threes, in a prickly, scaly involucre, forming a bur, which at length encloses 1–3 coriaceous nutlets, and, at maturity, opens by 4 valves. Leaves oblong, or oblong-lanceolate, serrate with pointed teeth. **Castanea.**

Sterile flowers in small heads, on drooping, axillary penduncles. Fertile flowers commonly in pairs at the apex of a short peduncle in a soft-prickly involucre, formed of numerous, united awl-shaped bractlets. Nut sharply 3-cornered. Leaves oblong-ovate, taper-pointed, toothed.
Fagus ferruginea, Ait.

789. Involucre 1-flowered, of many imbricated, small scales, forming a cup around the base of the hard and rounded nut. Sterile flowers clustered in slender and naked, drooping aments, without bracts. Stamens 6–12. **Quercus.**

Involucre 1- to 2-flowered, of 2–3 confluent scales, which become leafy-coriaceous, much enlarged and cut or torn at the apex, enclosing a bony nut. Sterile flowers in drooping cylindric aments; the concave bracts and the 2-cleft calyx united into 3-lobed scales. Stamens 8; anthers 1-celled. **Corylus.**

790. Sterile flowers in long aments with imbricate scales.
791

Sterile aments none of this sort...............795

791. Anthers bearded at the apex. Sterile flowers in aments, their scales simple, with 12 stamens. Fertile aments flaccid. Fruit a small nut. Leaves oblong-ovate, serrate.
792

Anthers not bearded at the apex............,.....793

792. Involucral scales of the fertile aments large, 3-lobed, the middle lobe much largest, unequally serrate-toothed on one side; the scales in pairs, each 2-flowered.

Sterile aments drooping from the sides of the small twigs near the end. Flowers preceding the leaves. Trunk ridged.
Carpinus Americana, Mx.
(*Water-Beech.*)

Involucral scales of the fertile aments lance-oblong, bag-like, 1-flowered, forming a sort of strobile, resembling that of the hop. Sterile aments 1–2 inches long, with the entire scales ciliate. Flowers appearing with the leaves. Trunk even. **Ostrya Virginica**, Willd.
(*Hop Hornbeam. Iron-wood.*)

793. Fertile flowers single under each scale of the globular, bur-like aments; ovary surrounded by 5 or 6, long, linear-awl-shaped scales, persistent around the nut. Scales of the sterile, cylindric aments kidney-shaped; stamens 3–6. Leaves linear-lanceolate, pinnatifid with rounded lobes, preceded by the flowers.
Comptonia asplenifolia, Ait.

Fertile flowers 2 or 3 under each scale. Nutlets naked, winged, or woody..........................794

794. Calyx of the fertile flowers of 4 little scales, adherent to the scales of the ovoid or oblong ament; each 2-flowered. Sterile aments elongated and drooping, with 5 bractlets, and 1–3 flowers under each scale; each flower usually with a 4-parted calyx and 4 stamens. Aments (sterile and fertile ones) often racemed. **Alnus.**
(*Alder.*)

Calyx of the fertile flowers none. Three of the fertile flowers under each 3-lobed bract, each consisting of a naked ovary with 2 filiform stigmas. Fruit a broadly winged nutlet. Sterile aments long and drooping, terminal and lateral; flowers golden-yellow, 3 of them and 2 bractlets under each scale. **Betula.**
(*Birch.*)

795. Flowers naked. Stamens numerous..........796

Flowers with a perianth—that is, a 4-lobed calyx. Stamens 4. Both sorts of flowers in spikes. Calyx of the fertile flowers becoming succulent in fruit; and the whole fertile spike transformed into a juicy (almost blackberry-like) aggregate fruit. Ovary 2-celled and becoming, by the disap-

pearance of one of the cells, an achenium. Leaves heart-ovate, serrate, often-lobed. **Morus** (870).
(*Mulberry.*)

796. Ovary 1-celled, 1-seeded. Fruit a nutlet. Both sorts of flowers in spherical heads, the fertile heads long-peduncled, pendulous. Leaves angularly sinuate-lobed or toothed, large; lobes sharp-pointed.
Platanus occidentalis, L.
(*Buttonwood. Sycamore.*)

Ovary 2-celled, consisting of 2 pistils united below; the cells many-seeded. Fruit a 2-celled, 2-beaked pod. Fertile flowers in spherical, long-peduncled, drooping heads (in fruit a globose strobile with the pods immersed into its scales). Sterile flowers in a conical, erect ament. Leaves deeply 5- to 7-lobed; lobes long, acuminate, serrate; veins villous at their bases. **Liquidambar stryaciflua**, L.
(*Sweet-gum Tree.*)

797. (775.) Staminate and pistillate flowers, each sort in a common (calyx-like) involucre of distinct, or united scales, composing what is called a head; both kinds of heads on the same plant—the pistillate heads only 1- or 2-flowered, the staminate (open-cup-shaped involucres) with several flowers, intermixed with chaff............................798

No common, calyx-like involucre to several flowers.
799

798. Fertile heads, each in a 1-leaved, entire, or 5-toothed, 1-flowered involucre (the flower naked—that is, a 2-styled ovary), 1–3 together in the axils. Sterile heads (about as large as a pea) composed of 7–12 scales, united into a cup, 5- to 20-flowered; flowers funnel-form; heads in cylindric spikes or panicled racemes. Herbs, 1–12 feet high.
Ambrosia (935).

Fertile heads, each a prickly, 2-leaved, 2-flowered, 2-beaked, coriaceous, ovoid or oblong involucre, forming a bur; its flowers consisting of a pistil with a slender, filiform corolla. Sterile heads: involucres of distinct scales, containing several staminate, tubular flowers. Racemes of flowers compact, in the axils, the sterile flowers above. Herbs sometimes spiny, with the spines 3-parted. **Xanthium** (932).

799. Flowers, at least the sterile, with both calyx and corolla...800

Flowers with only 1 set of perianth, sometimes with hood-shaped, petal-like staminodia, or glumaceous, or naked. 809

800. Herbs with leaves............................801

Herbs leafless, beset with small, scattered scales, purplish- or yellowish-brown, much branched; the simple or spicate branches flower-bearing their whole length—the upper flowers sterile, the lower fertile (with scattered perfect ones). Parasite on the roots of the beech.

(323) **Epiphegus Virginica**, Bart.

801. I. Corolla in both sorts of flowers of 2 or 3 petals, or tubular, and 2- to 3-lobed. Stemless, scape-bearing marsh or aquatic herbs..................................802

II. Corolla (at least of the staminate flowers) 4-petalled or parted......................................804

III. Corolla (at least of the staminate flowers) 5- to 6-petalled or parted. Calyx 5-toothed. Stamens 5......805

802. Both kinds of flowers on the same receptacle (forming a head). Ovary 2- to 3-celled, its cells 1-seeded. Fruit a pod. Leaves grass-like, flat, clustered at the base of the slender, simple, 1-headed, fluted scape..................803

Each kind separate, and not in heads. Flowers of 3 sepals and 3 petals, commonly whorled in threes; the sterile occupying the upper part of the scape. Leaves, if present (the blade is sometimes obscure), arrow-shaped, or lanceolate, with longitudinal nerves and cross-veinlets.

Sagittaria (884).

803. Stamens, petals and sepals 3. Stigmas in the fertile flowers 3. Scape 5-ribbed, puberulent. Leaves linear, bristle-like. **Pæpalanthus flavidus**, Kunth.

Stamens 4 or 6. Petals 2 or 3 (distinct in the fertile, united in the sterile flowers). Sepals 3. Fertile flowers in the margin; style 1; stigmas 2 or 3. Scape 7- to 12-ribbed. Leaves sword-shaped or subulate. **Eriocaulon**.

804. Submersed aquatics. Leaves, usually whorled in 3's, 4's, or 5's, parted into capillary segments. Sessile stigmas 4, pubescent. Some of the flowers perfect. Calyx 4-lobed. Petals 4 (sometimes wanting). Stamens 4–8. Fruit of 4 nut-like carpels, cohering by their angles. Upper flowers usually sterile, middle ones perfect, and the lowest fertile.

Myriophyllum (828).

Non-aquatic. Sterile flowers with a 4-parted calyx, 4 petals, a 4-rayed disk, and eight stamens. Fertile flowers with a 5-parted calyx and very minute, awl-shaped rudimentary petals; the 3 styles 2-cleft. The latter sort capitate-clustered at the base of the sterile spike, sessile in the forks and terminal. Herb rough-hairy. Leaves oblong or linear-oblong, obtusely toothed. **Croton glandulosum**, L.
(§ *Geisleria*, Klotzsch.)

805. Ovary superior. Stamens distinct. Fertile flowers without a corolla. Herbs downy, woolly, or scurfy. Leaves entire. Styles or stigmas 3, forked or 3-cleft...........808

Ovary inferior. Stamens 3–5, united more or less by their (often tortuous) anthers, as well as by their filaments. Succulent, tendril-bearing, climbing vines. Fruit a sort of berry, called pepo.................................806

806. Corolla of the sterile and fertile flowers 5-lobed. Leaves roundish-heart-shaped, 5-angled, or lobed. Fruit indehiscent...807

Corolla of the sterile and fertile flowers flat and spreading, 6-parted. Pepo prickly, 2-celled, 4-seeded, bursting at the top. Leaves sharply 5-lobed. Sterile flowers in compound racemes; fertile ones in small clusters or solitary; from the same axil, all greenish-white.

Echinocystis lobata, Torr. & Gr.

807. Flowers white, strictly monœcious, the sterile and fertile ones mostly from the same axils—the former corymbed, the latter in globose clusters, long-peduncled. Corolla of the sterile flowers flat, spreading, 5-lobed. Berry ovate, 1-seeded, covered with barbed, prickly bristles.

Sicyos angulatus, L.

Flowers greenish or yellowish, monœciously-polygamous; the sterile with a bell-shaped, 5-lobed corolla, few in small racemes, the fertile solitary. Berry ($\frac{1}{2}$–1 inch long), green, smooth, many-seeded.

(507) **Melothria pendula**, L

808. Flowers spiked or glomerate. Styles 3, twice- or thrice-cleft, or 2, and 2-parted. Ovary 3- (or 2-) celled. Pod

dehiscent into 3 or 2 carpels. Sterile flowers with a 5-parted calyx, 5 petals, and 5-12 stamens, on a 5-tubercled (glandular) disk. Fertile flowers without petals and with a 5- or 8-lobed calyx. Leaves entire. (804) **Croton.**
(§ *Pilinophytum*, Klotzsch, et
§ *Gynamblosis*, Torr.)

Flowers scattered on the branches, axillary on them, minute, the upper sterile. Stigmas 3, 2-cleft. Ovary 1-celled, 1-seeded. Pod indehiscent. The sterile flowers with 5 perfect, the fertile with 5 rudimentary scale-like stamens—the calyx of the former 5-, of the latter 3- to 5-parted. Petals none in the fertile flowers. Leaves silvery-hoary.
Crotonopsis linearis, Mx.

809. (799.) A maritime herb. Flowers naked, the sterile alternating with the fertile on the inner side of the linear, flat spadix, enclosed by a spathe. Anther 1, sessile. Ovary with a bifid style. Leaves grass-like, entire, obtuse.
Zostera marina, L.
(*Sea-Wrack.*)

Terrestrial, or fresh-water herbs..............810

810. Herbs with terminal, cylindric, club-shaped, or globular heads, bearing a multitude of crowded flowers, with not any or a very doubtful perianth (bristles, scales). Stamens innumerable, or rarely whorls of them attributable to the several sterile flowers........................811

Herbs with flowers having, at least the sterile ones, a true, simple perianth, or some obvious substitute of it; or herbs with glumaceous (grass-like) blossoms. Stamens definite. Flowers sometimes many on a common receptacle.
815

811. Spadix enclosed in a spathe..................812
Spadix without a spathe....................814

812. Flowers covering the base of the spadix.
(550) **Arisæma** (903).

Flowers covering the whole spadix...........813

813. The lowest flowers of the spadix perfect. Spathe open. (550) **Calla.**

The upper flowers of the spadix staminate, the

lower pistillate ; perfect ones none. Spathe convolute, wavy at the margins. Anthers 8-12, attached to the margin of a shield-shaped, oblong connectile, and opening by a terminal pore. Leaves arrow-shaped. **Peltandra Virginica**, Raf.

814. Flowers in a cylindric, club-shaped spike (that is, on a spadix), sterile ones above, the fertile below (sometimes separated by an interval, which is about 1 inch long). Each stamen surrounded by 3 bristles. Fertile flowers: ovary stalked; stalk beset with long hairs; style simple, long. In marshes and ditches. Leaves sword-shaped. **Typha.**

Flowers in spherical heads, the sterile above. Stamens mixed with minute scales, irregularly interposed. Ovaries ovoid, surrounded by 3–6 scales, sessile. Stigma oblique. Leaves linear, sheathing below. Borders of ponds. **Sparganium.**

815. I. Sterile flowers with 1 stamen. Flowers axillary. Aquatics..816

II. Sterile flowers with 2 or 3 stamens........819

III. Sterile flowers with 4 or 5 stamens........828

IV. Sterile flowers with more than 5 stamens. Styles 1, 2, 3, or 4 sessile stigmas....................837

816. Leaves whorled, linear, awl-shaped, 8–10. Flowers (very often perfect), the upper sterile, the lower fertile.
(545) **Hippuris vulgaris**, L.

Leaves not whorled, but alternate, or opposite. Herbs submersed...................................817

817. The naked flowers subtended by 2 small, opposite bracts (supplying the wanting perianth), axillary, diclinous, or perfect; the sterile ones consisting of 1 stamen, the fertile of 1, 2-styled ovary. Fruit indehiscent, nut-like, 4-lobed and 4-celled. Flowers expanding above the surface of the water. (564) **Callitriche.**

Flowers not subtended by two opposite bracts. 818

818. Leaves thread-form, entire. Stamen 1 ; filament filiform ; anther sagittate at the base. Fertile flower ; pistils 4 (2–5) in a cup-shaped involucre ; stigma peltate.
(549) **Zanichellia palustris**, L.

Leaves sparingly and very minutely repand-toothed (under a lens). Sterile flowers : stamen sessile, enclosed in a

little cowl-shaped involucre (spathe). Fertile flowers: ovary oblong, sessile; style short; stigmas 2–4.
Najas flexilis, Rostk (903).

819. Leaves parallel-veined, grass- or rush-like......820
 Leaves not grass- nor rush-like; or netted-veined. Flowers with a regular calyx.........................822

820. Stem a culm. Flowers glumaceous. Leaves grass-like..821
 Scape slender, simple, bearing a single, hairy head of flowers. Calyx of 3 sepals. Sterile flowers: stamens 3; filaments united below into a club-shaped tube around a rudimentary pistil. Fertile flowers: ovary 3-celled, surrounded by 3 tufts of hairs; stigmas 3, 2-cleft. Leaves linear-sword-shaped, tufted.
Lachnocaulon Michauxii, Kunth.

821. I. Grasses. Culms hollow..................565
 II. Sedges. Scales 1-flowered, if not empty.
 a. Ovary surrounded by a ventricose bag, open above and growing with the fruit. Stamens 3. Stigmas 2 or 3. Scales of the spikes 1-flowered, equally imbricated around the axis. Staminate flowers separated, borne either on the same spike, or in separate spikes. (551) **Carex** (902).
 b. Ovary not enclosed in a sac. Stamens 1-3. Style 3-cleft. Scales loosely imbricated, the lower ones empty. Fertile spikes 1-flowered, usually intermixed with clusters of few-flowered staminate spikes. Perianth not any. Achenium white or whitish. (551) **Scleria**.

822. Ovary 3-celled; styles or stigmas 3, often forked, or style 1, 3-cleft, and stigmas 3......................823
 Ovary 1-celled; styles (or stigmas) 1, 2 or 3, never forked..825

823. Flowers in racemes, or in a terminal spike. Ovules only 1 in each cell of the ovary......................824
 Flowers axillary, commonly 2 in each axil. Calyx 5- to 6-parted. Stamens 3, their filaments united in a column and surrounded by 5 or 6 glands. Cells of the ovary 2-ovuled. Styles 3, each 2-cleft; stigmas 6. Leaves 2-ranked.
Phyllanthus Caroliniensis, Walt.

824. Flowers in racemes. Stamens 2 or 3. Style 3-cleft; stigmas 3, simple. Calyx of the sterile flowers 3-, that of the fertile 6- (5- to 8-) parted. **Tragia.**

Flowers in a terminal spike. Stamens 2. Stigmas 3, simple. Calyx of the sterile flowers a 2-cleft, or crenulate cup, that of the fertile ones 3-toothed.
Stillingia sylvatica, L.

825. Leaves alternate (at least most of them), exstipulate...826

Leaves opposite, somewhat 3-nerved, long-petioled-stipulate; stipules united. Flowers, the sterile of 3 (or 4) sepals and as many stamens; the fertile ones of 3 sepals, 1 pistil, and 3 rudimentary cucullate stamens; both kinds mixed in dense clusters. **Pilea pumila,** Gray (833).

826. Flowers 3-bracted. Sepals 3. Stamens 3 or 2.
827

Flowers not 3-bracted; the sterile with a 3- (to 5-) sepalled calyx and 3 (to 5) stamens; the fertile consisting of a 2-styled ovary between 2 leaf-like bracts. Leaves triangular-hastate, sinuate-toothed, or nearly entire (upper ones lanceolate), sometimes meally; most of them alternate. Fruit-bracts deltoid. (666) **Atriplex hastata,** L. (836).

827. Fruit (utricle) opening transversely. Stamens 3 or 2. Stems whitish, mostly spreading next the ground. Leaves spatulate-oblong, long-petioled, very obtuse, or retuse. Flowers greenish, in close axillary clusters.
(649) **Amarantus albus,** L.

Fruit (utricle) indehiscent, or bursting irregularly. Stamens 3. Stems erect, ascending, or decumbent. Spikes or heads axillary and terminal. Leaves petioled, elliptic, or rhombic-lanceolate. (649) **Euxolus.**

828. (815.) I. Ovary 1-celled......................829

II. Ovary 3-celled. Styles 3, thick, subulate and recurved. Pod 3-horned. Calyx 4-parted. Stamens (of the sterile flowers) 4, surrounding a rudimentary ovary. Flowers, each 1- to 3-bracted, in a few, many-flowered spikes, from near the base of the stem, the fertile below the sterile. Leaves few, oval, crenate-toothed, long-petioled.
Pachysandra procumbens, Mx.

III. Ovary 4-celled, inferior. Leaves pinnately dissected into linear divisions, commonly whorled. Floating aquatics. **Myriophyllum** (838).

829. Style or stigma simple. Leaves stipulate, on (mostly long) petioles..830

Styles or stigmas 2 or 3. Leaves alternate, exstipulate...834

830. Leaves opposite..832

Leaves alternate..831

831. Herb beset with stinging bristles. Sterile flowers with 5 sepals, 5 stamens, and a rudimentary ovary; the fertile with 4 (apparently only 2) sepals and an ovary with a long subulate stigma. Achenium very oblique, and at length reflexed on the winged pedicel. Flowers in axillary, loosely and divaricately branched panicles, with the sterile flowers below. Leaves long-petioled, serrate, large.

Laportea Canadensis, Gaud. (899).

Herbs destitute of stinging hairs. Calyx of the sterile, 4-androus flowers 4-sepalled, that of the fertile tubular-bell-shaped, 4-lobed; stigma tufted. Achenium polished. Involucrate-bracted clusters of green flowers axillary. Petioled leaves entire.

(667) **Parietaria Pennsylvanica**, Muhl.

832. Herbs without stinging bristles...............833

Herbs beset with stinging bristles. Sepals 4 in both sterile and fertile flowers. Stigma capitate-tufted. Achenium straight, enclosed by the 2 inner and larger sepals. Flowers in branching, panicled spikes, or in peduncled capitate clusters. **Urtica** (898).

833. Sepals 3 or 4, those of the fertile flowers all, or all but one small. Stigma pencil-tufted.

(825) **Pilea pumila**, Gray (898).
(*Richweed*)

Calyx of the fertile flowers 4- (or 2-) toothed, or nearly entire, tubular or urn-shaped. Style long and thread-form. Calyx of the sterile flowers 4-parted. Flowers glomerate, the clusters spiked.

Bœhmeria cylindrica, Willd. (897).
(*False Nettle.*)

834. Flowers all with three scarious bracts.........835
　　　Flowers not 3-bracted. Herbs scurfy, sometimes silvery-mealy.......................................836

835. Utricle opening transversely. (827) **Amarantus.**
　　　Utricle indehiscent, or bursting irregularly. Stamens and sepals 5, the latter half the length of the 5-ribbed, thickish utricle. Herb prostrate. Leaves obovate, or rhombic-ovate, emarginate and petioled, often purple-veined. Flowers in axillary clusters.
　　　　　　　　(649) **Euxolus pumilus,** Raf.

836. Fruit-bearing, broadly wedge-shaped bracts united. Leaves oblong. **Obione arenaria,** Moq. (900).
　　　Fruit-bearing, ovate, or halberd-shaped bracts distinct or barely united at the base. Leaves triangular and halberd-form, alternate, or partly opposite. Flowers in spiked clusters. (826) **Atriplex hastata,** L. (893).

837. (815.) Styles 3, each 3- to many-cleft. Ovary 3-celled. Pod separating into 3 carpels..................839
　　　Styles 2, feathery (sometimes partly united). Stamens 6. Spikelets 1-flowered, the flower either staminate or pistillate. Glumes wanting. Stout water-grasses, 6–10 feet high. (567) **Zizania.**
　　　Style 1, long and filiform, or sessile stigmas 4. Leaves whorled (or rarely scattered). Aquatic weeds...838

838. Leaves pinnately parted into capillary divisions, often whorled. Calyx of the sterile flowers 4-parted, of the fertile 4-toothed. Petals caducous, or not developed. Stamens 8 (or 4). Herbs partly submersed.
　　　　　　　　(828) **Myriophyllum.**
　　　Leaves cut into thrice-forked, thread-like divisions. Staminate flower: stamens 12, or more, with large, oblong, sessile anthers. Pistillate flower: ovary ovoid, 1-ovuled; style filiform. A submersed herb.
　　　　　　Ceratophyllum demersum, L.

839. Flowers cymose. Calyx 5-lobed, or parted, corolla-like, white, in the staminate flowers salver-form. Stamens 10–15. Pod bristly-hairy. Herbs with stinging bristles.
　　　　　　Cnidoscolus stimulosa, Gray.

Sterile flowers spiked, and fertile ones glomerate; the sterile very small. Stamens 8–16. Fertile flowers surrounded by a large, leaf-like, cut-lobed, persistent bract. Calyx of the sterile flowers 4 parted, of the fertile 3-parted. Leaves alternate, long-petioled. Fruit rarely echinate.
Acalypha.

---

840. (774.) Trees, shrubs and other woody plants. Leaves deciduous, or evergreen 841
 Herbs 875

841. A yellowish-green, woody parasite on the branches of trees. Stems jointed and much branched. Leaves thick, obovate, or oval, somewhat petioled, longer than the axillary spikes, yellowish. Calyx 3- (2- to 4-) lobed. Sterile flowers; anthers 3, sessile. Fertile flowers; ovary inferior, stigma sessile. Berry 1-seeded, pulpy, white.
Phoradendron flavescens, Nutt.
(*American Mistletoe.*)
 No parasites 842

842. Leaves pinnate 843
 Leaves simple, often lobed, sometimes linear, needle-shaped, or evergreen 848

843. Trees or shrubs with thorns or prickles. Stamens 3–5. Pistils 3–5, or only 1 844
 Trees or shrubs unarmed 845

844. Leaves odd-pinnate; leaflets 7–11. Pistils of the fertile flowers 3–5. Flowers yellowish-green, in axillary clusters, or in a terminal cyme. **Xanthoxylum.**
 Leaves abruptly once- or twice-pinnate; leaflets (on the particular petiole) 18. Pistil 1 in the fertile flowers. Thorns above the axils, simple or branched.
(92) **Gleditschia.**

845. Petals present. Stamens 5, or 10 847
 Petals none. Calyx small, or wanting. Trees 846

846. Stamens 4, or 5, inserted into a fleshy disk. Styles 2, long and slender. Ovary 2-lobed. Fruit a double samara. Sterile flowers, in clusters on capillary pedicels;

fertile ones in long, gracefully drooping racemes, from lateral buds. Leaflets 3 or 5. **Negundo aceroides**, Mœnch.

Stamens usually 2 (rarely 3 or 4). Disk none. Style single; stigma 2-cleft. Ovary not lobed. Fruit a simple samara, winged from the apex only, or all around. The small flowers in crowded panicles or racemes, from the axils of last year's leaves. Leaflets 3–15. **Fraxinus.**
(*Ash.*)

847. Leaves trifoliolate; leaflets rhombic-ovate, sweet-scented, when bruised. Stamens 5, on a large, 5-parted disk. Styles 3. A low straggling shrub. Flowers (polygamous) yellow, in clustered, scaly-bracted, ament-like spikes, preceding the leaves. Fruit a dry drupe.
Rhus aromatica, Ait.

Leaves twice-pinnate, the several partial leaf-stalks bearing 7–13 leaflets, the lowest pair with single leaflets. Stamens 10. No disk. Style 1. A tall, large tree. Flowers (strictly diœcious) whitish, in axillary racemes. Fruit a legume. **Gymnocladus Canadensis**, Lam.

848. Leaves ample, at least lanceolate, often lobed, usually deciduous..................................852

Leaves linear, awl-shaped, needle-form, evergreen.
849

849. Pistil an open scale, or none (and the ovules naked).
850

Pistil closed. Stamens 3, rarely 4. Perianth of about 6 scales, the 3 inner ones petaloid. Low undershrubs.
851

850. Fertile flowers consisting of carpellary scales, without bracts, bearing 1–3 erect ovules on their base. Flowers all in very small, axillary, or lateral catkins. Anther-cells of the sterile flowers 3–6, attached to the lower edge of the shield-shaped scale. Leaves whorled in threes, or 4-ranked and much crowded. **Juniperus.**
(*Juniper and Red Cedar.*)

Fertile flowers solitary, bud-like, consisting merely of an erect, sessile ovule, surrounded by 6–8 imbricated scales, forming at length a cup-shaped disk, which becomes pulpy and berry-like, globular and red in fruit, partly en-

closing the seed. Sterile flowers: aments globular, surrounded by scales at the base; stamens monadelphous (anther-cells 3-8 under a peltate connectile). Tree.
Taxus baccata, L.

851. Stigma 6- to 9-rayed. Anthers pendulous on long filaments. Berries 6- to 9-seeded, black. Flowers very small, reddish, crowded in the axils of the upper leaves. Leaves imbricated. Shrub procumbent, 1-4 feet long.
(96) **Empetrum nigrum**, L.

Stigmas 3 or 4; style slender. Stamens exserted. Drupe 3- or 4-seeded, globular, minute (dry, when ripe). Flowers in terminal clusters of 10-15, with brownish scales and purple stamens and styles. Shrub diffuse, very slender. Stems 1 foot high. Leaves revolute on the margin.
(96) **Corema Conradii**, Torr.

852. Leaves alternate..........................855

Leaves opposite, or fascicled. Calyx sometimes early deciduous................................853

853. Corolla none............................854

Corolla present, 4-parted. Calyx 4-toothed. Stamens 2. (329) **Olea Americana**, L.

854. Leaves silvery-scurfy, elliptical, or ovate, entire. Calyx 4-parted in the sterile, 4-cleft and urn-shaped in the fertile flowers, becoming here pulpy and berry-like in fruit, enclosing the achenium. Fruit yellowish-red. Stamens 8. Ovary 1-celled; style slender; stigma 1-sided. Flowers small, clustered in the axils, or the fertile solitary.
Shepherdia Canadensis, Nutt.

Leaves not silvery-scurfy, oblong- or lance-ovate, acuminate at both ends, often serrulate. Calyx of 4 minute sepals, early deciduous. Ovary 2-celled; style slender; stigma somewhat 2-lobed. Drupe small, 1-seeded, fleshy, glaucous-purple. Flowers small, crowded in ament-like clusters, from the axils of last year's leaves, imbricated with scales. **Forestiera acuminata**, Poir.

855. Flowers, at least the fertile, with a complete perianth, calyx and corolla.............................856

Flowers with only one set of perianth, or else in aments 864

856. Woody climbers. Leaves palmately-veined, and commonly lobed..................................857

Trees or shrubs, not climbing...............859

857. Tendrils none. Stamens 6-20 (not on a disk); anthers 4-celled. Both sepals and petals conspicuous. Pistils 2-6. Fruit a 1-seeded drupe. Flowers in axillary racemes, or panicles..................................858

Tendrils present, with the panicled, or thyrsoid flower-clusters opposite the leaves. Stamens 5 on a 5-lobed disk. Calyx truncate, minute. Petals 5, cohering at the top, while they separate at the base, and usually falling off, resembling a mitre. Berry 2-celled, usually 4-seeded.

(158) **Vitis.**

858. Stamens, petals and sepals 6.

Cocculus Carolinus, D.C.

Stamens 12-20 in the sterile flowers, as long as the sepals. Sepals 4-8. Petals 6-8, short.

Menispermum Canadense, L.

859. Petals present in both sorts of flowers.........860

Petals only in the pistillate flowers present, small and fleshy, deciduous (often wanting). Stamens 10 (5-12), inserted on the margin of a convex disk. Fertile flowers 1 or 3-8 at the summit of a slender peduncle. Staminate flowers many in a simple or compound dense cluster of fascicles. Style long, revolute, stigmatic down one side. Drupe ovoid, with a striate, 1-seeded stone, blue, or bluish-black. Trees. (657) **Nyssa.**

860. Stamens 4-6..................................861

Stamens usually 16 in the sterile flowers, and 8 imperfect ones in the fertile. Calyx and corolla 4- to 6-lobed. Styles 4, 2-lobed at the apex. Ovary 8-celled. Berry plum-like, yellow, when ripe, 1 inch in diameter, 4- to 8-seeded, surrounded at the base by the thickish calyx. Fertile flowers axillary and solitary, the sterile smaller and often clustered. (421) **Diospyros Virginiana,** L.

861. Stamens alternate with the petals. Leaves sometimes armed with spiny teeth. Flowers axillary, in sessile or peduncled clusters, sometimes also solitary. Fruit a drupe with 2-6 stones..................................862

Stamens opposite the petals. Calyx-lobes, petals and stamens 4. Petals wrapped around the stamens. Drupe berry-like, with 3-4 nutlets, black. Branchlets thorny.
(661) **Rhamnus catharticus**, L.

862. Parts of the flower, calyx-teeth (or lobes), petals, stamens and stigmas 4 or 5.......................863
Parts of the flower habitually 6. (389) **Prinos**.

863. Petals oblong-linear, widely-spreading. Drupe (light red) with 4 horny, smooth nutlets. Flowers on long and slender, axillary peduncles, solitary or sparingly clustered. (130) **Nemopanthes Canadensis**, D. C.
Petals oval or obovate, obtuse. Drupe (red or purple) with 4 bony, furrowed nutlets. Fertile flowers mostly solitary, and the partly sterile ones habitually clustered in the axils. Leaves sometimes armed with spiny teeth.
(130) **Ilex**.

864. (855.) Flowers without any perianth, in aments or ament-like spikes..873
Flowers with a calyx........................865

865. Woody climbers. Perianth 6-cleft, or parted. Stamens 6 (5-7)...866
Trees or shrubs, not climbing................868

866. Shrubs (sometimes prickly) climbing by a pair of *tendrils* on the petioles. Ovary free from the calyx. Berry globular, 1- to 3-seeded, black or red. **Smilax** (890).
Tendrils none...............................867

867. Pistils 3, spindle-shaped, tipped with a radiate, many-cleft stigma. Sepals 6. Stamens twelve in the sterile flowers, short. Fruit a drupe, with the thin crustaceous putamen hollowed out like a cup on one side. Leaves large, thin, deeply 3- to 5-lobed, with a heart-shaped base; lobes acuminate. **Calycocarpum Lyoni**, Nutt.
Pistil 1 (adherent to the tube of the 6-cleft perianth). Pod 3-celled, 3-winged, loculicidal, 3-valved. Leaves broad-ovate, cordate, acuminate, 9- to 11-nerved; petioles long, the lowest whorled in fours. Sterile plant with the spikes panicled; the fertile with the spikes simple.
Dioscorea villosa, L.

868. Ovary superior..............................870
Ovary inferior.............................869

869. A low, straggling shrub. Calyx 5-cleft, with the lobes recurved, and a 5-glandular disk in the bottom. Stamens 5. Style short and thick. Fruit fleshy, drupe-like, pear-shaped. **Pyrularia oleifera**, Gray.

Large trees. Calyx small, 5-parted. Stamens 5–10, etc...859

870. Stamens 9, in 3 rows, the inner set usually gland-bearing at the base, in the sterile flowers; and 6, or 12–18 rudimentary ones in the fertile kind. Calyx 6-parted. Trees or shrubs......................................871

Stamens 4. Calyx 4-parted. Ovary 2-celled, one of the cells smaller and disappearing. Achenium ovate, compressed, covered by the succulent calyx, the whole fertile spike, therefore, converted in fruit into an oblong, juicy, aggregate (blackberry-like, dark purple) berry, or sorosis. Leaves ovate, heart-shaped, serrate, often 3-lobed. Shrubs or small trees. (795) **Morus.**

871. The axillary clusters or umbels of flowers surrounded by a 2- to 4-leaved involucre. Fertile flowers with 12–18 rudiments of stamens. Flowers preceding the leaves. 872

Involucre none. Anthers of the sterile kind 4-celled, 4-valved. Fertile flowers with 6 short rudiments of stamens and an ovoid ovary. Leaves many of them ovate and entire, and some 3-lobed. Flowers appearing with the leaves. Drupes blue, on red pedicels.

(702) **Sassafras officinale**, Nees.

872. Anthers 2-celled, 2-valved. Involucre 4-leaved.
(703) **Benzoin.**

Anthers 4-celled, 4-valved. Involucre 2- to 4-leaved. Branchlets zigzag.
(703) **Tetranthera geniculata**, Nees.

873. (864.) Scales of the aments irregularly cut-lobed at the apex. Stamens (8–12), as well as the ovary, surrounded below by a cup-shaped, perianth-like disk. Styles 2, bifid. Pod 2-valved. Seeds clothed with woolly down.

Populus, Tourn.

Scales of the aments entire. No cup-shaped, perianth-like disk 874

874. Sterile flowers; stamens 1, 2, 3–6, with 1 or 2 little glands at the base. Fertile flowers with a flat gland at the base of the ovary. Style short, or wanting; stigmas 2. Pod 2-valved, at length revolute at the summit. Seeds clothed with woolly down. **Salix,** Tourn.

Glands none at the base of the stamens or the ovary. Sterile aments oblong; stamens 2–8, shorter than the scales. Fertile aments ovoid. Ovary surrounded by 3 scales at the base. Stigmas 2, thread-form. Nutlet (dry drupe) 1-seeded, studded with resinous grains or wax. Leaves (and aments) resinous-dotted. **Myrica,** L.

875. (840.) Many flowers in a common calyx-like involucre, or head. Anthers united into a ring.
Baccharis (927) and **Antennaria** (930).

No common, calyx-like involucre to several flowers. Anthers not united 876

876. Flowers, at least one sort of them, with a double perianth, a calyx and a (usually colored) corolla 877

Flowers with a simple, herbaceous, or colored perianth (which rarely consists of 2 pales only), or naked (and then sometimes covered by a glume) 887

877. Corolla polypetalous 879

Corolla gamopetalous, funnel-form or salver-shaped. Stamens 3 or 4 878

878. Stamens 4. Calyx of 4 imbricated, scarious-margined sepals. Corolla salver-form, with the limb 4-parted. Sterile flowers with the anthers on long capillary filaments, and the corolla-lobes reflexed, or spreading. Fertile flowers with minute anthers on short filaments, and the corolla closed over the fruit in form of a beak. Stamens 4. Style 1. Pod opening by a transverse line. Flowers whitish, small, in a bracted, dense spike, or in clusters, or scattered.
(354) **Plantago.**

Stamens 3. Style slender; stigmas 1–3. Limb of the calyx of several plumose bristles (like a pappus), at first involute, at length unrolled. Corolla funnel-form, gibbous above the base, with the limb 5-lobed. Fruit indehiscent.

Leaves thickish, minutely ciliate, the radical ones lanceolate and spatulate, the cauline pinnately parted into 3-7 narrow segments. Root spindle-shaped (eaten by the Oregon Indians). Panicle interrupted. **Valeriana edulis,** Nutt.

879. Petals 3, at least in one sort of flowers........883

Petals 5 (or 4)..............................880

880. Leaves compound.........................881

Leaves heart-kidney-shaped, somewhat 5-lobed, serrate, wrinkled. Stem simple, 2- or 3-leaved, 1-flowered. Calyx-lobes 5, pointless. Petals obovate, white. Pistils several or few. Fruit a collective mass of few drupes. **Rubus Chamæmorus,** L.

Leaves palmately 9- to 11-parted. Stamens 15-20, monadelphous in a column. Styles 8-10. A tall, roughish herb. **Napæa dioica,** L.

881. Flowers in a spiked raceme, or racemed panicle. Leaves twice- or thrice-ternate or pinnate..............882

Flowers in a simple umbel. A whorl of 3, palmately 3- to 7-foliolate leaves at the summit of the stem. Calyx-tube coherent with the ovary, its limb obscurely toothed or obsolete. Stamens 5. Root spindle-shaped, or globular. **Panax,** L. (*Aralia,* Gray.)

882. Pistils 5. Stamens 10-50. Calyx 5-cleft. Petals obovate, whitish. Raceme spiked. Flowers strictly diœcious. **Spiræa Aruncus,** L.

Pistil 1; styles 2. Stamens 10. Calyx 4- to 5-parted. Petals 4 or 5, spatulate. Panicle. Leaflets heart-shaped. **Astilbe decandra,** Don.

883. Herbs submersed. Flowers on scape-like peduncles, from a spathe. Perianths 6- (or 3-) parted, the inner 3 of the 6 segments petaloid. Stamens 3-12. Ovary inferior..885

Herbs not submersed.......................884

884. Petals white. Flowers pretty large, usually whorled in threes. Stamens indefinite. Pistils numerous on a globular receptacle. Leaves commonly arrow-shaped.

(802) **Sagittaria.**

Perianth greenish, the 3 inner sepals larger, somewhat colored. Flowers small, in slender, panicled racemes, which on the fertile plant become reddish in summer. Stamens 6. **Rumex Acetosella**, L.

885. Filaments of the sterile flowers united into a solid column; anthers 6–12, linear. Ovary of the fertile flowers 6- to 9-celled; stigmas 6–9. Sterile spathe 1-leaved, with about 3 long-peduncled flowers; the fertile 2-leaved, with a solitary, short-peduncled flower. Leaves round-heart-shaped, long-petioled, spongy-reticulated and purplish beneath. **Limnobium Spongia**, Rich.

Filaments not united, or scarcely any. Ovary 1-celled, with 3 parietal placentæ. Stigmas 3. Leaves linear, or oval-oblong..886

886. Stems leafy and elongated. Both sorts of flowers solitary and sessile, from a sessile and tubular, 2-cleft spathe. Sterile flowers small or minute, with 3 sepals, as many narrower petals, and 9 sessile anthers. Fertile flowers with a tubular, exceedingly long and thread-form, 6-lobed, colored perianth, and 3–6 antherless filaments, the tube rising to the surface of the water, to be fertilized by the sterile flowers, which break off under water and then float on its surface. Leaves linear, or oval-oblong, whorled or opposite. (707) **Anacharis Canadensis**, Planch.

Stems none. Only the fertile flowers solitary. Sterile ones crowded on a conical spadix in a 3-parted spathe, which is borne on a short scape, and consisting of a 3-sepalled calyx only and 3 stamens; they break away and rise to the surface of the water, to fertilize the pistillate flowers. The latter have a 3-lobed calyx-tube, which is not extended beyond the elongated ovary, and 3 linear, small petals; they are sessile in a tubular, bifid spathe, borne on an exceedingly long, peduncle-like scape. The latter coils up spirally, to draw the fertilized ovary under water, to ripen. Leaves linear, thin, long and ribbon-like. **Vallisneria spiralis**, L. (*Eel-grass.*)

887. (876.) Flowers (at least the sterile) with a regular perianth of 3 to 6 segments or sepals..................888

Flowers naked (the sterile, in one genus, in a little, cowl-shaped spathe), or glumaceous...................902

888. Perianth of 6 segments or sepals. Stamens 6..**889**
Perianth of fewer segments or sepals.........**891**

889. Perianth-segments, or sepals unequal. Fruit a berry...**890**

Sepals equal, spatulate-linear, spreading, white, withering-persistent. Fertile flowers with rudimentary stamens. Styles 3, linear, club-shaped. Pod 3-valved from the apex. Stem wand-like, terminated by a virgate, long, spiked raceme of small, bractless flowers. Fertile plant more leafy than the sterile. Leaves spatulate and lanceolate, petioled. (736) **Chamælirium luteum**, Gray.

890. The 3 inner segments of the perianth broader. Sterile flowers with a small ovary without a stigma. Leaves awl-shaped. Stem panicled. Berries red.
Asparagus officinalis, L.

The 3 outer sepals broader. Sterile flowers without an ovary. Leaves heart-shaped, halberd-form, or roundish-ovate. Herbs usually climbing, mostly by tendrils. Flowers in axillary, long-peduncled umbels. (Peduncles 3–8 inches long.) Berries bluish-black, with a bloom.
Smilax herbacea, L.
(*Carrion Flower.*)

891. Leaves simple............................**893**
Leaves compound..........................**892**

892. Leaves 3 to 4 times ternately compound. Sepals 4. Stamens numerous. Pistils several.
Thalictrum dioicum, L.

Leaves digitate, of 5–7 linear-lanceolate, coarsely toothed leaflets. Sterile flowers in axillary, compound, racemes or panicles, with 5 sepals and 5 drooping stamens. Fertile flowers spiked-clustered, 1-bracted, with only 1 sepal, folded around the 2-styled ovary.
Cannabis sativa, L.

893. Leaves halberd-shaped.
(836) **Atriplex hastata, L.**
Leaves not halberd-shaped..................**894**

894. Leaves alternate...........................**899**
Leaves opposite...........................**895**

895. Herbs erect, straight, not twining............896

Herbs twining. Leaves heart-shaped and 3- to 5-lobed, with persistent, ovate stipules between the petioles. Sterile flowers in panicles; calyx of 5 sepals; stamens 5. Fertile flowers: aments with several-flowered scales; each flower (pistil) with 1 sepal and embraced by it; styles 2, long. The fertile aments forming in fruit a strobile with large bracts; achenia invested by the enlarged scale-like calyx. **Humulus lupulus,** L.
(*Hop.*)

896. Stigmas 2 or 3. Flowers with 3 scarious bracts. Bracts and calyx silvery-white, the latter woolly at the base. Calyx of 5 sepals. Stamens 5; filaments slender, united into a short cup at the base; anthers 1-celled. Leaves petioled, ovate-lanceolate. Panicles narrow.
(667) **Iresine celosioides,** L.

Style or stigma 1. Flowers not 3-bracted.....897

897. Calyx of the fertile flowers of 2–4 separate or almost separate sepals...............................898

Calyx of the fertile flowers urn-shaped, entire, or 2- to 4-toothed, enclosing the ovary; style long, awl-shaped, stigmatic and hairy down one side. Stinging hairs none. Leaves oblong-ovate, or ovate-lanceolate, pointed, serrate, 3-nerved. Flowers clustered in axillary spikes.
(833) **Bœhmeria cylindrica,** Willd.

898. Herb beset with stinging bristles. Sepals 4 in both sterile and fertile flowers; the 2 inner larger ones enclosing the achenium in fruit. Stigma capitate-tufted. The much-branched spikes longer than the petioles, drooping. Leaves ovate-heart-shaped, pointed, deeply serrate.
Urtica dioica, L.

Stinging bristles none. Sterile flowers with 3 or 4 sepals and as many stamens. Fertile flowers: sepals 3, lanceolate, slightly unequal, a rudiment of a scale-like stamen usually before each. Stigma sessile, pencil-tufted. Clustered axillary, scarcely branched spikes much shorter than the petioles, spreading. Leaves ovate, coarsely toothed, pointed, 3-nerved. Stems whitish- or bluish-green, translucent and shining. (833) **Pilea pumila,** Gr.
(*Richweed.*)

899. Styles or stigmas 2, 3, or 5 900
  Stigma 1, subulate, elongated. Sterile flowers: calyx 5-parted; stamens 5; ovary rudimentary. Fertile flowers: calyx 4-sepalled, the 2 outer sepals minute, the 2 inner foliaceous in fruit; achenium finally reflexed on the winged pedicel. Hairs stinging. Leaves broad-ovate, long-petioled, rounded or slightly cordate at the base, acuminate. Panicles axillary, solitary, or in spikes.
  (831) **Laportea Canadensis**, Gaud.
900. Stigmas 3-5. Flowers all 3-bracted. Sterile flowers with 5 sepals and 5 stamens. Fertile flowers without a calyx... 901
  Styles 2. Sterile flowers bractless; calyx of 4 or 5 sepals. Stamens 4 or 5. Fertile flowers 2-bracted; bracts more or less united, at length hardened and connivent; calyx none. Flowers densely glomerate. Herb mealy-canescent. Leaves oval or oblong, short-petioled.
  (836) **Obione arenaria**, Moq.
  (*Sand Orache.*)
901. Utricle indehiscent and valveless. Stigmas 3-5, shorter than the ovary, and linear-awl-shaped. Stem nearly terete. Flowers in slender, axillary, and terminal spikes. Leaves ovate, lanceolate, acuminate, wavy, wedge-shaped at the base, petioled. **Acnida cannabina**, L.
  Utricle dehiscent, circumscissile. Stigmas long, bristle-form, plumose-hairy. Stem angular. Flowers in axillary spikes. Leaves lanceolate or oblong-ovate, long-petioled. **Montelia tamariscina**, Gray.
902. (887.) A grass and a sedge.
  Brizopyrum (613) and **Carex** (821).
  No grasses, nor sedges 903
903. Herb submersed, slender, branching, with linear, minutely repand-toothed, opposite leaves, and axillary flowers. Stigmas usually 3 or 4. (818) **Najas flexilis**, Rostk.
  A non-aquatic, scape-bearing herb. Flowers naked, on a spadix, covering its base and leaving its upper part free. The subulate spadix in some plants with wholly sterile flowers, in others with sterile and fertile ones, the latter below. Spathe oblong, acuminate, convolute. Leaf pedate, with 7-11 segments.
  Arisæma Dracontium, Schott.

E. COMPOSITES.

904. (3.) *Tubuliflorœ.* Heads presenting tubular flowers—alone or with ligulate ones.....................905

Liguliflorœ. Heads with all the flowers ligulate. Plants with milky juice............................987

905. Heads discoid—that is, without ligulate flowers (rays), at least apparently so.......................906

Heads radiate—that is, with the outer flowers ligulate (rays)...943

906. I. Receptacle (at least in the centre) naked—that is, with no pales or bristles among the flowers; very rarely hairy...907

II. Receptacle chaffy, having pales or little scales among the flowers.......................................931

III. Receptacle bearing bristles, or deeply honey-combed..938

907. Pappus of scales, or bristles present..........912

Pappus none, or a short-toothed margin.......908

908. Heads alike. Leaves alternate...............909

Heads of two sorts on the same plant, the sterile top-shaped, and 5- to 20-flowered, the fertile 1-flowered. Leaves lobed or dissected, opposite or alternate.
Ambrosia.

909. Heads beset with glands, whitish, loosely panicled. Leaves triangular, or somewhat heart-shaped, white-woolly beneath; petioles margined. Stem leafless above. Achenia long, club-shaped, glandular.
Adenocaulon bicolor, Hook.

Heads glandless. Leaves, at least the lower ones, more or less pinnately incised or (once- to thrice-) dissected. Heads yellowish-white, yellow, or purplish.............910

910. Receptacle conical or convex. Corollas yellow. Pappus a short, 5-lobed or toothed crown..............911

Receptacle flattish. Flowers mostly of 2 sorts, the central one usually sterile. Corollas yellow or purplish. Racemes or panicled spikes. Plants often hoary or whitened-woolly.
Artemisia.

911. Herbs 1-3 feet high. Pappus-crown 5-lobed or toothed. Receptacle convex. Heads corymbed.
Tanacetum.

Herb low, 6-8 inches high. Pappus-crown obsolete. Involucral scales oval, with white, scarious margins. Receptacle conical. Heads on simple peduncles.
Matricaria discoidea, D. C.

912. (907.) Pappus of many capillary bristles. Achenia sometimes 5-angled.................................915

Pappus a circle of 5-20 chaffy scales..........913

913. Leaves alternate.............................914

Leaves whorled in 5s or 6s, linear. Pappus a single row of almost horny, oval, obtuse scales. Flowers flesh-color. In shallow water. **Sclerolepis verticillata,** Cass.

914. Leaves ovate-oblong. Pappus of awned scales (stout bristles, chaffy dilated at the base). Heads 3- to 5-flowered in a compound head; flowers purplish.
Elephantopus Carolinianus, Willd.

Leaves once- to twice-pinnately parted into linear or oblong lobes. Pappus of 15-20 small, very thin, blunt scales in a single row. Heads many-flowered, separate, not in a compound head, corymbed.
Hymenopappus scabiosæus, L'Her.

915. Leaves alternate..........................918

Leaves opposite (or whorled)................916

916. Involucral scales 4 only. Heads 4-flowered. Climbing herbs. Leaves heart-shaped, or halberd-form. Heads in peduncled axillary corymbs.
Mikania scandens, Willd.

Involucral scales more than 4. No climbers...917

917. Receptacle flat. Involucral scales 8-20. Flowers white or purp.e. Heads generally corymbose.
Eupatorium (925).

Receptacle conical. Heads in terminal corymbs; flowers sky-blue, becoming purplish. Leaves triangular-ovate, slightly cordate, toothed.
Conoclinium cœlestinum, D. C.

918. Flowers in the heads all alike and perfect. (Heads homogamous)..................................919

Flowers not all, or none of them perfect. (Heads heterogamous, or diœcious) 926

919. Involucral scales imbricated................. 922
Involucral scales in one row (or nearly so)..... 920

920. Leaves ample. Heads rather large, whitish, or yellow-flowered 921
Leaves linear. Heads very small............ 953

921. Flowers whitish, or cream-color. Heads 5- to many-flowered, rather large, in flat corymbs. **Cacalia.**
Flowers yellow. Heads many-flowered, solitary, or corymbed. Leaves pinnatifid and toothed, clasping.
Senecio vulgaris, L. (958).

922. Flowers whitish, or purple.................. 923
Flowers yellow. Achenia inversely conical. Receptacle with an awl-shaped prolongation in the centre. Leaves oblanceolate, or linear, 1- to 3-nerved. Heads small, in a flat-topped corymb. **Bigelovia nudata,** D. C.

923. Flowers whitish........................... 925
Flowers purple............................ 924

924. Pappus double; the outer a row of very short, the inner of longer bristles. Heads of purple flowers, in fastigiate cymes. Involucral scales sometimes tipped with an awl-shaped appendage. **Vernonia.**
Pappus simple. Bristles of the pappus plumose, or at least conspicuously bearded, stiff. Achenia slender, tapering to the base, about 10-ribbed. Leaves entire. Heads spicate-racemose, or panicled-cymose. **Liatris.**

925. Pappus capillary. Achenia 4-cornered. Herb much branched. Leaves pinnately dissected. Heads in a pyramidal panicle, small, 3- to 5-flowered. Stem 3 to 10 feet high. **Eupatorium fœniculaceum,** Willd.
(*Dog-fennel.*)

Pappus very plumose. Achenia terete, many-striate. Leaves chiefly toothed and lanceolate; upper ones linear and entire. Heads panicled-corymbose.
Kuhnia eupatorioides, L.

926. Scales herbaceous, often deciduous..........927
Scales scarious, mostly whitish, persistent, sometimes yellowish or brown..........................929

927. Herbs. Leaves serrate. Heads heterogamous..928
Scurfy shrubs. Flowers all tubular, diœcious—that is, pistillate and staminate flowers in separate heads, borne by different plants. Heads in loose, terminal panicles, or sessile in axillary glomerules. Pappus very long and white. Leaves wedge-form or spatulate-oblong, toothed.
Baccharis.

928. Heads purplish, exhaling a camphor-like odor. Herbs somewhat glandular. Corymbs compound, dense. Involucre imbricated. **Pluchea.**

Heads whitish, corymbed. Involucre cylindrical; scales in a single row, with a few bractlets at the base. Pappus bright white and very soft. Leaves large, lanceolate or oblong, unequally and deeply toothed.
Erechthites hieracifolia, Raf.

929. Receptable perfectly naked.................930
Receptacle (elongated, columnar, or top-shaped) naked at the summit only, and bearing broad chaff toward its base. Heads, few-flowered, in dense, dichotomously arranged clusters, woolly. **Filago Germanica,** L.

930. Heads diœcious, or nearly so, corymbose, rarely single. Pappus of the staminate flowers club-shaped at the summit. Involucre usually pearl-white. Herbs white-woolly. Leaves linear lanceolate. **Antennaria,** Gærtn.
(*Everlasting.*)

Heads containing both perfect (central) and pistillate (marginal) flowers. Bristles of the pappus all slender. Herbs woolly. Leaves oblong-spatulate, lanceolate, or linear, sessile, often decurrent. Involucral scales white, yellowish, or brown. **Gnaphalium.**
(*Cudweed.*)

931. (906.) Leaves opposite, at least the lower.....936
Leaves alternate............................932

932. Heads all alike, some of the flowers sometimes sterile...933

Heads of two sorts, the sterile with many, the fertile with 1 or 2 flowers..............................935

933. Heads corymbed. Pappus of bristles........934

Heads solitary, terminal. Pappus of 5 or 6 scales. Anthers blue. Leaves entire, ovate-lanceolate, 3-nerved, sessile. **Marshallia latifolia,** Pursh.

934. Receptacle chaffy throughout, with short, rigid, narrow, 3-nerved pales between the flowers. Flowers purple. **Carphephorus tomentosus,** Torr. and Gr.

The columnar receptacle naked at the top, chaffy on the margins and toward the base. Heads woolly. Herb downy-canescent......................................929

935. Fertile involucres 1-flowered, oblong and 4-angled. Sterile heads crowded in a cylindrical spike, the cup-shaped involucres produced on one side into a recurved hispid tooth. Leaves lanceolate, partly clasping.
Ambrosia bidentata, Mx.

Fertile involucres 2-flowered, forming in fruit an oblong prickly bur. Sterile involucres globose, of distinct scales. (798) **Xanthium.**

936. Flowers of the margin pistillate and fertile, 1–5, those of the centre staminate, sterile. Shrubs or herbs. Involucral scales roundish, few. Heads small, greenish-white, on short, recurved peduncles, forming leafy, panicled racemes or spikes. **Iva.**
(*Marsh-Elder.*)

Flowers in the heads all of one sort..........937

937. Pappus of 2 or few retrorsely barbed awns.
Bidens (985).

Pappus of 2 upwardly barbed, or naked awns.
Coreopsis (985).

938. (906.) Pappus composed of many bristles.....939

Pappus none, or a few scabrous bristles. Involucral scales fringed-margined, or ciliate, the middle ones sometimes spiny. Heads single. Flowers purple, rarely whitish, the marginal often falsely radiate, large, blue. **Centaurea.**

939. Receptacle deeply honeycombed. Stems winged by

the decurrent base of the lobed and toothed, somewhat prickly leaves. Heads large; involucral scales spiny.
Onopordon Acanthium, L.
Receptacle not honeycombed................940

940. Involucral scales hooked. Leaves large, petioled, upper ones ovate, the lower heart-shaped. Heads solitary, or clustered; flowers purplish. **Lappa Major**, Gærtn.
(*Burdock.*)
Involucral scales not hooked................941

941. Pappus threefold—namely, 10 short, horny teeth, 10 slender, rigid bristles, and as many shorter ones. Flowers yellow. **Cnicus benedictus**, L.
Pappus of one sort; its bristles united at the base in a ring. Leaves sinuate, or pinnatifid, spiny, usually decurrent. Flowers purplish................................942

942. Pappus plumose. **Cirsium.**
Pappus capillary. Heads solitary, drooping.
Carduus nutans, L.

943. (905.) Receptacle chaffy—that is, with pales among the flowers..964
Receptacle naked, or deeply honeycombed.....944

944. Receptable deeply honeycombed. Heads solitary, globular, at the summit of each stem. Rays neutral, long, narrowly wedge-shaped, yellow; disk-flowers at length purple. Receptacle strongly convex. Achenia silky-villous. Pappus of 7–9 oblong-lanceolate chaffy scales. Leaves oblanceolate. **Baldwinia uniflora**, Nutt.
Receptacle not honeycombed................945

945. Pappus of 5–12 scales, or of many capillary bristles
946
Pappus none, or of a few short awns or a crown. Leaves alternate.......................................961

946. Pappus of 5–12 scales......................947
Pappus of many capillary bristles............949

947. Leaves alternate...........................948
Leaves opposite. Pappus-scales deeply cleft into capillary bristles. Leaves dotted, pinnately parted into nar-

row, bristly-toothed segments. Heads terminating the branches; rays few, yellow.
Dysodia chrysanthemoides, Lag.

948. Rays sterile, wedge-form, yellow. Agreeing in all other respects with Helenium.
Leptopoda brachypoda, Torr. & Gr.

Rays fertile, wedge-form, 3- to 5-cleft. Receptacle spherical; involucral scales linear, reflexed. Achenia top-shaped, ribbed. Pappus of 5–8, chaffy, 1-nerved, bristle-pointed scales. Leaves lanceolate, toothed, decurrent on the angled stem and branches. Heads yellow, at the end of the branches, or corymbed. **Helenium autumnale**, L.

949. Leaves alternate..................................952

Leaves opposite or radical..................950

950. Leaves radical.................................951

Leaves opposite. Scales of the involucre in 2 rows. Pappus of rough, denticulate bristles. Heads large, solitary, or corymbed; flowers yellow. Stem simple. **Arnica.**

951. Heads purplish; flowers fragrant. Herb somewhat diœcious. Leaves roundish, or kidney-shaped, deeply 5- to 7-lobed, the lobes toothed and cut, white-woolly beneath, very large. Scape with sheathing scaly bracts. Corymb.
Nardosmia palmata, Hook.
(*Sweet Coltsfoot.*)

Heads yellow, solitary, with fertile flowers in the ray and sterile ones in the disk. Scape simple, with purple scales, appearing with its flower before the leaves. Leaves heart-shaped, angular, toothed, woolly beneath, long-petioled.
Tussilago farfara, L.

952. Rays whitish or purple.....................953

Rays yellow, in about one row (very rarely cream-color) 957

953. Rays in a single row, pistillate and fertile.....954

Rays in several rows. **Erigeron** (956)

954. Pappus simple 955

Pappus double. Rays 8–12. Heads corymbose or solitary. **Diplopappus**, Cass.

955. Achenia very silky. Rays about 5, white. Leaves linear or ample, somewhat serrate sometimes. Heads small, in little clusters, forming a flat corymb.
Sericocarpus.

Achenia not silky..........................956

956. Rays 5–75. Scales of the involucre (more or less) loosely or closely imbricated, usually with leaf-like tips. Achenia flattish.
Aster.

Rays 40–200, always white. Involucral scales nearly equal, almost in a single row. Achenia flattened, usually pubescent and 2-nerved. (953) **Erigeron.**

957. Pappus simple; the bristles equal or nearly so.
958

Pappus double, the outer set of bristles very small. Rays numerous, yellow. Herbs woolly or hairy, usually low. Heads rather large, often corymbed at the end of the branches.
Chrysopsis.

958. Scales of the involucre imbricated, the outer ones shorter..959

Scales of the involucre not imbricated, equal. Pappus bright white, very soft. Heads solitary, or corymbed; flowers chiefly yellow. (921) **Senecio.**

959. Either rounded-heart-shaped (lower) leaves, and small heads, or very large, showy heads, and anthers tailed at the base. Flowers always yellow. Pappus never unequal...960

Leaves never heart-shaped. Heads small. Pappus sometimes slightly and irregularly unequal. Rays (1–16) usually yellow, rarely cream-color, or nearly white. Achenia many-ribbed, nearly terete. Stems mostly wand-like. Heads clustered, or racemed.
Solidago.

960. Heads corymbed, small, 8- to 10-flowered, racemed, or spiked along the branches. Lower leaves rounded-heart-shaped, upper ones ovate, all serrate. Pappus of minute, scale-like bristles. **Brachychæta cordata**, Torr. & Gr.

Heads solitary, large showy. Rays numerous. Involucral scales ovate. Pappus scabrous. Anthers with two bristles at the base. **Inula Helenium**, L.

961. (945.) Pappus none. Flowers all perfect......963
    Pappus of a few short awns or bristles, or a membranaceous crown.................................962

962. Pappus of a few short awns or bristles. Achenia wing-margined. Heads loosely corymbed, or panicled; rays white or purplish. Leaves thickish, chiefly entire. Herbs Aster-like. **Boltonia.**
    Pappus a membranaceous margin, rather cup-like. Receptacle conical. Achenia angular. Heads corymbed; rays white, disk yellow. Involucral scales scarious-margined. Leaves twice-pinnately divided.
              Matricaria Parthenium, L.
                    (*Feverfew*.)

963. Rays pale violet-purple. Leaves lanceolate or oblong, or the lower spatulate-obovate. Heads solitary.
              Bellis integrifolia, Mx.
    Rays white. Heads large, terminating the stem and branches. Leaves pinnatifid-toothed.
              Leucanthemum vulgare, Lam.
                  (*Ox-eye Daisy*.)

964. (943.) One sort of the flowers in the heads, either the disk- or ray-flowers sterile......................973
    Disk- and ray-flowers both fertile, the latter pistillate....................................965

965. Leaves opposite..........................968
    Leaves alternate. Rays white..............966

966. Pappus none, or a minute crown. Leaves once- or twice-pinnately parted, or entire and lance-linear. Rays over 4...967
    Pappus of 2 awns. Leaves ovate-lanceolate. Stem narrowly, or interruptedly winged. Rays 3-4, oval. Achenia narrowly winged. **Verbesina Virginica**, L. (968).

967. Achenia terete, striate or smooth. Leaves once- to twice-pinnately divided. Branches terminated by single heads. **Anthemis arvensis**, L.
    Achenia oblong, flattened, margined. Leaves twice-pinnately parted, or simple and lance-linear. Rays 4-5 or 8-12. Heads corymbose. **Achillea.**

968. Pappus of cut-fringed chaffy scales. Involucral scales in one row. Leaves triple-nerved, short-petioled, heart-ovate, toothed. Heads small, rays whitish.
       Galinsoga parviflora, Cav.
 Pappus-scales not cut-fringed................969

969. Heads solitary on the stem or branches.......970
 Heads in compound corymbs. Flowers yellow. Pappus of 2 awns. Stem 4-winged. Leaves ovate, triple-nerved, serrate. Rays 1–5. **Verbesina Siegesbeckia,** Mx.

970. Receptacle convex or conical 971
 Receptacle flat............................972

971. Achenia 4-cornered. Involucre of separate scales. Heads showy, peduncled, terminating the stem or branches; rays yellow, about 10, fertile. Leaves lance-ovate, or oblong-ovate, triple-nerved, serrate. **Heliopsis lævis,** Pers.
 Achenia inversely ovoid. Involucre a leafy cup. Heads large, single, on terminal peduncles; flowers pale yellow. Leaves oval or oblong, their bases usually connate.
     Tetragonotheca helianthoides, L.

972. Pappus obsolete, or none. Chaff of the receptacle oristle-form. Heads solitary, small, long-peduncled; flowers whitish; anthers brown. Achenia 3- to 4-sided, or those of the disk laterally flattened, hairy at the top. Stems procumbent. Leaves oblong-lanceolate, slightly serrate, sessile.
      Eclipta procumbens, Mx.
 Pappus an obscure crown. Chaff of the receptacle scale-like and rigid. Low shrubby maritime herb. Heads solitary, peduncled, terminal; flowers yellow; anthers blackish. Achenia rather wedge-shaped, 3- to 4-cornered. Leaves spatulate-oblong or lanceolate, silky-downy.
      Borrichia frutescens, D. C.

973. (964.) Leaves opposite, at least the lower, or sometimes whorled. Rays yellow 981
 Leaves alternate...........................974

974. Disk-flowers fertile, rays sterile.............976
 Disk-flowers sterile, rays fertile..............975

975. Heads yellow-flowered; rays numerous. Achenia winged, sometimes with 2 teeth, which are confluent with the winged margin. Corymbs or panicles. **Silphium** (983).

Heads white-flowered; rays 5 only, their ligules very short and broad, inversely heart-shaped, not projecting beyond the woolly disk. Pappus of 2 chaffy scales. Heads small, in a dense, flat corymb. Leaves oblong, or ovate, crenate-toothed. **Parthenium integrifolium**, L.

976. Rays yellow.................................978
Rays white or purple.....................977

977. Rays white, spreading. Pappus none. Receptacle chaffy only at the summit; the chaff deciduous. Leaves finely thrice-pinnately divided. Branches terminated by single heads; rays at length reflexed. Involucral scales white-margined. **Maruta Cotula**, D. C.

Rays purple, drooping. Chaffy scales sharply pointed, elongated. Disk purplish. Head solitary, showy, at the summit of the stem. Leaves serrate, or entire, 3- to 5-nerved (sometimes bristly-hairy). **Echinacea.**

978. Pappus of 2 awns.........................980
Pappus none...............................979

979. Achenia 4-cornered, flat at the top. Rays generally drooping. Heads showy, terminating the stem and branches. Disk greenish-yellow, or dark purple, sometimes columnar. **Rudbeckia.**

Achenia compressed laterally and margined. Rays few, large (2 inches long), drooping. Disk grayish. Leaves pinnate, of 3–7 leaflets. Heads single, showy, at the end of the branches. **Lepachys pinnata**, Torr. & Gr.

980. Achenia broadly winged. Awns persistent. Heads corymbed. Leaves serrate, feather-veined, tapering to the base and mostly decurrent on the stem.
Actinomeris (986).

Achenia wingless. Awns deciduous. Heads solitary, or corymbed. **Helianthus** (986).

981. (973.) Disk-flowers fertile; rays sterile; achenia with awns.................................984
Disk-flowers sterile; rays fertile............ 982

982. Pappus 2- or 3-toothed....................983

Pappus none. Heads in panicled corymbs; flowers light yellow. Leaves large, thin, lobed, and with stipule-like appendages at the base. **Polymnia.**

983. Achenia winged and sometimes also 2-toothed. Heads corymbed, or nearly solitary. Tall rough herbs. Lower leaves sometimes whorled in threes. **Silphium.**

Achenia flattened. Heads always single, long-peduncled. A hairy, low herb (2–6 inches high). Flowers yellow. Leaves ovate or spatulate, crenate, long-petioled.
Chrysogonum Virginianum, L.

984. Achenia compressed laterally, or not at all.....986

Achenia compressed back and front..........985

985. Achenia with upwardly barbed awns, or awnless, never beaked. Rays yellow, or partly colored, rarely purple. **Coreopsis.**
(*Tickseed.*)

Achenia with retrorsely hispid awns, often attenuated above. **Bidens.**
(*Bur-Marigold.*)

986. Achenia wingless. Awns deciduous. Heads solitary or corymbed. (980) **Helianthus.**

Achenia broadly winged. Awns persistent.
(980) **Actinomeris.**

987. (904.) Pappus present......................988

Pappus none. Involucre cylindric, of 8 scales, in a single row, 8- to 12-flowered. Heads small, loosely panicled; flowers yellow. Leaves angled, or toothed, ovate, or sometimes lyre-shaped. **Lampsana communis,** L.
(*Nipple-wort.*)

988. Pappus of plumose bristles, which are chaffy-dilated at the base, either on a short beak, or sessile. Scapes bearing one or more yellow heads. The radical leaves pinnatifid. **Leontodon autumnale,** L.

Pappus not plumose......................989

989. Pappus chaffy, or of both chaff and bristles...990

Pappus composed exclusively of capillary bristles.
992

990. Rays yellow. Pappus double. Heads terminating the naked scapes, or branches. Leaves chiefly radical...991

Rays blue. Pappus a small crown of little, bristle-form scales. Involucre double. Heads large, 2 or 3 together, axillary and terminal. Stem-leaves oblong or lanceolate, partly clasping, the uppermost rigid, minute, the lowest and radical ones runcinate. **Cichorium Intybus,** L.

991. Outer pappus of 5 broad chaffy scales, the inner of as many alternate, slender bristles. Heads small, terminating the scapes and branches. Leaves chiefly radical, lyrate and toothed, sometimes roundish; the other leaves narrower. **Krigia Virginica,** Willd.

Outer pappus of numerous, very small, bristle-chaffy scales, the inner of many long, capillary bristles. Heads rather showy, single at the summit of the scapes or naked peduncles. **Cynthia.**

992. Pappus bright white........................997
Pappus reddish, tawny, or dirty white........993

993. Flowers yellow............................994
Flowers bluish, or cream-color turning bluish. Achenia contracted above into a short neck or beak. Leaves irregularly pinnatifid, sometimes, or rather often, runcinate, coarsely toothed (the upper ones undivided).. Heads in a large compound panicle. Herb tall and stout, 3–12 feet high. **Mulgedium leucophæum,** D. C. (998).

994. Achenia not beaked........................995
Achenia long-beaked, nearly terete. Pappus soft. Heads solitary, terminal on both stem and branches, rather large. Leaves oblong or lanceolate, entire, cut, or pinnatifid; those of the stem sessile and somewhat clasping. **Pyrrhopappus Carolinianus,** D. C.

995. Flowers whitish, rose-purple, or purplish......996
Flowers yellow. Stems sometimes scape-like. Leaves entire, or toothed. Heads single, or panicled; flowers yellow. **Hieracium.**
(*Hawk-weed.*)

996. Flowers rose-purple, about 5 in each head. Stem almost leafless, scape-like, erect, slender, once or twice forked above. Leaves linear-subulate. Heads few, erect, on long peduncles. Radical leaves 6–10 inches long. **Lygodesmia juncea,** Don.

Flowers whitish or purplish. Involucral scales often purplish. Heads racemose-panicled, mostly nodding. Stem leafy. **Nabalus.**

997. Achenia beaked..............................998

Achenia beakless. Pappus copious. Heads large. 1000

998. Flowers yellow. Beak of the achenia long...899

Flowers blue. Beak of the achenia short and thick. .. **Mulgedium.**

999. Heads large. A scape-bearing herb. Scapes naked, hollow, bearing solitary large heads. Leaves pinnatifid, or runcinate, in a tuft. Outer involucre reflexed.
Taraxacum Dens-leonis, Desf.
(*Dandelion.*)

Heads small. A leafy-stemmed herb. Lower leaves runcinate, upper ones lanceolate and entire. Heads in a long and narrow, naked panicle.
Lactuca elongata, Muhl.

1000. Leaves all radical, lance-linear, long, woolly on the margins, tufted. Scape simple, naked. Heads solitary, large. Pappus rather rigid; some of the bristles gradually thickened toward the base.
Troximon cuspidatum, Pursh.

Leafy-stemmed weeds. Heads corymbed or umbellate. Pappus exceedingly soft. Leaves spiny-toothed, usually runcinate-pinnatifid. **Sonchus.**
(*Sow-thistle.*)

CONSPECTUS

OF THE

EXOGENS AND ENDOGENS

CONSTITUTING

THE PHÆNOGAMOUS FLORA

OF THE

NORTHERN AND MIDDLE STATES.

(Gray's Arrangement.)

DIVISION I

EXOGENS, OR DICOTYLEDONS.

CLASS I. *of the Exogens:* **Angiosperms**, consisting of three sub-classes—that is, the Polypetalous, Gamopetalous, and Apetalous Exogens.

SUB-CLASS I

Polypetalous Exogens.

ORDER I.

RANUNCULACEÆ, OR CROWFOOTS.

GENERA:

1. *Atragene,* Sims, represented by the species:
    Atragene Americana, Sims.
    (*Clematis verticillaris,* D.C.)
2. *Clematis,* L. Virgin's Bower. *Species* 5.
3. *Pulsatilla,* Tourn. Pasque-Flower. *Species:*
    Pulsatilla Nuttalliana, Gray.
    (*Anemone,* D.C.)
4. *Anemone,* L. Anemone. Wind-Flower. *Species* 6.
5. *Hepatica,* Dill. Liver-Leaf. *Species* 2.
6. *Thalictrum,* Tourn. Meadow-Rue. *Species* 3.

7. *Trautvetteria,* Fischer and Meyer. False Bugbane.
   Species: *Trautvetteria palmata,* Fisch. and M.
8. *Ranunculus,* L. Crowfoot. Buttercup. *Species* 17.
9. *Myosurus,* Dill. Mouse-Tail. *Species:*
   Myosurus minimus, L.
10. *Isopyrum,* L. *Species:*
    Isopyrum biternatum, Torr. and Gr.
11. *Caltha,* L. Marsh-Marigold. *Species:*
    Caltha palustris, L. Marsh-Marigold.
12. *Trollius,* L. Globe-Flower. *Species:*
    Trollius laxus, Salisb. Spreading Globe-Flower.
13. *Coptis,* Salisb. Goldthread. *Species:*
    Coptis trifolia, Salisb. Three-leaved Goldthread.
14. *Helleborus,* L. Hellebore. *Species:*
    Helleborus viridis, L. Green Hellebore.
15. *Aquilegia,* Tourn. Columbine. *Species:*
    Aquilegia Canadensis, L. Wild Columbine.
16. *Delphinium,* Tourn. Larkspur. *Species* 4.
17. *Aconitum,* Tourn. Aconite. Monkshood. Wolfsbane. *Species* 2.
18. *Xanthorrhiza,* Marshall. Shrub Yellow-root. *Species:*   *Xanthorrhiza apiifolia,* L'Her.
19. *Hydrastis,* L. Orange-root. Yellow Puccoon. *Species:*   *Hydrastis Canadensis,* L.
20. *Actæa,* L. Baneberry. Cohosh. *Species:*
    Actæa spicata, L. (Varieties 2: *rubra* and *alba.*)
21. *Cimicifuga,* L. Bugbane. *Species* 2.

ORDER II.

MAGNOLIACEÆ, OR MAGNOLIADS.

GENERA:

1. *Magnolia,* L. *Species* 5.
2. *Liriodendron,* L. Tulip-Tree. *Species:*
   Liriodendron tulipifera, L.

ORDER III.

ANONACEÆ, OR ANONADS.

GENUS:
1. *Asimina,* Adans. North-American Papaw. *Speci*
   Asimina triloba, Dunal. Common Pap:

ORDER IV.

MENISPERMACEÆ, OR MENISPERMADS.

GENERA:
1. *Cocculus,* D.C. *Species: Cocculus Carolinus,* D
2. *Menispermum,* L. Moonseed. *Species:*
   Menispermum Canadense, L. Canadian Moonse
3. *Calycocarpum,* Nutt. Cupseed. *Species:*
   Calycocarpum Lyoni, Nu

ORDER V.

BERBERIDACEÆ, OR BERBERIDS.

GENERA:
1. *Berberis,* L. Barberry. *Species* 2.
2. *Caulophyllum,* Mx. Blue Cohosh. *Species:*
   Caulophyllum thalictroides, I
3. *Diphylleia,* Mx. Umbrella-leaf. *Species:*
   Diphylleia cymosa, I
4. *Jeffersonia,* Bart. Twin-leaf. *Species:*
   Jeffersonia diphylla, P
5. *Podophyllum,* L. May-Apple. Mandrake. *Speci*
   Podophyllum peltatum,

ORDER VI.

NELUMBIACEÆ, OR WATER-BEANS.

GENUS:
1. *Nelumbium,* Juss. Nelumbo. Sacred Bean. *Speci*
   Nelumbium luteum, Willd. Yellow Nelumbo.

ORDER VII.

CABOMBACEÆ, OR WATER-SHIELDS.

GENUS:

1. *Brasenia*, Schreber. Water-Shield. *Species:*
Brasenia peltata, Pursh.

ORDER VIII.

NYMPHÆACEÆ, OR WATER-LILIES.

GENERA:

1. *Nymphœa*, Tourn. Water-Lily. Water-Nymph. *Species:* *Nymphœa odorata*, Ait. Sweet-scented Water-Lily.
2. *Nuphar*, Smith. Yellow Pond-Lily. Spatter-Dock. *Species* 2.

ORDER IX.

SARRACENIACEÆ, OR PITCHER PLANTS.

GENUS:

1. *Sarracenia*, Tourn. Side-saddle-Flower. *Species* 2.

ORDER X.

PAPAVERACEÆ, OR POPPY-WORTS.

GENERA:

1. *Papaver*, L. Poppy. *Species* 2.
2. *Argemone*, L. Prickly Poppy. *Species:*
Argemone Mexicana, L.
3. *Stylophorum*, Nutt. Celandine Poppy. *Species:*
Stylophorum diphyllum, Nutt.
(*Meconopsis diphylla*, D.C.)
4. *Chelidonium*, L. Celandine. *Species:*
Chelidonium majus, L.
5. *Glaucium*, Tourn. Horn-Poppy. *Species:*
Glaucium luteum, Scop.
6. *Sanguinaria*, Dill. Blood-Root. *Species:*
Sanguinaria Canadensis, L.

ORDER XI.

FUMARIACEÆ, OR FUMEWORTS.

GENERA:
1. *Adlumia*, Raf. Climbing Fumitory. *Species:*
   Adlumia cirrhosa, Raf.
2. *Dicentra*, Bork. Dutchman's Breeches. *Species* 3.
3. *Corydalis*, Vent. *Species* 2.
4. *Fumaria*, L. Fumitory. *Species:*
   Fumaria officinalis, L. Common Fumitory.

ORDER XII.

CRUCIFERÆ, OR CRUCIFERS.

GENERA:
1. *Nasturtium*, R. Br. Water-Cress. *Species* 6.
2. *Iodanthus*, Torr. & Gr. False Rocket. *Species:*
   Iodanthus hesperidoides, Torr. & Gr.
   (*Iodanthus pinnatifida*, Wood.)
3. *Leavenworthia*, Torr. *Species* 2.
4. *Dentaria*, L. Toothwort. Pepper-root. *Species* 4.
5. *Cardamine*, L. Bitter-Cress. *Species* 5.
6. *Arabis*, L. Rock-Cress. *Species* 7.
7. *Turritis*, Dill. Tower Mustard. *Species* 3.
8. *Barbarea*, R. Br. Winter-Cress. *Species:*
   Barbarea vulgaris, R. Br. Common Winter-Cress.
9. *Erysimum*, L. Treacle Mustard. *Species* 2.
10. *Sisymbrium*, L. Hedge Mustard. *Species* 3.
11. *Sinapis*, Tourn. Mustard. *Species* 3.
12. *Draba*, L. Whitlow-Grass. *Species* 8.
13. *Armoracia*, Rupp. Horseradish. (Nasturtium, Tries, Gray.) *Species:*
    Armoracia Americana, Arn.
14. *Vesicaria*, Lam. Bladder-Pod. *Species:*
    Vesicaria Shortii, Torr. & Gr.
15. *Camelina*, Crantz. False Flax. *Species:*
    Camelina sativa, Crantz.

16. *Lepidium*, L. Pepperwort. Peppergrass. *Species 4.*
17. *Capsella*, Vent. Shepherd's Purse. *Species :*
   Capsella Bursa-Pastoris, Mœnch.
18. *Subularia*, L. Awlwort. *Species :*
   Subularia aquatica, L.
19. *Senebiera*, D.C. Wart-Cress. Swine-Cress. *Species 2.*
20. *Cakile*, Tourn. Sea-rocket. *Species :*
   Cakile Americana, Nutt.
21. *Raphanus*, L. Radish. *Species :*
   Raphanus Raphanistrum, L.
   (*Wild Radish. Jointed Charlock.*)

ORDER XIII.

CAPPARIDACEÆ, OR CAPPARIDS.

GENUS :

1. *Polanisia*, Raf. *Species :*
   Polanisia graveolens, Raf.

ORDER XIV.

RESEDACÆ, OR MIGNONETTES.

GENUS :

1. *Reseda*, L. Mignonette. Dyer's Rocket. *Species :*
   Reseda luteola, L. Dyer's Weed, or Weld.

ORDER XV.

VIOLACEÆ, OR VIOLETS.

GENERA :

1. *Solea*, Ging. Green Violet. *Species :*
   Solea concolor, Ging.
2. *Viola*, L. Violet. Heart's-ease. *Species 18.*

ORDER XVI.

CISTACEÆ, OR ROCK-ROSES.

GENERA :

1. *Helianthemum*, Tourn. Rock-Rose. *Species 2.*
2. *Hudsonia*, L. *Species 2.*
3. *Lechea*, L. Pinweed. *Species 3.*

ORDER XVII.

DROSERACEÆ, OR SUNDEWS.

GENUS:
1. *Drosera*, L. Sundew. *Species* 4.

ORDER XVIII.

PARNASSIACEÆ, OR PARNASSIADS.

GENUS:
1. *Parnassia*, Tourn. Grass of Parnássus. *Species* 3.

ORDER XIX.

HYPERICACEÆ, OR ST. JOHN'S-WORTS.

GENERA:
1. *Ascyrum*, L. St. Peter's-wort. *Species* 2.
2. *Hypericum*, L. St. John's-wort. *Species* 15.
3. *Elodea*, Pursh. Marsh St. John's-wort. *Species* 2.

ORDER XX.

ELATINACEÆ, OR WATER-PEPPERS.

GENUS:
1. *Elatine*, L. Water-wort. *Species:*
   Elatine Americana, Arnott.

ORDER XXI.

CARYOPHYLLACEÆ, OR PINKWORTS.

GENERA:
1. *Dianthus*, L. Pink. Carnation. *Species:*
   Dianthus Armeria, L. Deptford Pink.
2. *Saponaria*, L. Soapwort. *Species:*
   Saponaria officinalis, L. Common Soapwort.
3. *Vaccaria*, Medik. Cow-herb. *Species:*
   Vaccaria vulgaris, Host.
4. *Silene*, L. Catchfly. Campion. *Species* 12.

5. *Agrostemma,* L. Corn-cockle. *Species:*
   Agrostemma Githago, L.
6. *Honkenya,* Ehrh. Sea-sandwort. *Species:*
   Honkenya peploides, Ehrh.
7. *Alsine,* Wahl. Grove Sandwort. *Species* 4.
8. *Arenaria,* L. Sandwort. *Species:*
   Arenaria serphyllifolia, L. Thyme-leaved Sandwort.
9. *Mœhringia,* L. *Species:*
   Mœhringia lateriflora, L.
10. *Stellaria,* L. Chickweed. Starwort. *Species* 7.
11. *Holosteum,* L. Jagged Chickweed. *Species:*
    Holosteum umbellatum, L.
12. *Cerastium,* L. Mouse-ear Chickweed. *Species* 5.
13. *Mœnchia,* Ehrh. (Cerastium, Fenzl.) *Species:*
    Mœnchia quaternella, Ehrh.
14. *Sagina,* L. Pearlwort. *Species* 3.
15. *Spergularia,* Pers. Spurrey-Sandwort. *Species:*
    Spergularia rubra, Pers.
16. *Spergula,* L. Spurrey. *Species:*
    Spergula arvensis, L.
17. *Anychia,* Mx. Forked Chickweed. (Sessile stigmas 2. The stem-part of the embryo or radicle turned downward.) *Species:* *Anychia dichotoma,* Mx.
18. *Paronychia,* Tourn. Whitlow-wort. (Radicle of the embryo ascending.) *Species* 2.
19. *Scleranthus,* L. Knawel. *Species:*
    Scleranthus annuus, L.
20. *Mollugo,* L. Indian Chickweed. *Species:*
    Mollugo verticillata, L. Carpet-weed.

ORDER XXII.

PORTULACACEÆ, OR PURSLANES.

GENERA:
1. *Sesuvium,* L. Sea-Purslane. *Species:*
   Sesuvium Portulacastrum, L.
2. *Portulaca,* Tourn. Purslane. *Species:*
   Portulaca oleracea, L. Common Purslane.

3. *Talinum*, Adans. *Species:*
Talinum teretifolium, Pursh.
4. *Claytonia*, L. Spring-Beauty. *Species 2.*

ORDER XXIII.

MALVACEÆ, OR MALLOWS.

GENERA:
1. *Althœa*, L. Marsh-Mallow. *Species:*
Althœa officinalis, L. Common Marsh-Mallow.
2. *Malva*, L. Mallow. *Species 2.*
3. *Callirrhoë*, Nutt. *Species 2.* (*Malva* and *Sida*, L.)
4. *Napœa*, Clayt. Glade Mallow. *Species:*
Napœa dioica, L.
5. *Sida*, L. *Species 3.*
6. *Abutilon*, Tourn. Indian Mallow. *Species:*
Abutilon Avicennœ, Gærtn. Velvet-Leaf.
7. *Modiola*, Mœnch. *Species:*
Modiola multifida, Mœnch.
8. *Kosteletzkya*, Presl. *Species:*
Kosteletzkya Virginica, Presl.
9. *Hibiscus*, L. Rose-Mallow. *Species 3.*

ORDER XXIV.

TILIACEÆ, OR LINDENBLOOMS.

GENUS:
1. *Tilia*, L. Linden. Basswood. *Species 2.*

ORDER XXV.

CAMELLIACEÆ, OR TEAWORTS.

GENERA:
1. *Stuartia*, Catesby. *Species:*
Stuartia Virginica, Cav.
2. *Gordonia*, Ellis. Loblolly Bay. *Species:*
Gordonia Lasianthus, L.

ORDER XXVI.

LINACEÆ, OR FLAXWORTS.

GENUS:
1. *Linum*, L. Flax. *Species 2.*

ORDER XXVII.

OXALIDACEÆ, OR WOOD-SORRELS.

GENUS:
1. *Oxalis*, L. Wood-Sorrel. *Species 3.*

ORDER XXVIII.

GERANIACEÆ, OR CRANESBILLS.

GENERA:
1. *Geranium*, L. Cranesbill. *Species 4.*
2. *Erodium*, L'Her. Storksbill. *Species:*
    Erodium cicutarium, L'Her.

ORDER XXIX.

BALSAMINACEÆ, OR JEWEL-WEEDS.

GENUS:
1. *Impatiens*, L. Jewel-Weed. Balsam. *Species 2.*

ORDER XXX.

LIMNANTHACEÆ, OR LIMNANTHS.

GENUS:
1. *Flœrkea*, Willd. False Mermaid. *Species:*
    Flœrkea proserpinacoides, Willd.

ORDER XXXI.

RUTACEÆ, OR RUEWORTS.

GENERA:
1. *Xanthoxylum*, Colden. Prickly Ash. *Species 2.*
2. *Ptelea*, L. Shrubby trefoil. Hop-tree. *Species:*
    Ptelea trifoliata, L.

ORDER XXXII.

ANACARDIACEÆ, OR SUMACHS.

GENUS:
1. *Rhus*, L. Sumach. *Species* 6.

ORDER XXXIII.

VITACEÆ, OR VINES.

GENERA:
1. *Vitis*, Tourn. Grape. *Species* 6 (including the genus *Cissus* as a sub-genus).
2. *Ampelopsis*, Mx. Virginian Creeper. *Species:*
Ampelopsis quinquefolia, Mx.

ORDER XXXIV.

RHAMNACEÆ, OR BUCKTHORNS.

GENERA:
1. *Berchemia*, Necker. Supple-Jack. *Species:*
Berchemia volubilis, D. C.
2. *Rhamnus*, L. Buckthorn, including *Frangula Caroliniana*, Gray. *Species* 4.
3. *Ceanothus*, L. New-Jersey Tea. Red-root. *Species* 2.

ORDER XXXV.

CELASTRACEÆ, OR STAFF-TREES.

GENERA:
1. *Celastrus*, L. Staff-Tree. Shrubby Bitter-sweet. *Species:*
Celastrus scandens, L. Wax-work.
2. *Euonymus*, Tourn. Spindle-Tree. *Species* 2.

ORDER XXXVI.

SAPINDACEÆ, OR SOAPBERRY FAMILY.

SUB-ORDER I.—*Staphyleaceæ*. Bladdernuts.

GENUS:
   1. *Staphylea*, L. Bladdernut. *Species:*
                                  Staphylea trifolia, L.

SUB-ORDER II.—*Sapindaceæ* proper.
GENUS:
   2. *Æsculus*, L. Buckeye. Horse-chestnut. *Species* 4.

SUB-ORDER III.—*Acerineæ*, Maples.
GENERA:
   3. *Acer*, Tourn. Maple. *Species* 5.
   4. *Negundo*, Mœnch. Box-Elder. Ash-leaved Maple.
      Species: *Negundo aceroides*, Mœnch.

ORDER XXXVII.

POLYGALACEÆ, OR MILKWORTS.

GENUS:
   1. *Polygala*, Tourn. Milkwort. *Species* 13.

ORDER XXXVIII.

LEGUMINOSÆ, OR LEGUMINOUS PLANTS.

GENERA:
   1. *Lupinus*, Tourn. Lupine. *Species:*
                    Lupinus perennis, L. Wild Lupine.
   2. *Crotalaria*, L. Rattle-box. *Species:*
                    Crotalaria sagittalis, L.
   3. *Genista*, L. Woad-waxen. Whin. *Species:*
            Genista tinctoria, L. Dyer's Green-weed.
   4. *Trifolium*, L. Clover. Trefoil. *Species* 8.
   5. *Melilotus*, Tourn. Melilot. Sweet Clover. *Species* 2.
   6. *Medicago*, L. Medick. *Species* 4.
   7. *Psoralea*, L. *Species* 5.
   8. *Dalea*, L. *Species:*   *Dalea alopecuroides*, Willd.
   9. *Petalostemon*, Mx. Prairie-Clover. *Species* 2.
  10. *Amorpha*, L. False Indigo. *Species* 2.
  11. *Robinia*, L. Locust-Tree. *Species* 3.

12. *Wistaria*, Nutt. *Species:*
    Wistaria frutescens, D. C.
13. *Tephrosia*, Pers. Hoary Pea. *Species* 3.
14. *Astragalus*, L. Milk-vetch. *Species* 6.
15. *Æschyomene*, L. Sensitive Joint Vetch. *Species:*
    Æschyomene hispida, Willd.
16. *Hedysarum*, Tourn. *Species:*
    Hedysarum boreale, Nutt.
17. *Desmodium*, D. C. Tick-trefoil. *Species* 18.
18. *Lespedeza*, Mx. Bush-Clover. *Species* 6.
19. *Stylosanthes*, Swartz. Pencil-flower. *Species:*
    Stylosanthes elatior, Swartz.
20. *Vicia*, Tourn. Vetch. Tare. *Species* 6.
21. *Lathyrus*, L. Vetchling. Everlasting Pea. *Species* 4.
22. *Phaseolus*, L. Kidney-Bean. *Species* 4.
23. *Apios*, Bœrh. Ground-Nut. Wild Bean. *Species:*
    Apios tuberosa, Mœnch.
24. *Rhynchosia*, D. C. *Species:*
    Rhynchosia tomentosa, Torr. & Gr.
25. *Galactia*, P. Browne. Milk Pea. *Species* 2.
26. *Amphicarpæa*, Ell. Hog Peanut. *Species:*
    Amphicarpæa monoica, Nutt.
27. *Clitoria*, L. Butterfly Pea. *Species:*
    Clitoria Mariana, L.
28. *Centrosema*, D. C. Spurred Butterfly Pea. *Species:*
    Centrosema Virginianum, Benth.
29. *Baptisia*, Vent. False Indigo. *Species* 5.
30. *Cladrastis*, Raf. Yellow-Wood. *Species:*
    Cladrastis tinctoria, Raf.
31. *Cercis*, L. Judas-Tree. Red-Bud. *Species:*
    Cercis Canadensis, L. Red-Bud.
32. *Cassia*, L. Senna. *Species* 4.
33. *Gymnocladus*, Lam. Kentucky Coffee-tree. *Species:*
    Gymnocladus Canadensis, Lam.
34. *Gleditschia*, L. Honey-Locust. *Species* 2.
35. *Desmanthus*, Willd. *Species:*
    Desmanthus brachylobus, Benth.
36. *Schrankia*, Willd. Sensitive Brier. *Species* 2.

ORDER XXXIX.

ROSACEÆ, OR ROSEWORTS.

1. *Prunus*, L. Plum and Cherry—that is,
    § *Prunus* proper, with 4 *Species.*
    § *Cerasus*, with 4 *Species.*
2. *Spiræa*, L. Meadow-Sweet. *Species* 6.
3. *Gillenia*, Mœnch. Indian Physic. *Species* 2.
4. *Agrimonia*, Tourn. Agrimony. *Species* 2.
5. *Sanguisorba*, L. Great Burnet. *Species:*
    Sanguisorba Canadensis, L. Canadian Burnet.
6. *Alchemilla*, Tourn. Lady's Mantle. *Species:*
    Alchemilla arvensis, L. Parsley Piert.
7. *Sibbaldia*, L. *Species:* *Sibbaldia procumbens*, L.
8. *Dryas*, L. *Species:* *Dryas integrifolia*, Vahl.
9. *Geum*, L. Avens. *Species* 8.
10. *Waldsteinia*, Willd. *Species:*
    Waldsteinia fragarioides, Tratt. Barren Strawberry.
11. *Potentilla*, L. Cinquefoil. Five-finger. *Species* 11.
12. *Fragaria*, Tourn. Strawberry. *Species* 2.
13. *Dalibarda*, L. *Species:* *Dalibarda repens*, L.
14. *Rubus*, L. Bramble. *Species* 11.
15. *Rosa*, Tourn. Rose. *Species* 6.
16. *Cratægus*, L. Hawthorn. White Thorn. *Species* 9.
17. *Pyrus*, L. Pear and Apple. *Species* 4.
18. *Amelanchier*, Medic. June-berry. *Species:*
    Amelanchier Canadensis, Torr. & Gr.

ORDER XL.

CALYCANTHACEÆ, OR CALYCANTHS.

GENUS:

1. *Calycanthus*, L. Carolina Allspice. Sweet-scented shrub. *Species* 3.

ORDER XLI.

MELASTOMACEÆ, OR MELASTOMES.

GENUS:
1. *Rhexia*, L. Deer-Grass. Meadow Beauty. *Species* 3.

ORDER XLII.

LYTHRACEÆ, OR LOOSESTRIFES.

GENERA:
1. *Ammannia*, Houst. *Species* 2.
2. *Lythrum*, L. Loosestrife. *Species* 4.
3. *Nesœa*, Juss. Swamp Loosestrife. *Species:*
    Nesœa verticillata, H. B. K.
4. *Cuphea*, Jacq. *Species:*
    Cuphea viscosissima, Jacq. Clammy Cuphea.

ORDER XLIII.

ONAGRACEÆ, OR EVENING PRIMROSES..

GENERA:
1. *Epilobium*, L. Willow-Herb. *Species* 5.
2. *Œnothera*, L. Evening Primrose. *Species* 9.
3. *Gaura*, L. *Species* 2.
4. *Jussiœa*, L. *Species:* *Jussiœa decurrens*, D. C.
5. *Ludwigia*, L. False Loosestrife. *Species* 7.
6. *Circœa*, Tourn. Enchanter's Nightshade. *Species* 2.
7. *Proserpinaca*, L. Mermaid-weed. *Species* 2.
8. *Myriophyllum*, Vaill. Water-Milfoil. *Species* 6.
9. *Hippuris*, L. Mare's-Tail. *Species:*
    Hippuris vulgaris, L.

ORDER XLIV.

LOASACEÆ, OR LOASADS.

GENUS:
1. *Mentzelia*, Plum. *Species: Mentzelia oligosperma*, Nutt.

ORDER XLV.

CACTACEÆ, OR INDIAN FIGS.

GENUS:
1. *Opuntia*, Tourn. Prickly Pear. Indian Fig. *Species:*
Opuntia vulgaris, Mill.

ORDER XLVI.

GROSSULACEÆ, OR CURRANTS.

GENUS:
1. *Ribes*, L. Currant. Gooseberry.
  § *Grossularia*, Tourn. Gooseberry. *Species* 4.
  § *Ribesia*, Berl. Currant. *Species* 3.

ORDER XLVII.

PASSIFLORACEÆ, OR PASSION-FLOWERS.

GENUS:
1. *Passiflora*, L. Passion-flower. *Species* 2.

ORDER XLVIII.

CUCURBITACEÆ, OR CUCURBITS.

GENERA:
1. *Sicyos*, L. One-seeded Star-Cucumber. *Species:*
Sicyos angulatus, L.
2. *Echinocystis*, Torr. & Gr. Wild Balsam-Apple. *Species:*
Echinocystis lobata, Torr. & Gr.
3. *Melothria*, L. *Species:*   Melothria pendula, L.

ORDER XLIX.

CRASSULACEÆ, OR HOUSE-LEEKS.

GENERA:
1. *Tillæa*, L. *Species:*   Tillæa simplex, Nutt.
2. *Sedum*, L. Stone-crop. Orpine. *Species* 4.
3. *Penthorum*, Gron. Ditch-stone-crop. *Species:*
Penthorum sedoides, L.

ORDER L.

SAXIFRAGACEÆ, OR SAXIFRAGES.

GENERA :
1. *Astilbe*, Don.   False Goat's-beard. *Species:*
     Astilbe decandra, Don.
2. *Saxifraga*, L.   Saxifrage.   *Species* 8.
3. *Boykinia*, Nutt.   *Species:*
     Boykinia aconitifolia, Nutt.
4. *Sullivantia*, Torr. & Gr.   *Species:*
     Sullivantia Ohionis, Torr. & Gr.
5. *Heuchera*, L.   Alum-root.   *Species* 4.
6. *Mitella*, Tourn. Mitre-wort. Bishop's-cap. *Species* 2.
7. *Tiarella*, L.   False Mitre-wort.   *Species:*
     Tiarella cordifolia, L.
8. *Chrysosplenium*. Tourn. Golden Saxifrage. *Species:*
     Chrysosplenium Americanum, Schwein.
9. *Itea*, L.   *Species:*   *Itea Virginica*, L.
10. *Hydrangea*, Gron.   *Species:*
     Hydrangea arborescens, L. Wild Hydrangea.
11. *Philadelphus*, L.   Mock-Orange.   *Species:*
     Philadelphus inodorus, L.

ORDER LI.

HAMAMELACEÆ, OR WITCH-HAZEL-WORTS.

1. *Hamamelis*, Witch-Hazel.   *Species:*
     Hamamelis Virginica, L.
2. *Fothergilla*, L. f.   *Species:* *Fothergilla alnifolia*, L. f.
3. *Liquidambar*, L.   Sweet Gum-Tree.   *Species:*
     Liquidambar Styraciflua, L.

ORDER LII.

UMBELLIFERÆ, OR UMBELWORTS.

1. *Hydrocotyle*, Tourn.   Water Penny-wort.   *Species* 4.
2. *Crantzia*, Nutt.   *Species:*   *Crantzia lineata*, Nutt.
3. *Sanicula*, Tourn. Sanicle. Black Snakeroot. *Species* 2.

4. *Eryngium.* Tourn. Button-Snakeroot. *Species* 2.
5. *Daucus,* Tourn. Carrot. *Species: Daucus carota,* L.
6. *Polytænia,* D. C. *Species: Polytænia Nuttallii,* D. C.
7. *Heracleum,* L. Cow-Parsnip. *Species:*
Heracleum lanatum, Mx.
8. *Pastinaca,* Tourn. Parsnip. *Species:*
Pastinaca sativa, L.
9. *Archemora,* D. C. Cowbane. *Species:*
Archemora rigida, D. C.
10. *Tiedemannia,* D. C. False Water-dropwort. *Species:*
Tiedemannia teretifolia, D. C.
11. *Angelica,* L. *Species:* *Angelica Curtisii,* Buckley.
12. *Archangelica,* Hoffm. *Species* 3.
13. *Conioselinum,* Fischer. Hemlock-Parsley. *Species:*
Conioselinum Canadense, Torr. & Gr.
14. *Æthusa,* L. Fool's Parsley. *Species:*
Æthusa Cynapium, L.
15. *Ligusticum,* L. Lovage. *Species* 2.
16. *Thaspium,* Nutt. Meadow-Parsnip. *Species* 4.
17. *Zizia,* D. C. *Species:* *Zizia integerrima,* D. C.
18. *Bupleurum,* Tourn. Thorough-wax. *Species:*
Bupleurum rotundifolium, L.
19. *Discopleura,* D. C. Mock-Bishop-weed. *Species* 2.
20. *Cicuta,* L. Water-Hemlock. *Species* 2.
21. *Sium,* L. Water-Parsnip. *Species* 2:
Sium lineare, Mx., and *Sium angustifolium,* L.
22. *Cryptotænia,* D. C. Honewort. *Species:*
Cryptotænia Canadensis, D. C.
23. *Chærophyllum,* L. Chervil. *Species:*
Chærophyllum procumbens, Lam.
24. *Osmorrhiza,* Raf. Sweet Cicely. *Species* 2.
25. *Conium,* L. Poison-Hemlock. *Species:*
Conium maculatum, L.
26. *Eulophus,* Nutt. *Species:*
Eulophus Americanus, Nutt.
27. *Erigenia,* Nutt. Harbinger-of-Spring. *Species:*
Erigenia bulbosa, Nutt.

ORDER LIII.

ARALIACEÆ, OR ARALIADS.

GENERA:
1. *Aralia*, L. Wild Sarsaparilla. *Species* 4.
2. *Panax*, L. Ginseng. *Species* 2.

ORDER LIV.

CORNACEÆ, OR CORNELS.

GENERA:
1. *Cornus*, Tourn. Cornel. Dogwood. *Species* 9.
2. *Nyssa*, L. Tupelo. Pepperidge. Sour Gum-Tree. *Species* 2.

SUB-CLASS II.

Gamopetalous Exogens.

ORDER LV.

CAPRIFOLIACEÆ, OR HONEYSUCKLES.

1. *Linnæa*, Gron. Twin-flower. *Species:*
 Linnæa borealis, Gron.
2. *Symphoricarpus*, Dill. Snowberry. *Species* 3.
3. *Lonicera*, L. Honeysuckle. Woodbine. *Species* 8.
4. *Diervilla*, Tourn. Bush-Honeysuckle. *Species:*
 Diervilla trifida, Mœnch.
5. *Triosteum*, L. Fever-wort. Horse-Gentian. *Species* 2.
6. *Sambucus*, Tourn. Elder. *Species* 2.
7. *Viburnum*, L. Arrow-wood. Laurestinus. *Species* 10.

ORDER LVI.

RUBIACEÆ, OR MADDERWORTS.

SUB-ORDER I.—*Stellatæ.* True Madderworts.
1. *Galium*, L. Bedstraw. Cleavers. *Species* 14.

SUB-ORDER II.—*Cinchoneæ.* Cinchona-Plants.
2. *Spermacoce*, L. Button-weed. *Species:*
   Spermacoce glabra, Mx.
3. *Diodia*, L. Button-weed. *Species* 2.
4. *Cephalanthus*, L. Button-bush. *Species:*
   Cephalanthus occidentalis, L.
5. *Mitchella*, L. Partridge-berry. *Species:*
   Mitchella repens, L.
6. *Oldenlandia*, L. *Species:*
   Oldenlandia glomerata, Mx. Creeping Greenhead.
7. *Houstonia*, L. Bluets. *Species* 4.

SUB-ORDER III.—*Loganieæ.*
8. *Mitreola*, L. Mitre-wort. *Species:*
   Mitreola petiolata, Torr. & Gr.
9. *Spigelia*, L. Pink-root. Worm-Grass. *Species:*
   Spigelia Marilandica, L.
10. *Polypremum*, L. *Species:*
    Polypremum procumbens, L.

ORDER LVII.

VALERIANACEÆ, OR VALERIANS.

1. *Valeriana*, Tourn. Valerian. *Species* 3.
2. *Fedia*, Gærtn. (*Valerianella*, D. C.) Corn-Salad. Lamb-Lettuce. *Species* 5.

ORDER LVIII.

DIPSACEÆ, OR TEASELWORTS.

GENUS:
1. *Dipsacus*, Tourn. Teasel. *Species:*
   Dipsacus sylvestris, Mill. Wild Teasel.

CONSPECTUS. 309

ORDER LIX.

COMPOSITÆ, OR COMPOSITES.

SUB-ORDER I.—*Tubuliforæ.*

GENERA:
1. *Vernonia,* Schreb. Iron-weed. *Species* 2.
2. *Elephantopus,* L. Elephant's-foot. *Species:*
   Elephantopus Carolinianus, Willd.
3. *Sclerolepis,* Cass. *Species:*
   Sclerolepis verticillata, Cass.
4. *Carphephorus,* Cass. *Species:*
   Carphephorus tomentosus, Torr. & Gr.
5. *Liatris,* Schreb. Button-Snakeroot. Blazing Star. *Species* 10.
6. *Kuhnia,* L. *Species:* *Kuhnia eupatorioides,* L.
7. *Eupatorium,* Tourn. Thoroughwort. *Species* 16.
8. *Mikania,* Willd. Climbing Hemp-weed. *Species:*
   Mikania scandens, L.
9. *Conoclinium,* D. C. Mist-flower. *Species:*
   Conoclinium cœlestinum, D. C.
10. *Nardosmia,* Cass. Sweet Colt's-foot. *Species:*
    Nardosmia palmata, Hook.
11. *Tussilago,* Tourn. Colt's-foot. *Species:*
    Tussilago Farfara, L.
12. *Adenocaulon,* Hook. *Species:*
    Adenocaulon bicolor, Hook.
13. *Sericocarpus,* Nees. White-topped Aster. *Species* 3.
14. *Aster,* L. Starwort. Aster. *Species* 38.
15. *Erigeron,* L. Fleabane. *Species* 8.
16. *Diplopappus,* Cass. Double-bristled Aster. *Species* 4.
17. *Boltonia,* L'Her. *Species* 2.
18. *Bellis,* Tourn. Daisy. *Species:*
    Bellis integrifolia, Mx. Western Daisy.
19. *Brachychæta,* Torr. & Gr. False Golden-rod. *Species:*
    Brachychæta cordata, Torr. & Gr.
20. *Solidago,* L. Golden-rod. *Species* 35.

21. *Bigelovia*, D. C.   Rayless Golden-rod.   *Species:*
            Bigelovia nudata, D. C.
22. *Chrysopsis*, Nutt.   Golden Aster.   *Species* 5.
23. *Inula*, L.   Elecampane.   *Species:*
            Inula Helenium, L.
24. *Pluchea*, Cass.   Marsh Fleabane.   *Species* 2.
25. *Baccharis*, L.   Groundsel-Tree.   *Species* 2.
26. *Polymnia*, L.   Leaf-Cup.   *Species* 2.
27. *Chrysogonum*, L.   *Species:*
           Chrysogonum Virginianum, L.
28. *Silphium*, L.   Rosin-Plant.   *Species* 6.
29. *Parthenium*, L.   *Species:*
           Parthenium integrifolium, L.
30. *Iva*, L.   Marsh Elder.   Highwater-shrub.   *Species* 2.
31. *Ambrosia*, Tourn.   Ragweed.   *Species* 4.
32. *Xanthium*, Tourn.   Cocklebur.   Clotbur.   *Species* 2.
33. *Tetragonotheca*, Dill.   *Species:*
          Tetragonotheca helianthoides, L.
34. *Eclipta*, L.   *Species:*
           Eclipta procumbens, Mx.
35. *Borrichia*, Adans.   Sea-Ox-eye.   *Species:*
           Borrichia frutescens, D. C.
36. *Heliopsis*, Pers.   Ox-eye.   *Species:*
           Heliopsis lævis, Pers.
37. *Echinacea*, Mœnch.   Purple Cone-flower.   *Species* 2.
38. *Rudbeckia*, L.   Cone-flower.   *Species* 6.
39. *Lepachys*, Raf.   *Species:*
           Lepachys pinnata, Torr. & Gr.
40. *Helianthus*, L.   Sunflower.   *Species* 18.
41. *Actinomeris*, Nutt.   *Species* 2.
42. *Coreopsis*, L.   Tickseed.   *Species* 12.
43. *Bidens*, L.   Bur-Marigold.   *Species* 6.
44. *Verbesina*, L.   Crownbeard.   *Species* 2.
45. *Dysodia*, Cav.   Fetid Marigold.   *Species:*
           Dysodia chrysanthemoides, Lag.
46. *Hymenopappus*, L'Her.   *Species:*
          Hymenopappus scabiosæus, L'Her.

47. *Helenium*, L. False Sunflower. *Species:*
   Helenium autumnale, L. Sneeze-weed.
48. *Leptopoda*, Nutt. *Species:*
   Leptopoda brachypoda, Torr. & Gr.
49. *Baldwinia*, Nutt. *Species:*
   Baldwinia uniflora, Nutt.
50. *Marshallia*, Schreb. *Species:*
   Marshallia latifolia, Pursh.
51. *Galinsoga*, Ruiz & Pav. *Species:*
   Galinsoga parviflora, Cav.
52. *Maruta*, Cass. May-weed. *Species:*
   Maruta Cotula, D. C.
53. *Anthemis*, L. Chamomile. *Species:*
   Anthemis arvensis, L. Corn Chamomile.
54. *Achillea*, L. Yarrow. *Species* 2.
55. *Leucanthemum*, Tourn. Ox-eye-Daisy. *Species:*
   Leucanthemum vulgare, Lam.
56. *Matricaria*, Tourn. Wild Chamomile. Feverfew. *Species* 2.
57. *Tanacetum*, L. Tansy. *Species* 2.
58. *Artemisia*, L. Wormwood. *Species* 7.
59. *Gnaphalium*, L. Cudweed. *Species* 5.
60. *Antennaria*, Gærtn. Everlasting. *Species* 2.
61. *Filago*, Tourn. Cotton-Rose. *Species:*
   Filago Germanica, L.
62. *Erechthites*, Raf. Fireweed. *Species:*
   Erechthites hieracifolia, Raf.
63. *Cacalia*, L. Indian Plantain. *Species* 4.
64. *Senecio*, L. Groundsel. *Species* 4.
65. *Arnica*, L. *Species* 2.
66. *Centaurea*, L. Star-Thistle. *Species* 3.
67. *Cnicus*, Vaill. Blessed Thistle. *Species:*
   Cnicus benedictus, L.
68. *Cirsium*, Tourn. Plumed Thistle. *Species* 10.
69. *Carduus*, Tourn. Plumeless Thistle. *Species:*
   Carduus nutans, L. Musk Thistle.

70. *Onopordon*, Vaill.   Cotton Thistle.   *Species:*
Onopordon Acanthium, L.

71. *Lappa*, Tourn.   Burdock.   *Species:*
Lappa major, Gærtn.

Sub-order II.—*Ligulifloræ.*

72. *Lampsana*, Tourn.   Nipple-wort.   *Species:*
Lampsana communis, L.

73. *Cichorium*, Tourn.   Succory.   *Species:*
Cichorium Intybus, L.

74. *Krigia*, Schreb.   Dwarf Dandelion.   *Species:*
Krigia Virginica, Willd.

75. *Cynthia*, Don.   *Species* 2.

76. *Leontodon*, L., Juss.   Hawkbit.   Fall Dandelion.
Species: *Leontodon autumnale*, L.

77. *Hieracium*, Tourn.   Hawkweed.   *Species* 6.

78. *Nabalus*, Cass.   Rattlesnake-root.   *Species* 9.

79. *Troximon*, Nutt.   *Species:*
Troximon cuspidatum, Pursh.

80. *Taraxacum*, Haller.   Dandelion.   *Species:*
Taraxacum Dens-leonis, Desf.   Common Dandelion.

81. *Pyrrhopappus*, D. C.   False Dandelion.   *Species:*
Pyrrhopappus Carolinianus, D. C.

82. *Lygodesmia*, Don.   *Species : Lygodesmia junica*, Don.

83. *Lactuca*, Tourn.   Lettuce.   *Species:*
Lactuca elongata, Muhl.   Wild Lettuce.

84. *Mulgedium*, Cass.   False or Blue Lettuce.   *Species* 3.

85. *Sonchus*, L.   Sow-Thistle.   *Species* 3.

ORDER LX.

LOBELIACEÆ, OR LOBELIADS.

Genus:

1. *Lobelia*, L.   *Species* 12.

ORDER LXI.

CAMPANULACEÆ, OR BELLWORTS.

GA:
Campanula, Tourn. Bellflower. Species 4.
Specularia, Heist. Venus' Looking-glass. Species:
Specularia perfoliata, A. D. C.

ORDER LXII.

ERICACEÆ, OR HEATHWORTS.

RDER I.—*Vacciniecæ*. Whortleberries.

GA:
Gaylussacia, H. B. K. Huckleberry. Species 4.
Vaccinium, L. Cranberry. Blueberry. Bilberry.
Species 11, including § *Oxycoccus*.
Chiogenes, Salisb. Creeping Snowberry. Species:
Chiogenes hispidula, Torr. & Gr.

RDER II.—*Ericineæ*. The proper Heath Family.

Arctostaphylos, Adans. Bearberry. Species 2.
Epigæa, L. Ground Laurel. Trailing Arbutus.
Species: *Epigæa repens*, L.
Gaultheria, Kalm. Aromatic Wintergreen. Species:
Gaultheria procumbens, L.
Leucothöe, Don. Species 4.
Cassandra, Don. Leather-Leaf. Species:
Cassandra calyculata, Don.
Cassiope, Don. Species:
Cassiope hypnoides, Don.
Andromeda, L. Gray. Species 4.
Oxydendrum, D. C. Sorrel-Tree. Sour-wood.
Species: *Oxydendrum arboreum*, D. C.
Clethra, L. White Alder. Sweet Pepperbush.
Species 2.
Phyllodoce, Salisb. Species:
Phyllodoce taxifolia, Salisb.

14. *Kalmia*, L.   American Laurel.   Species 4.
15. *Menziesia*, Smith.   Species :
   Menziesia ferruginea, Smith ; var. *globularis*.
16. *Azalea*, L.   False Honeysuckle.   Species 4.
17. *Rhododendron*, L.   Rose-Bay.   Species 3.
18. *Rhodora*, Duham.   Species :
   Rhodora Canadensis, L.
19. *Ledum*, L.   Labrador Tea.   Species :
   Ledum latifolium, Ait.
20. *Loiseleuria*, Desv.   Alpine Azalea.   Species :
   Loiseleuria procumbens, Desv.
21. *Leiophyllum*, Pers.   Sand Myrtle.   Species :
   Leiophyllum buxifolium, Ell.

Sub-order III.—*Pyroleæ*.   The Pyrola Family.

22. *Pyrola*, L.   False Wintergreen.   Species 5.
23. *Monesis*, Salisb.   Species :
   Monesis uniflora, Gray.   One-flowered Pyrola.
24. *Chimaphila*, Pursh.   Pipsissewa.   Species 2.

Sub-order IV.—*Monotropeæ*.   Indian Pipe Family.

25. *Pterospora*, Nutt.   Pine-drops.   Species :
   Pterospora Andromedea, Nutt.
26. *Schweinitzia*, Ell.   Sweet Pine-sap.   Species :
   Schweinitzia odorata, Ell.
27. *Monotropa*, L.   Indian Pipe.   Pine-sap.   Species 2.

ORDER LXIII.
GALACINEÆ.

Genera :
1. *Galax*, L.   Beetle-weed.   Species :
   Galax aphylla, L.

ORDER LXIV.
AQUIFOLIACEÆ, OR HOLLYWORTS.

Genera :
1. *Ilex*, L.   Holly.   Species 6.

2. *Prinos*, L. Winterberry. *Species* 3.
3. *Nemopanthes*, Raf. Mountain Holly. *Species:*
   Nemopanthes Canadensis, D. C.

ORDER LXV.

STYRACACEÆ, OR THE STORAX FAMILY.

GENERA:
1. *Styrax*, Tourn. Storax. *Species* 3.
2. *Halesia*, Ellis. Snowdrop or Silver-bell Tree.
   Species: *Halesia tetraptera*, L.
3. *Symplocos*, Jacq. Sweet-Leaf. *Species:*
   Symplocos tinctoria, L'Her.

ORDER LXVI.

EBENACEÆ, OR EBONY FAMILY.

GENUS:
1. *Diospyros*, L. Date-Plum. Persimmon. *Species:*
   Diospyros Virginiana, L.

ORDER LXVII.

SAPOTACEÆ, OR SOAPWORTS.

GENUS:
1. *Bumelia*, Swartz. Southern Buckthorn. *Species* 2.

ORDER LXVIII.

PLANTAGINACEÆ, OR RIBWORTS.

GENUS:
1. *Plantago*, L. Plantain. Ribgrass. *Species* 8.

ORDER LXIX.

PLUMBAGINACEÆ, OR LEADWORTS.

GENUS:
1. *Statice*, Tourn. Sea-Lavender. Marsh-Rosemary.
   Species: Statice Limonium, L. *Var. Caroliniana.*

ORDER LXX.

PRIMULACEÆ, OR PRIMROSES.

GENERA :
1. *Primula*, L. Primrose. Cowslip. *Species* 2.
2. *Androsace*, Tourn. *Species:*
    Androsace occidentalis, Pursh.
3. *Dodecatheon*, L. American Cowslip. *Species:*
    Dodecatheon Meadia, L.
4. *Trientalis*, L. Chickweed-Wintergreen. *Species:*
    Trientalis Americana, Pursh.
5. *Lysimachia*, L. Loosestrife. *Species* 6.
6. *Naumburgia*, Mœnch. Tufted Loosestrife. *Species:*
    Naumburgia thyrsiflora, Reichenb.
7. *Glaux*, L. Sea-Milkwort. *Species:*
    Glaux maritima, L.
8. *Anagallis*, Tourn. Pimpernel. *Species:*
    Anagallis arvensis, L.
9. *Centunculus*, L. Chaffweed. *Species:*
    Centunculus minimus, L.
10. *Samolus*, L. Water-Pimpernel. Brookweed. *Species:* *Samolus Valerandi*, L. *Var. Americanus.*
11. *Hottonia*, L. Featherfoil. Water-Violet. *Species:*
    Hottonia inflata, Ell.

ORDER LXXI.

LENTIBULACEÆ, OR BLADDERWORTS.

GENERA:
1. *Utricularia*, L. Bladderwort. *Species* 11.
2. *Pinguicula*, L. Butterwort. *Species:*
    Pinguicula vulgaris, L.

ORDER LXXII.

BIGNONIACEÆ, OR TRUMPET-FLOWERS.

SUB-ORDER I.—*Bignonieœ.*

GENERA :
1. *Bignonia*, Tourn. *Species:*
    Bignonia capreolata, L. Cross-Vine.

2. *Tecoma*, Juss. Trumpet-flower. *Species:*
   Tecoma radicans, Juss. Trumpet-Creeper.
3. *Catalpa*, Walt. Indian Bean. *Species:*
   Catalpa bignonioides, Walt.

SUB-ORDER II.—*Sesameæ.*

4. *Martynia*, L. Unicorn-plant. *Species:*
   Martynia proboscidea, Glox.

ORDER LXXIII.
OROBANCHACEÆ, OR BROOM-RAPES.

GENERA:
1. *Epiphegus*, Nutt. Beech-drops. Cancer-root. *Species:*
   Epiphegus Virginiana, Bart.
2. *Conopholis*, Wallr. Squaw-root. *Species:*
   Conopholis Americana, Wallr.
3. *Aphyllon*, Mitchell. Naked Broom-rape. *Species 2.*
4. *Phelipæa*, Tourn. Broom-rape. *Species:*
   Phelipæa Ludoviciana, Don.

ORDER LXXIV.
SCROPHULARIACEÆ, OR FIGWORTS.

GENERA:
1. *Verbascum*, L. Mullein. *Species 3.*
2. *Linaria*, Tourn. Toad-flax. *Species 4.*
3. *Antirrhinum*, L. Snapdragon. *Species:*
   Antirrhinum Orontium, L.
4. *Scrophularia*, Tourn. Figwort. *Species:*
   Scrophularia nodosa, L.
5. *Collinsia*, Nutt. *Species 2.*
6. *Chelone*, Tourn. Turtle-head. Snake-head. *Species:*
   Chelone glabra, L.
7. *Pentstemon*, Mitchell. Beard-Tongue. *Species 3.*
8. *Mimulus*, L. Monkey-flower. *Species 3.*
9. *Conobea*, Aublet. *Species: Conobea multifida*, Benth.
10. *Herpestis*, Gærtn. *Species 3.*
11. *Gratiola*, L. Hedge-Hyssop. *Species 5.*

12. *Ilysanthes*, Raf. Species: *Ilysanthes gratioloides*, Benth.
13. *Micranthemum*, Rich.
   Micranthemum Micrantha, Rich. (*Hemianthus micranthemoides*, Nutt.)
14. *Limosella*, L. Mudwort. Species:
   Limosella aquatica, L. Var. *Tenuifolia*.
15. *Synthyris*, Benth. Species:
   Synthyris Houghtoniana, Benth.
16. *Veronica*, L. Speedwell. Species 12.
17. *Buchnera*, L. Blue-hearts. Species:
   Buchnera Americana, L.
18. *Seymeria*, Pursh. Mullein-Foxglove. Species:
   Seymeria macrophylla, Nutt.
19. *Gerardia*, L. Species 6.
20. *Dasystoma*, Raf. Yellow Foxglove. Species 5.
21. *Castilleja*, Mutis. Painted-cup. Species 3.
22. *Schwalbea*, Gron. Chaff-seed. Species:
   Schwalbea Americana, L.
23. *Euphrasia*, Tourn. Eye-bright. Species:
   Euphrasia officinalis, L.
24. *Rhinanthus*, L. Yellow-Rattle. Species:
   Rhinanthus Crista-galli, L.
25. *Pedicularis*, Tourn. Lousewort. Species 2.
26. *Melampyrum*, Tourn. Cow-wheat. Species:
   Melampyrum Americanum, Mx.
   (*Melampyrum pratense*. Var. *Americanum*, Benth.)
27. *Gelsemium*, Juss. Yellow Jessamine. Species:
   Gelsemium sempervirens, Ait.

ORDER LXXV.

ACANTHACEÆ, OR ACANTHADS.

GENERA:

1. *Dianthera*, Gron. Water-Willow. Species:
   Dianthera Americana, L.
   (*Rhytiglossa pedunculosa*, Nees.)
2. *Dipteracanthus*, Nees. Species 2.

ORDER LXXVI.
VERBENACEÆ, OR VERVAINS.
GENERA:
1. *Verbena*, L. Vervain. *Species* 7.
2. *Lippia*, L. Fog-fruit. *Species:*
 Lippia lanceolata, Mx.
3. *Callicarpa*, L. French Mulberry. *Species:*
 Callicarpa Americana, L.
4. *Phryma*, L. Lopseed. *Species:*
 Phryma Leptostachya, L.

ORDER LXXVII.
LABIATÆ, OR LABIATES.
GENERA:
1. *Teucrium*, L. Germander. *Species:*
 Teucrium Canadense, L.
2. *Trichostema*, L. Blue Curls. *Species* 2.
3. *Isanthus*, Mx. False Pennyroyal. *Species:*
 Isanthus cœruleus, Mx.
4. *Mentha*, L. Mint. *Species* 4.
5. *Lycopus*, L. Water-Horehound. *Species* 2.
6. *Cunila*, L. Dittany. *Species:* *Cunila Mariana*, L.
7. *Hyssopus*, L. *Species:* *Hyssopus officinalis*, L.
8. *Pycnanthemum*, Mx. Mountain Mint. Basil. *Species* 9.
9. *Origanum*, L. Wild Marjoram. *Species:*
 Origanum vulgare, L.
10. *Thymus*, L. Thyme. *Species:*
 Thymus Serpyllum, L. Creeping Thyme.
11. *Satureja*, L. Savory. *Species:*
 Satureja hortensis, L.
12 *Calamintha*, Mœnch. Calaminth.
 § *Calamintha Proper*, Benth. *Species:*
 Calamintha Nepeta, Link. Basil Thyme.
 § *Calomelissa*, Benth. *Species:*
 Calamintha glabella, Benth.
 § *Clinopodium*, L. *Species:*
 Calamintha Clinopodium, Benth. Basil.

13. *Melissa*, L. Balm. *Species:* Melissa officinalis, L.
14. *Hedeoma*, Pers. Mock Pennyroyal. *Species* 2.
15. *Collinsonia*, L. Horse-Balm. *Species:*
   Collinsonia Canadensis, L. Stone-root.
16. *Salvia*, L. Sage. *Species* 2.
17. *Monarda*, L. Horse-Mint. *Species* 4.
18. *Blephilia*, Raf. *Species* 2.
19. *Lophanthus*, Benth. Giant Hyssop. *Species* 3.
20. *Nepeta*, L. Cat-Mint.
   § *Nepeta Proper.* *Species:* Nepeta Cataria, L. Catnip.
   § *Glechoma*, L. *Species:*
   Nepeta Glechoma, Benth. Ground-Ivy. Gill.
21. *Dracocephalum*, L. Dragon-head. *Species:*
   Dracocephalum parviflorum, Nutt.
22. *Cedronella*, Mœnch. *Species:*
   Cedronella cordata, Benth.
23. *Synandra*, Nutt. *Species:*
   Synandra grandiflora, Nutt.
24. *Physostegia*, Benth. False Dragon-Head. *Species:*
   Physostegia Virginiana, Benth.
25. *Brunella*, Tourn. Self-heal. *Species:*
   Brunella vulgaris, L. Self-heal. Heal-all.
26. *Scutellaria*, L. Skullcap. *Species* 10.
27. *Marrubium*, L. Horehound. *Species:*
   Marrubium vulgare, L.
28. *Galeopsis*, L. Hemp-Nettle. *Species* 2.
29. *Stachys*, L. Hedge-Nettle. *Species* 3.
30. *Leonurus*, L. Motherwort. *Species* 2.
31. *Lamium*, L. Dead-Nettle. *Species* 2.
32. *Ballota*, L. Fetid Horehound. *Species:*
   Ballota nigra, L. Black Horehound.
33. *Phlomis*, L. Jerusalem Sage. *Species:*
   Phlomis tuberosa, L.

ORDER LXXVIII.
BORRAGINACEÆ, OR BORRAGEWORTS.

GENERA:

1. *Echium*, Tourn. Viper's Bugloss. *Species:*
   Echium vulgare, L. Blue-weed.

2. *Lycopsis*, L. Bugloss. *Species:*
Lycopsis arvensis, L. Small Bugloss.
3. *Symphytum*, Tourn. Comfrey. *Species:*
Symphytum officinale, L.
4. *Onosmodium*, Mx. False Gromwell. *Species* 3.
5. *Lithospermum*, Tourn. Gromwell. Puccoon. *Species* 7.
6. *Mertensia*, Roth. Smooth Lungwort. *Species* 3.
7. *Myosotis*, L. Scorpion-Grass. Forget-me-not. *Species* 3.
8. *Echinospermum*, Swartz. Stickseed. *Species:*
Echinospermum Lappula, Lehm.
9. *Cynoglossum*, Tourn. Hound's Tongue. *Species* 3.
10. *Heliotropium*, Tourn. Heliotrope. *Species:*
Heliotropium Europœum, L.
11. *Heliophytum*, D.C. Indian Heliotrope. *Species:*
Heliophytum Indicum, D.C.

ORDER LXXIX.

HYDROPHYLLACEÆ, OR HYDROPHYLLS.

GENERA:

1. *Hydrophyllum*, L. Water-leaf. *Species* 4.
2. *Nemophila*, Nutt. *Species:*
Nemophila microcalyx, Fish. & Meyer.
3. *Ellisia*, L. *Species:* *Ellisia Nyctelea*, L.
4. *Phacelia*, Juss. *Species* 5 (including § *Cosmanthus*).

ORDER LXXX.

POLEMONIACEÆ, OR PHLOXWORTS.

GENERA:

1. *Polemonium*, Tourn. Greek Valerian. *Species:*
Polemonium reptans, L. Jacob's-Ladder.
2. *Phlox*, L. *Species* 9.
3. *Diapensia*, L. *Species:* *Diapensia Lapponica*, L.
4. *Pyxidanthera*, Mx. *Species:*
Pyxidanthera barbulata, Mx.

ORDER LXXXI.

CONVOLVULACEÆ, OR BINDWEEDS.

GENERA:
1. *Quamoclit*, Tourn. Cypress-Vine. *Species:*
   Quamoclit coccinea, Mœnch.
2. *Ipomœa*, L. Morning-Glory. *Species 4.*
   including the sub-genus *Pharbitis*.
3. *Convolvulus*, L. Bindweed. *Species:*
   Convolvulus arvensis, L.
4. *Calystegia*, R. Br. Bracted Bindweed. *Species 2.*
5. *Stylisma*, Raf. *Species 2.*
6. *Dichondra*, Forst. *Species:*
   Dichondra repens, Forst. *Var. Carolinensis.*
7. *Cuscuta*, Tourn. Dodder. *Species 9.*

ORDER LXXXII.

SOLANACEÆ, OR NIGHTSHADES.

GENERA:
1. *Solanum*, L. Nightshade. *Species 3.*
2. *Physalis*, L. Ground Cherry. *Species 3.*
3. *Nicandra*, Adans. Apple of Peru. *Species:*
   Nicandra physaloides, Gærtn.
4. *Hyosciamus*, Tourn. Henbane. *Species:*
   Hyosciamus niger, L. Black Henbane.
5. *Datura*, L. Jamestown-weed. Thorn-Apple. *Species:*
   Datura Stramonium, L.
6. *Nicotiana*, L. Tobacco. *Species:*
   Nicotiana rustica, L. Wild Tobacco.

ORDER LXXXIII.

GENTIANACEÆ, OR GENTIANWORTS.

GENERA:
1. *Sabbatia*, Adans. American Centaury. *Species 8.*
2. *Erythræa*, Pers. Centaury. *Species 3.*

3. *Frasera,* Walt. American Columbo. *Species:*
   Frasera Carolinensis, Walt.
4. *Halenia,* Borkh. Spurred Gentian. *Species:*
   Halenia deflexa, Griseb.
5. *Gentiana,* L. Gentian. *Species* 9.
6. *Bartonia,* Muhl. *Species* 2.
7. *Obolaria,* L. *Species:* *Obolaria Virginica,* L.
8. *Menyanthes,* Tourn. Buckbean. *Species:*
   Menyanthes trifoliata, L.
9. *Limnanthemum,* Gmel. Floating Heart. *Species:*
   Limnanthemum lacunosum, Griseb.

ORDER LXXXIV.

APOCYNACEÆ, OR DOGBANES.

GENERA:

1. *Amsonia,* Walt. *Species:*
   Amsonia Tabernæmontana, Walt.
2. *Forsteronia,* Meyer. *Species:*
   Forsteronia difformis, A. D. C.
3. *Apocynum,* Tourn. Dogbane. Indian Hemp. *Species* 2.

ORDER LXXXV.

ASCLEPIADACEÆ, OR ASCLEPIADS.

GENERA:

1. *Asclepias,* L. Milkweed. Silkweed. *Species* 14.
2. *Acerates,* Ell. Green Milkweed. *Species* 2.
3. *Enslenia,* Nutt. *Species:* *Enslenia albida,* Nutt.
4. *Gonolobus,* Mx. *Species* 2.
5. *Periploca,* L. *Species:* *Periploca Græca,* L.

ORDER LXXXVI.

OLEACEÆ, OR OLIVES.

GENERA:

1. *Ligustrum,* Tourn. Privet. *Species:*
   Ligustrum vulgare, L.

2. *Olea*, Tourn. Olive. *Species:*
   Olea Americana, L. Devil-wood.
3. *Chionanthus*, L. Fringe-Tree. *Species:*
   Chionanthus Virginica, L.
4. *Fraxinus*, Tourn. Ash. *Species* 6.
5. *Forestiera*, Poir. *Species: Forestiera acuminata*, Poir.

SUB-CLASS III.

Apetalous Exogens.

ORDER LXXXVII.

ARISTOLOCHIACEÆ, OR BIRTHWORTS.

GENERA:

1. *Asarum*, Tourn. Asarabacca. Wild Ginger. *Species* 3.
2. *Aristolochia*, Tourn. Birthwort. *Species* 3.

ORDER LXXXVIII.

NYCTAGINACEÆ, OR MARVELWORTS.

GENERA:

1. *Oxybaphus*, Vahl. *Species:*
   Oxybaphus nyctagineus, Sweet.

ORDER LXXXIX.

PHYTOLACCACEÆ, OR POKEWORTS.

GENERA:

1. *Phytolacca*, Tourn. Pokeweed. *Species:*
   Phytolacca decandra, L. Common Poke or Scoke.
   Garget. Pigeon-berry.

ORDER XC.

CHENOPODIACEÆ, OR GOOSEFOOTS.

GENERA:

1. *Cycloloma*, Moq.   Winged Pigweed. *Species:*
   Cycloloma platyphyllum, Moq.
2. *Chenopodium*, L. Goosefoot. Pigweed. *Species* 9.
3. *Roubieva*, Moq. *Species:*   *Roubieva multifida*, Moq.
4. *Blitum*, Tourn. Blite. *Species* 3.
5. *Atriplex*, Tourn. Orache. *Species:*
   Atriplex hastata, L.
6. *Obione*, Gærtn. *Species:*
   Obione arenaria, Moq. Sand-Orache.
7. *Salicornia*, Tourn. Glasswort. Samphire. *Species* 3.
8. *Chenopodina*, Moq. Sea-Goosefoot. *Species:*
   Chenopodina maritima, Moq.
9. *Salsola*, L. Saltwort. *Species:*   *Salsola Kali*, L.

ORDER XCI.

AMARANTACEÆ, OR AMARANTHS.

GENERA:

1. *Amarantus*, Tourn. Amaranth. *Species* 7.
2. *Euxolus*, Raf. False Amaranth. *Species* 3.
3. *Montelia*, Moq. *Montelia tamariscina*, Gray.
4. *Acnida*, L. Water-Hemp. *Species:*
   Acnida cannabina, L.
5. *Iresine*, P. Browne. *Species:*   *Iresine celosioides*, L.
6. *Frœlichia*, Mœnch. *Species:*   *F. Floridiana*, Moq.

ORDER XCII.

POLYGONACEÆ, OR SORRELWORTS.

GENERA:

1. *Polygonum*, L. Knotweed. *Species* 19.
2. *Polygonella articulata*, Meisn. (*Polygonum articulatum*, L., Gray.)

3. *Fagopyrum*, Tourn. Buckwheat. *Species:*
   F. esculentum, Mœnch.
4. *Oxyria*, Hill. Mountain Sorrel. *Species:*
   O. digyna, Campd. (*O. reniformis*, Hook.)
5. *Rumex*, L. Dock. Sorrel. *Species* 10.

ORDER XCIII.

LAURACEÆ, OR LAURELS.

GENERA:

1. *Persea*, Gærtn. Alligator Pear. *Species:*
   P. Carolinensis, Nees. Red Bay.
2. *Sassafras*, Nees. *Species: Sassafras officinale*, Nees.
3. *Benzoin*, Nees. Wild Allspice. Fever-bush. *Species* 2.
4. *Tetranthera*, Jacq. *Species:*
   T. geniculata, Nees. Pond-Spice.

ORDER XCIV.

THYMELACEÆ, OR DAPHNADS.

GENUS:

1. *Dirca*, L. Leatherwood. Moose-wood. *Species:*
   Dirca palustris, L.

ORDER XCV.

ELÆAGNACEÆ, OR OLEASTERS.

GEN.:

1. *Shepherdia*, Nutt. *Species:*   *Sh. Canadensis*, Nutt.

ORDER XCVI.

SANTALACEÆ, OR SANDALWORTS.

GENERA:

1. *Comandra*, Nutt. Bastard Toad-flax. *Species* 2.
2. *Pyrularia*, Mx. Oil-nut. Buffalo-nut. *Species:*
   Pyrularia oleifera, Gray.
   (*P. pubera*, Mx.)

ORDER XCVII.

LORANTHACEÆ, OR LORANTHS.

GENUS:
1. *Phoradendron,* Nutt. False Mistletoe. *Species:*
Phoradendron flavescens, Nutt.

ORDER XCVIII.

SAURURACEÆ, OR SAURURADS.

GENUS:
1. *Saururus,* L. Lizard's-tail. *Species:*
Saururus cernuus, L.

ORDER XCIX.

CERATOPHYLLACEÆ, OR HORNWORTS.

GEN.:
1. *Ceratophyllum,* L. Hornwort. *Species:*
Ceratophyllum demersum, L.

ORDER C.

CALLITRICHACEÆ, OR WATER-STARWORTS.

GEN.:
1. *Callitriche,* L. Water-Starwort. *Species* 3.

ORDER CI.

PODOSTEMACEÆ, OR THREADFOOTS.

GEN.:
1. *Podostemon,* Mx. River-weed. *Species:*
Podostemon ceratophyllum, Mx.

ORDER CII.

EUPHORBIACEÆ, OR SPURGEWORTS.

1. *Euphorbia,* L. Spurge. *Species* 19.

2. *Cnidoscolus*, Pohl. Spurge-Nettle. *Species:*
      Cnidoscolus stimulosa, Gray.
       (*Tread-Softly*.)
3. *Acalypha*, L. Three-seeded Mercury. *Species* 3.
4. *Tragia*, Plum. *Species* 3.
5. *Stillingia*, Garden. *Species:*   *S. sylvatica*, L.
6. *Croton*, L. *Species* 3.
7. *Crotonopsis*, Mx. *Species:*   *C. linearis*, Mx.
8. *Phyllanthus*, L. *Species:*   *Ph. Carolinensis*, Walt.
9. *Pachysandra*, Mx. *Species:*   *P. procumbens*, Mx.

ORDER CIII.

EMPETRACEÆ, OR CROWBERRIES.

GENERA:
1. *Empetrum*, Tourn. Crowberry. *Species:*
    Empetrum nigrum, L. Black Crowberry.
2. *Corema*, Don. *Species:*   *Corema Conradii*, Torr.

ORDER CIV.

URTICACEÆ, OR NETTLEWORTS.

SUB-ORDER I.—*Ulmaceæ*. Elmworts.

GENERA:
1. *Ulmus*, L. Elm. *Species* 4.
2. *Planera*, Gmel. Planer-Tree. *Species:*
       Pl. aquatica, Gmel.
3. *Celtis*, Tourn. Nettle-Tree. Hackberry. *Species* 2.

SUB-ORDER II.—*Artocarpeæ*. Artocarps.

GENUS:
4. *Morus*, Tourn. Mulberry. *Species* 2.

SUB-ORDER III.—*Urticeæ*. The true Nettles.
5. *Urtica*, Tourn. Nettle. *Species* 4.
6. *Laportea*, Gaud. Wood-Nettle. *Species:*
      Laportea Canadensis, Gaud.

7. *Pilea*, Lindl. Richweed. Clearweed. *Species:*
   Pilea pumila, Gray.
8. *Bœhmeria*, Jacq. False Nettle. *Species:*
   Bœhmeria cylindrica, Willd.
9. *Parietaria*, Tourn. Pellitory. *Species:*
   Parietaria Pennsylvanica, Muhl.

Sub-order IV.—*Cannabineæ.* Hempworts.

10. *Cannabis*, Tourn. Hemp. *Species:* *C. sativa*, L.
11. *Humulus*, L. Hop. *Species:* *Humulus lupulus*, L.

ORDER CV.

PLATANACEÆ, OR SYCAMORES.

Genus:
1. *Platanus*, L. Plane-Tree. Buttonwood. *Species:*
Platanus occidentalis, L. American Plane-Tree or Sycamore.

ORDER CVI.

JUGLANDACEÆ, OR THE WALNUT FAMILY.

1. *Juglans*, L. Walnut. *Species* 2.
2. *Carya*, Nutt. Hickory. *Species* 7.

ORDER CVII.

CUPULIFERÆ, OR MASTWORTS.

Genera:
1. *Quercus*, L. Oak. *Species* 18.
2. *Castanea*, Tourn. Chestnut. *Species* 2.
3. *Fagus*, Tourn. Beech. *Species:*
   Fagus ferruginea, Ait.
4. *Corylus*, Tourn. Hazelnut. Filbert. *Species* 2.
5. *Carpinus*, L. Hornbeam. Iron-wood. *Species:*
   Carpinus Americana, Mx. Blue or Water-beach.
6. *Ostrya*, Mich. Hop-Hornbeam. Iron-wood. *Species:*
   Ostrya Virginica, Willd. Lever-wood.

ORDER CVIII.

MYRICACEÆ, OR GALEWORTS.

GENERA:
1. *Myrica*, L. Bayberry. Wax-Myrtle. *Species* 2.
2. *Comptonia*, Solander. Sweet Fern. *Species:*
   C. asplenifolia, Ait.

ORDER CIX.

BETULACEÆ, OR BIRCHWORTS.

1. *Betula*, Tourn. Birch. *Species* 7.
2. *Alnus*, Tourn. Alder. *Species* 3.

ORDER CX.

SALICACEÆ, OR WILLOW-WORTS.

GENERA:
1. *Salix*, Tourn. Willow. Osier. *Species* 22.
2. *Populus*, Tourn. Poplar. Aspen. *Species* 6.

CLASS II. *of the Exogens:* **Gymnosperms.**

ORDER CXI.

CONIFERÆ, OR CONIFERS.

SUB-ORDER I.—*Abietineæ.* The proper Pine Family.

GENERA:
1. *Pinus*, Tourn. Pine. *Species* 8.
2. *Abies*, Tourn. Spruce. Fir. *Species* 5.
3. *Larix*, Tourn. Larch. *Species:*
   Larix Americana, Mx. Black Larch. Tamarack. Hackmatack.

SUB-ORDER II.—*Cupressineæ.* The Cypress Family.

4. *Thuja*, Tourn. Arbor-vitæ. *Species:*
   Th. occidentalis, L.

5. *Cupressus*, Tourn. Cypress. *Species:*
   C. thyoides, L. White Cedar.
6. *Taxodium*, Rich. Bald Cypress. *Species:*
   T. distichum, Rich.
7. *Juniperus*, L. Juniper. *Species* 2.

Sub-order III—*Taxineæ*. The Yew Family.

8. *Taxus*, Tourn. Yew. *Species:*
   Taxus baccata, L. *Var. Canadensis*, Gray.

DIVISION II.

ENDOGENS, OR MONOCOTYLEDONS.

ORDER CXII.

ARACEÆ, OR AROIDS.

Genera:
1. *Arisæma*, Mart. Indian Turnip. Dragon Arum. *Species* 2.
2. *Peltandra*, Raf. Arrow Arum. *Species:*
   Peltandra Virginica, Raf.
3. *Calla*, L. Water-Arum. *Species:* *Calla palustris*, L.
4. *Symplocarpus*, Salisb. Skunk Cabbage. *Species:*
   Symplocarpus fœtidus, Salisb.
5. *Orontium*, L. Golden-Club. *Species:*
   Orontium aquaticum, L.
6. *Acorus*, L. Sweet Flag. Calamus. *Species:*
   Acorus Calamus L.

ORDER CXIII.

TYPHACEÆ, OR TYPHADS.

Genera:
1. *Typha*, Tourn. Cat-tail Flag. *Species* 2.
2. *Sparganium*, Tourn. Bur-reed. *Species* 5.

ORDER CXIV.

LEMNACEÆ, OR DUCKMEATS.

GENUS:
1. *Lemna,* L.   Duckweed.   Duck's-meat.   *Species* 5.

ORDER CXV.

NAJADACEÆ, OR NAJADS.

GENERA:
1. *Najas,* L.   Najad.   *Species:*   *Najas flexilis,* Rostk.
2. *Zanichellia,* Micheli.   Horned Pondweed.   *Species:*
    Zanichellia palustris, L.
3. *Zostera,* L.   Grass-wrack.   Eel-grass.   *Species:*
    Zostera marina, L.
4. *Ruppia,* L.   Ditch-grass.   *Species:*
    Ruppia maritima, L.
5. *Potamogeton,* Tourn.   Pondweed.   *Species* 12.

ORDER CXVI.

ALISMACEÆ, OR WATER-PLANTAINS.

1. *Triglochin,* L.   Arrow-grass.   *Species* 2.
2. *Scheuchzeria,* L.   *Species:*   *Scheuchzeria palustris,* L.
3. *Alisma,* L.   Water-Plantain.   *Species:*
    Alisma Plantago, L.   Var. *Americanum.*
4. *Echinodorus,* Rich.   *Species* 3.
5. *Sagittaria,* L.   Arrow-head.   *Species* 5.

ORDER CXVII.

HYDROCHARIDACEÆ, OR FROGBITS.

GENERA:
1. *Limnobium,* Richard.   American Frog's-bit.   *Species:*
    Limnobium Spongia, Rich.
2. *Anacharis,* Rich.   Water-weed.   *Species:*
    Anacharis Canadensis, Planchon.
3. *Vallisneria,* Micheli.   Tape-grass.   Eel-grass.   *Species:*
    Vallisneria spiralis, L.

ORDER CXVIII.

BURMANNIACEÆ.

GENUS:
1. *Burmannia*, L.   Species: *Burmannia biflora*, L.

ORDER CXIX.

ORCHIDACEÆ, OR ORCHIDS.

GENERA:
1. *Orchis*, L. Orchis. Species:
   Orchis spectabilis, L.   Showy Orchis.
2. *Gymnadenia*, R. Brown. Naked-Gland Orchis. Species 2.
3. *Platanthera*, Rich. False Orchis. Species 16.
4. *Goodyera*, R. Brown. Rattlesnake-Plantain. Species 2.
5. *Spiranthes*, Rich. Ladies' Tresses. Species 3.
6. *Listera*, R. Br. Twayblade. Species 3.
7. *Arethusa*, Gron. Species: *Arethusa bulbosa*, L.
8. *Pogonia*, Juss. Species 4.
9. *Calopogon*, R. Brown. Species:
   Calopogon pulchellus, R. Br.
10. *Calypso*, Salisb. Species: *Calypso borealis*, Salisb.
11. *Tipularia*, Nutt. Crane-fly Orchis. Species:
    Tipularia discolor, Nutt.
12. *Bletia*, Ruiz and Pavon. Species:
    Bletia aphylla, Nutt.
13. *Microstylis*, Nutt. Adder's-Mouth. Species 2.
14. *Liparis*, Rich. Twayblade. Species 2.
15. *Corallorhiza*, Haller. Coral-root. Species 4.
16. *Aplectrum*, Nutt. Putty-root. Adam and Eve. Species: *Aplectrum hyemale*, Nutt.
17. *Cypripedium*, L. Lady's-Slipper. Species 6.

ORDER CXX.

AMARYLLIDACEÆ, OR AMARYLLIDS.

GENERA
1. *Zephyranthus*, Herb. (*Amaryllis*, L.) Species:
   Zephyranthus Atamasco, Herb. Atamosco-Lily.

2. *Pancratium.* *Species:*   *P. rotatum,* Ker
3. *Agave,* L. American Aloe. *Species:*
    Agave Virginica, L
4. *Hypoxis,* L. Star-grass. *Species:* *Hypoxis erecta,* L

ORDER CXXI.

HÆMODORACEÆ, OR BLOODWORTS.

GENERA:

1. *Lachnanthes,* Ell. Red-root. *Species:*
    Lachnanthes tinctoria, Ell
2. *Lophiola,* Ker. *Species:*   *Lophiola aurea,* Ker
3. *Aletris,* L. Colic-root. Star-grass. *Species* 2.

ORDER CXXII.

BROMELIACEÆ, OR BROMELIADS.

GENUS:

1. *Tillandsia,* L. Long-Moss. *Species:*
    Tillandsia usneoides, L

ORDER CXXIII.

IRIDACEÆ, OR IRIDS.

GENERA:

1. *Iris,* L. Flower-de-Luce. *Species* 5.
2. *Sisyrinchium,* L. Blue-eyed Grass. *Species:*
    Sisyrinchium Bermudiana, L

ORDER CXXIV.

DIOSCOREACEÆ, OR YAM-ROOTS.

GENUS:

1. *Dioscorea,* Plum. Yam. *Species: Dioscorea villosa,* L

ORDER CXXV.

SMILACEÆ, OR SARSAPARILLAS.

GENERA:
1. *Smilax*, Tourn. Greenbrier. Catbrier. *Species* 10.
2. *Trillium*, L. Three-leaved Nightshade. *Species* 7.
3. *Medeola*, Gron. Indian Cucumber-root. *Species:*
Medeola Virginica, L.

ORDER CXXVI.

LILIACEÆ, OR LILYWORTS.

GENERA:.
1. *Asparagus*, L. *Species:*   *A. officinalis*, L.
2. *Polygonatum*, Tourn. Solomon's Seal. *Species* 3.
3. *Smilacina*, Desf. False Solomon's Seal. *Species* 3.
4. *Majanthemum*, Mœnch. *Species:*
Majanthemum bifolium, D.C.
(*Smilacina bifolia*, Ker.)
5. *Convallaria*, L. *Species:*   *C. majalis*, L.
6. *Clintonia*, Raf. *Species* 2.
7. *Hemerocallis*, L. Day-Lily. *Species:*
Hemerocallis fulva, L.
8. *Ornithogalum*, Tourn. Star of Bethlehem. *Species:*
Ornithogalum umbellatum, L.
9. *Scilla*, L. Squill. *Species:*   *S. Fraseri*, Gray.
10. *Allium*, L. Onion, Garlic. *Species* 7.
11. *Lilium*, L. Lily. *Species* 4.
12. *Erythronium*, L. Dog's-tooth Violet. *Species* 2.
13. *Yucca*, L. Bear-grass. Spanish Bayonet. *Species:*
Yucca filamentosa, L. Adam's Needle.

ORDER CXXVII.

MELANTHACEÆ, OR MELANTHS.

SUB-ORDER I.—*Uvularieæ*.
1. *Uvularia*, L. Bellwort. *Species* 4.
2. *Prosartes*, Don. *Species:*   *P. lanuginosa*, Don.
3. *Streptopus*, Mx. Twisted-Stalk. *Species* 2.

SUB-ORDER II.—*Melanthieæ.*

4. *Melanthium*, Gron. *Species:*
   Melanthium Virginicum, L. Bunch-Flower.
5. *Zygadenus*, Mx. *Species* 3.
6. *Veratrum*, Tourn. False Hellebore. *Species* 4. Including *Stenanthium angustifolium*, Gray. (*Veratrum angustifolium*, Pursh.)
7. *Amianthium*, Gray. Fly-Poison. *Species:*
   Amianthium muscœtoxicum, Gray.
8. *Xerophyllum*, Mx. *Species:*
   Xerophyllum asphodeloides, Nutt.
9. *Helonias*, L. *Species:* *Helonias bullata*, L.
10. *Chamælirium*, Willd. Devil's-Bit. *Species:*
    Chamælirium luteum, Gray. Blazing-Star.
11. *Tofieldia*, Huds. False Asphodel. *Species:*
    Tofieldia glutinosa, Willd., and
    T. pubens, Ait.

ORDER CXXVIII.

JUNCACEÆ, OR RUSHES.

GENERA:

1. *Narthecium*, Mœhring. Bog-Asphodel. *Species:*
   Narthecium Americanum, Ker.
2. *Luzula*, D.C. Wood-Rush. *Species* 5.
3. *Juncus*, L. Rush. Bog-rush. *Species* 20.

ORDER CXXIX.

PONTEDERIACEÆ, OR PICKEREL-WEEDS.

GENERA:

1. *Pontederia*, L. Pickerel-weed. *Species:*
   Pontederia cordata, L.
2. *Heteranthera*, Ruiz and Pav. Mud-Plantain. *Species* 2.
3. *Schollera*, Schreb. Water Star-Grass. *Species:*
   Schollera graminea, Willd.

ORDER CXXX.

COMMELYNACEÆ, OR SPIDERWORTS.

1. *Commelyna*, Dill. Day-flower. *Species 3.*
2. *Tradescantia*, L. Spiderwort. *Species 3.*

ORDER CXXXI.

XYRIDACEÆ, OR XYRIDS.

GENERA:
1. *Mayaca*, Aublet. *Species:* *Mayaca Michauxii*, Schott & Endl.
2. *Xyris*, L. Yellow-eyed Grass. *Species 3.*

ORDER CXXXII.

ERIOCAULONACEÆ, OR PIPEWORTS.

GENERA:
1. *Eriocaulon*, L. Pipewort. *Species 3.*
2. *Pœpalanthus*, Mart. *Species:* *P. flavidus*, Kunth.
3. *Lachnocaulon*, Kunth. Hairy Pipewort. *Species:* *L. Michauxii*, Kunth.

ORDER CXXXIII.

CYPERACEÆ, OR SEDGES.

GENERA:
1. *Cyperus*, L. Galingale. *Species 19.*
2. *Kyllingia*, L. *Species:* *K. pumila*, Mx.
3. *Dulichium*, Rich. *Species:* *D. spathaceum*, Pers.
4. *Hemicarpha*, Nees. *Species:* *H. subsquarrosa*, Nees.
5. *Eleocharis*, R. Br. Spike-Rush. *Species 16.*
6. *Scirpus*, L. Bulrush. Club-Rush. *Species 14.*
7. *Eriophorum*, L. Cotton-Grass. *Species 5.*
8. *Fimbristylis*, Vahl. *Species 4.* (Including § *Trichelostylis*, Lestib.)

9. *Fuirena*, Rottbœll. Umbrella-Grass. *Species:*
   Fuirena squarrosa, Mx
10. *Psilocarya*, Torr. Bald-Rush. *Species:*
    Psilocarya scirpoides, Torr
11. *Dichromena*, Rich. *Species:*
    Dichromena leucocephala, Mx
12. *Ceratoschœnus*, Nees. Horned Rush. *Species* 2.
13. *Rhynchospora*, Vahl. Beak-Rush. *Species* 10.
14. *Cladium*, P. Browne. Twig-Rush. *Species:*
    Cladium mariscoides, Torr.
15. *Scleria*, L. Nut-Rush. *Species* 5.
16. *Carex*, Sedge. *Species* 132.

ORDER CXXXIV.
GRAMINEÆ, OR GRASSES.

TRIBE I. *Poaceæ*, R. Br.

1. *Leersia*, Solander. False Rice. White Grass. *Species* 3.
2. *Zizania*, Gron. Water- or Indian-Rice. *Species* 2.
3. *Alopecurus*, L. Fox-tail Grass. *Species* 3.
4. *Phleum*, L. Cat's-tail Grass. *Species* 2.
5. *Vilfa*, Adans, Beauv. Rush-Grass. *Species* 3.
6. *Sporolobus*, R. Br. Drop-seed Grass. *Species* 5.
7. *Agrostis*, L. Bent-Grass. *Species* 6.
8. *Polypogon*, Desf. Beard-Grass. *Species:*
   Polypogon Monspeliensis, Desf.
9. *Cinna*, L. Wood Reed-Grass. *Species:*
   Cinna arundinacea, L. *Var. pendula*, Gray.
10. *Muhlenbergia*, Schreb. *Species* 7.
11. *Brachyelytrum*, Beauv. *Species:*
    B. aristatum, Beauv. (*Muhlenbergia aristata*, Pers.)
12. *Calamagrostis*, Adans. Reed Bent-Grass.
    (§ *Calamagrostis*, § *Calamovilfa*, and § *Ammophila*.)
    Species 7.
13. *Oryzopsis*, Mx. Mountain Rice. *Species* 3.
14. *Stipa*, L. Feather-Grass. *Species* 3.
15. *Aristida*, L. Triple-awned Grass. *Species* 7.

16. *Spartina*, Schreb.   Cord or Marsh Grass.   *Species* 4.
17. *Ctenium*, Panzer.   Toothache-Grass.   *Species:*
    Ctenium Americanum, Spreng.
18. *Boutelona*, Lag.   Muskit-grass.   *Species* 3.
19. *Gymnopogon*, Beauv.   Naked Beard-Grass.   *Species* 2.
20. *Cynodon*, Rich.   Bermuda Grass.   Scutch-Grass.
    Species: Cynodon Dactylon, Pers.
21. *Dactyloctenium*, Willd.   Egyptian Grass.   *Species:*
    D. Ægyptiacum, Willd.
22. *Eleusine*, Gærtn.   Crab-Grass.   Yard-Grass.   *Species:*
    E. Indica, Gærtn.   Dog's-tail, or Wire-Grass.
23. *Leptochloa*, Beauv.   *Species* 2.
24. *Tricuspis*, Beauv.   *Species:*
    Tr. sesleroides, Torr.   Tall Red-top.
25. *Uralepis*, Nutt.   (*Tricuspis*, Gray.)   *Species:*
    U. purpurea, Nutt.   Sand-Grass.
26. *Diarrhena*, Raf.   *Species:*   *D. Americana*, Beauv.
27. *Dactylis*, L.   Cock's-foot.   Orchard-Grass.   *Species:*
    D. glomerata, L.
28. *Kœleria*, Pers.   *Species:*   *K. cristata*, Pers.
29. *Eatonia*, Raf.   *Species* 2.
30. *Melica*, L.   Melic-Grass.   *Species:*
    Melica mutica, Walt.
31. *Glyceria*, R. Br., Trin.   Manna-Grass.   *Species* 10.
32. *Brizopyrum*, Link.   Spike-Grass.   *Species:*
    Br. spicatum, Hook.
33. *Poa*, L.   Meadow-Grass.   Spear-Grass.   *Species* 12.
34. *Eragrostis*, Beauv.   *Species* 8.
35. *Briza*, L.   Quaking Grass.   *Species: Briza media*, L.
36. *Festuca*, L.   Fescue-Grass.   *Species* 4.
37. *Bromus*, L.   Brome-Grass.   *Species* 6.
38. *Uniola*, L.   Spike-Grass.   *Species* 3.
39. *Phragmites*, Trin.   Reed.   *Species:*
    Phr. communis, Trin.
40. *Arundinaria*, Mx.   Cane.   *Species:*
    A. macrosperma, Mx.
41. *Lepturus*, R. Br.   *Species:*   *L. paniculatus*, Nutt.

42. *Lolium*, L.   Darnel.   *Species* 2.
43. *Triticum*, L.   Wheat.   *Species* 3.
44. *Hordeum*, L.   Barley.   *Species* 2.
45. *Elymus*, L.   Lyme-Grass.   Wild Rye.   *Species* 4.
46. *Gymnostichum*, Schreb.   Bottle-brush Grass.   *Species :*
Gymnostichum Hystrix, Schreb.
(*Elymus Hystrix*, L.)
47. *Aira*, L.   Hair-Grass.   *Species* 3.
48. *Danthonia*, D.C.   Wild Oat-Grass.   *Species :*
D. spicata, Beauv.
49. *Trisetum*, Pers.   *Species* 2.
50. *Avena*, L.   Oat.   *Species* 2.
51. *Arrhenatherum*, Beauv.   Oat-Grass.   *Species :*
A. avenaceum, Beauv.
(*Avena elatior*, L.)
52. *Holcus*, L.   Meadow Soft-Grass.   *Species :*
Holcus lanatus, L.

Tribe II.—*Phalarideæ.*

53. *Hierochloa*, Gmel.   Holy-Grass.   *Species* 2.
54. *Anthoxanthum*, L.   Sweet-scented Vernal-grass.   *Species :*   *A. odoratum*, L.
55. *Phalaris*, L.   Canary-grass.   *Species* 2.

Tribe III.—*Paniceæ.*

56. *Milium*, L.   Millet-Grass.   *Species :*
Milium effusum, L.
57. *Amphicarpum*, Kunth.   *Species :*   *A. Purshii*, Kunth.
58. *Paspalum*, L.   *Species* 5.
59. *Panicum*, L.   Panic-Grass.   *Species* 20.
60. *Setaria*, Beauv.   Bristly Foxtail-Grass.   *Species* 4.
61. *Cenchrus*, L.   Hedgehog- or Bur-Grass.   *Species :*
C. tribuloides, L.
62. *Tripsacum*, L.   Gama-Grass.   Sesame-Grass.   *Species :*
Tripsacum dactyloides, L.
63. *Erianthus*, Mx.   Woolly Beard-Grass.   *Species* 2.

64. *Andropogon*, L.   Beard-Grass.   *Species* 5.
65. *Sorghum*, Pers.   Broom-Corn.   *Species:*
   Sorghum nutans, Gray.   Indian Grass.
   (*Andropogon*, Wood.)

In the § § of the Key, pointing out the genera of Grasses, the student will notice quotations on the left hand. The Roman numerals indicate the engravings in the respective tables appended to Gray's and Wood's Manuals, and the Arabic ones the place of the genus in the order of grasses, as it is given by each author.

INDEX.

The numerals of the first column indicate each order and genus in the Conspectus, preceding this Index; those of the second one the § § of the Key. In the first column 28. 6 means Order **XXIII.,** *genus 6.*

A

Abies, Spruce, Fir,	111.	2	779
Abutilon, Velvet-leaf,	28.	6	72
Acalypha, three-seeded Mercury,	102.	3	839
Acer, Maple,	36.	3	167
Acerates, Green Milkweed,	85.	2	417
Achillea, Milfoil, Yarrow,	59.	54	967
Acnida, Water-Hemp,	91.	4	901
Aconitum, Aconite, Monk's-hood,	1.	17	43
Acorus, Sweet Flag, Calamus,	112.	6	715
Actæa, Baneberry, Cohosh,	1.	20	18
Actinomeris,	59.	41	980
Adam and Eve, Aplectrum,	119.	16	766
Adam's Needle, sp. of Yucca,			
Adder's Mouth, Microstylis,	119.	13	772
Adder's Tongue, sp. of Erythronium,	126.	12	743
Adenocaulon,	59.	12	909
Adlumia, Climbing Fumitory,	11.	1	222
Æschyomene, Sensitive Joint Vetch,	38.	15	245
Æsculus, Buckeye,	36.	2	230
Æthusa, Fool's Parsley,	42.	14	309
Agrimonia, Agrimony,	39.	4	319
Agave, False Aloe,	120.	3	752
Agrostemma, Corn-Cockle,	21.	5	207
Agrostis, Bent-Grass,	134.	7	585
Aira, Hair-Grass,	134.	47	620
Airopsis,	134.	50	622
Alchemilla, Lady's Mantle,	39.	6	754
Alder, Alnus,	109.	2	794
Alder-Buckthorn, Subgenus of Rhamnus,			

Aletris, Star-Grass,	121.	3	749
Alisma, Water-Plantain,	116.	3	267
Alligator Pear, Persea,	93.	1	704
Allium, Garlic,	126.	10	722
Alnus, Alder,	109.	2	794
Alopecurus, Fox-tail Grass,	134.	3	569
Alpine Azalea, Loiseleuria,	62.	20	409
Alsine, Grove Sandwort,	21.	7	205
Althæa, Marsh-Mallow,	23.	1	70
Alum-root, Heuchera,	50.	5	288
Amaranth, Amarantus,	91.	1	827
Amarantus, Amaranth,	91.	1	827
Amaryllis, Atamasco Lily,	120.	1	752
Ambrosia, Ragweed,	59.	31	798
Amelanchier, June-berry,	39.	18	55
American Aloe, Agave,	120.	3	752
American Centaury, Sabbatia,	83.	1	375
American Columbo, Frasera,	83.	3	351
American Cowslip, Dodecatheon,	70.	3	361
American Ipecac, sp. of Gillenia,			
Amianthium, Fly-Poison,	127.	7	734
Ammannia,	42.	1	104
Amorpha, False Indigo,	38.	10	83
Ampelopsis, Virginian Creeper,	33.	2	157
Amphicarpæa, Hog-Peanut,	38.	26	252
Amphicarpum,	134.	57	588
Amsonia,	84.	1	412
Anacharis, Water-weed,	117.	2	886
Anagallis, Pimpernel,	70.	8	371
Andromeda, Staggerbush,	62.	10	430
Andropogon, Beard-Grass,	134.	64	602
Androsace,	70.	2	362
Anemone, Windflower,	1.	4	36
Angelica,	52.	11	302
Angelica-Tree, sp. of Aralia,			
Angelico, sp. of Ligusticum,			
Anise-Hyssop, sp. of Lophanthus,			
Antennaria, Everlasting,	59.	60	980
Anthemis, Corn-Chamomile,	59.	53	967
Anthoxanthum, Sweet-scented Vernal Grass,	134.	54	598
Antirrhinum, Snapdragon,	74.	3	462
Anychia, Forked Chickweed,	21.	17	691
Aphyllon, Naked Broom-Rape,	78.	3	454
Apios, Ground-nut, Wild Bean,	38.	23	249
Aplectrum, Adam and Eve,	119.	16	766
Apocynum, Dogbane,	84.	3	413
Apple, Pyrus in part,	39.	17	56
Apple of Peru, Nicandra,	82.	3	391

INDEX. 345

Aquilegia, Columbine,		1. 15	42
Arabis, Rock-Cress,		12. 6	119
Aralia, Wild Sarsaparilla,		53. 1	287
Arbor-Vitæ, Thuja,		111. 4	781
Arbutus, Trailing A., Epigæa,		62. 5	425
Archangelica,		52. 12	301
Archemora, Cow-bane,		52. 9	300
Arctostaphylos, Bearberry,		62. 4	421
Arenaria, Sandwort,		21. 8	203
Arethusa,		119. 7	771
Argemone, Prickly Poppy,		10. 2	13
Arisæma, Indian Turnip and Green Dragon,		112. 1	550
Aristida, Triple-awned Grass,		134. 15	582
Aristolochia, Birthwort,		87. 2	757
Armoracia, Horseradish,		12. 13	111
Arnica, American Leopard's-bane,		59. 64	950
Aromatic Wintergreen, Gaultheria,		62. 6	422
Arrhenatherum, Oat-Grass,		134. 51	618
Arrow-Arum, Peltandra,		112. 2	813
Arrow-Grass, Triglochin,		116. 1	728
Arrow-Head, Sagittaria,		116. 5	802
Arrow-Wood, Viburnum,		55. 7	520
Artemisia, Wormwood,		59. 57	910
Arundinaria, Cane,		134. 40	606
Asarum, Wild Ginger,		87. 1	644
Asclepias, Milkweed, Silkweed,		85. 1	417
Ascyrum, St. Peter's-wort,		19. 1	23
Ash, Fraxinus,		86. 4	846
Ash-leaved Maple, Negundo,		36. 3	846
Asimina, Papaw,		3. 1	28
Asparagus,		126. 1	717
Aspen, Populus,		110. 2	873
Aster,		59. 14	956
Astilbe, False Goatsbeard,		50. 1	190
Astragalus, Milk-Vetch,		38. 14	261
Atamasco Lily, Amaryllis,		120. 1	752
Atragene,		1. 1	31
Atriplex, Orache,		90. 5	826
Avena, Oat,		134. 50	625
Avens, Geum,		39. 9	65
Awlwort, Subularia,		12. 18	111
Azalea, False Honeysuckle,		62. 16	395

B

Baccharis, Groundsel-Tree,		59. 25	927
Bald Cypress, Taxodium,		111. 6	780
Bald-Rush, Psilocarya,		133. 10	562

Baldwinia,	59. 49	944
Ballota, Black Horehound,	77. 32	504
Balm, Melissa,	77. 13	495
Balmony, Chelone,	74. 6	476
Balsam, Impatiens,	29. 1	224
Baneberry, Actæa,	1. 20	18
Baptisia, False Indigo,	38. 29	232
Barbarea, Winter-Cress,	12. 8	124
Barberry, Berberis,	5. 1	139
Barley, Hordeum,	134. 44	577
Barnyard-Grass, sp. of Panicum,		
Barren Strawberry, Waldsteinia,	39. 10	63
Bartonia,	83. 6	353
Basil, Clinopodium,	77. 12	496
Basil-Thyme, Calamintha,	77. 12	496
Basswood, Tilia,	24. 1	75
Bastard-Toad-Flax, Comandra,	96. 1	699
Bayberry, Myrica,	108. 1	874
Beach-Pea, sp. of Lathyrus,		
Beak-Rush, Rhynchospora,	133. 13	563
Bearded Darnel, sp. of Lolium,		
Beard-Grass, Andropogon and Polypogon,		
Beard-Tongue, Pentstemon,	74. 7	476
Bear-Grass, Yucca,	126. 13	727
Beaver-Poison, Cicuta,	52. 20	307
Bed-Straw, Galium,	56. 1	529
Beech, Fagus,	107. 3	788
Beech-Drops, Epiphegus,	78. 1	323
Beggars-Lice, sp. of Cynoglossum,		
Beggar-Ticks, sp. of Bidens,		
Bellflower, Campanula,	61. 1	524
Bellis, Daisy,	59. 18	963
Bellwort, Uvularia,	127. 1	740
Bengal-Grass, sp. of Setaria,		
Benjamin-Bush, Benzoin,	93. 3	708
Bent-Grass, Agrostis,	134. 7	585
Benzoin, Benjamin-Bush,	93. 3	708
Berberis, Barberry,	5. 1	139
Berchemia, Supple Jack,	34. 1	160
Bermuda-Grass, Cynodon,	134. 20	573
Betula, Birch,	109. 1	794
Bidens, Bur-Marigold,	59. 43	937
Bigelovia, Rayless Golden-rod,	59. 21	922
Bignonia,	72. 1	460
Bilberry, Vaccinium,	62. 2	514
Bindweed, Covolvulus,	81. 3	404
Birch, Betula,	109. 1	794
Birthroot, sp. of Trillium,		

Birthwort, Aristolochia,		87. 2	757
Bishop's Cap, Mitella,		50. 6	192
Bishop's-Weed, Discopleura,		52. 19	308
Bitter-Cress, Cardamine,		12. 5	121
Bitter-Nut, sp. of Carya,			
Bittersweet, sp. of Solanum,			
Bitterweed, sp. of Ambrosia,			
Black Alder, sp. of Ilex,			
Blackberry, some sp. of Rubus,			
Black Bindweed, sp. of Polygonum,			
Black Grass, sp. of Juncus,			
Black Haw, Sloe, sp. of Viburnum,			
Black Horehound, Ballota,		77. 32	504
Black Jack, sp. of Quercus,			
Black-Moss, Tillandsia,		122. 1	86
Black Oat-Grass, sp. of Stipa,			
Black Snake-Root, Sanicula,		52. 3	293
Black Thorn, Sloe, sp. of Prunus,			
Bladder Ketmia, sp. of Hibiscus,			
Bladder-nut, Staphylea,		36. 1	178
Bladder-pod, Vesicaria,		12. 14	113
Bladder-wort, Utricularia,		71. 1	439
Blazing Star, Liatris and Chamælirium,			
Blephilia,		77. 18	443
Blessed Thistle, Cnicus,		59. 67	941
Bletia,		119. 12	765
Blite, Blitum,		90. 4	682
Blitum, Blite,		90. 4	682
Blood-root, Sanguinaria,		10. 6	11
Blue Beech, Carpinus,		107. 5	792
Blueberry, Vaccinium,		62. 2	514
Blue Bottle, sp. of Centaurea,			
Blue Cohosh, Caulophyllum,		5. 2	141
Blue Curls, Trichostema,		77. 2	340
Blue False Indigo, sp. of Baptisia,			
Blue Flag, sp. of Iris,			
Bluets, Oldenlandia and Houstonia,			
Blue-eyed Grass, Sisyrinchium,		123. 2	709
Blue-Grass, sp. of Poa,			
Blue Hearts, Buchnera,		74. 17	337
Blue Joint-Grass, sp. of Calamagrostis,			
Blue Lettuce, Mulgedium,		59. 84	993
Blue Tangle, sp. of Gaylussacia,			
Blue Weed, Echium,		78. 1	377
Bœhmeria, False Nettle,		104. 8	833
Bog-Asphodel, Narthecium,		128. 1	742
Bog-Rush, Juncus,		128. 3	741
Boltonia,		59. 17	962

Boneset, sp. of Eupatorium,
Borrichia, Sea Ox-eye, ... 59. 35 972
Bottle-brush Grass, Gymnostichum, . . . 134. 46 612
Bottle-Grass, sp. of Setaria,
Bouncing Bet, Saponaria, ... 21. 2 195
Bouteloua, Muskit-Grass, ... 134. 18 596
Bowman's Root, sp. of Gillenia,
Boxberry, Gaultheria, ... 62. 6 422
Box-Elder, Negundo, ... 36. 4 846
Boykinia, ... 50. 3 289
Brachychæta, False Golden-rod, . . . 59. 19 960
Brachyelytrum, ... 134. 11 587
Bracted Bindweed, Calystegia, . . . 81. 4 402
Bramble, Rubus, ... 39. 14 62
Brasenia, Watershield, ... 7. 1 264
Bristly Foxtail-Grass, Setaria, . . . 134. 60 601
Briza, Quaking Grass, ... 134. 35 629
Brizopyrum, Spike-Grass, ... 134. 32 613
Bromus, Brome-Grass, ... 134. 37 625
Brome-Grass, Bromus, ... 134. 37 625
Brook-weed, Samolus, ... 70. 10 365
Brook-lime, sp. of Veronica,
Broom-corn, Sorghum, ... 134. 65 602
Broom-crowberry, Corema, ... 103. 2 851
Broom-rape, Aphyllon and Phelipæa, .
Brunella, Self-heal, ... 77. 25 498
Buchnera, Blue Hearts, ... 74. 17 337
Buckbean, Menyanthes, ... 88. 8 360
Buckeye, Æsculus, ... 36. 2 230
Buckthorn, Rhamnus, ... 34. 2 103
Buckwheat, Fagopyrum, ... 92. 3 687
Buffalo-nut, Pyrularia, ... 96. 2 869
Bugle-weed, sp. of Lycopus,
Bugloss, Lycopsis, ... 78. 2 881
Bugbane, Cimicifuga, ... 1. 21 18
Bulrush, Scirpus, ... 133. 6 558
Bumelia, Southern Buckthorn, . . . 67. 1 389
Bunch-berry, sp. of Cornus,
Bunch-flower, Melanthium, ... 127. 4 733
Bupleurum, Thorough-wax 52. 18 305
Burmannia, ... 118. 1 708
Burdock, Lappa, ... 59. 71 940
Bur-Grass, Cenchrus, ... 134. 61 605
Bur-Marigold, Bidens, ... 59. 43 937
Burnet, Sanguisorba, ... 39. 5 700
Burning-Bush, sp. of Euonymus,
Bur-Reed, Sparganium, ... 113. 2 814
Bush-Clover, Lespedeza, ... 38. 18 246

INDEX. 349

Bush-Honeysuckle, Diervilla,	55.	4	518
Butter-and-Eggs, Linaria,	74.	2	462
Buttercup, Ranunculus,	1.	8	31
Butterfly-Pea, Clitoria,	38.	27	254
Butterfly-Weed, sp. of Asclepias,			
Butternut, sp. of Juglans,			
Butter-Weed, sp. of Erigeron,			
Butterwort, Pinguicula,	71.	2	439
Button-Bush, Cephalanthus,	56.	4	527
Button-Snakeroot, Eryngium,	52.	4	293
Button-Weed, Diodia and Spermacoce,			
Buttonwood, Platanus,	103.	1	796

C

Cacalia, Indian Plantain,	59.	63	921
Cactus Opuntia, Prickly Pear,	45.	1	51
Cakile, Sea-Rocket,	12.	20	106
Calamagrostis, Reed Bent-Grass,	134.	12	575
Calaminth, Calamintha,	77.	12	496
Calamintha, Basil-Thyme,	77.	12	496
Calamovilfa, subgenus of Calamagrostis,	134.	12	575
Calamus, Acorus,	112.	6	715
Calico-Bush, sp. of Kalmia,			
Calla, Water-Arum,	112.	3	813
Callicarpa, French Mulberry,	76.	3	339
Callirrhoë,	23.	3	71
Callitriche, Water-Starwort,	100.	1	817
Calopogon,	119.	9	768
Caltha, Marsh-Marigold,	1.	11	40
Calycanthus, Carolina Allspice,	40.	1	57
Calycocarpum, Cupseed,	4.	3	867
Calypso,	119.	10	767
Calystegia, Bracted Bindweed,	81.	4	402
Camelina, False Flax,	12.	15	113
Campanula, Bell-flower,	61.	1	524
Campion, Silene,	21.	4	197
Canary-Grass, Phalaris,	134.	55	594
Cancer-root, Conopholis and Epiphegus,			
Cane, Arundinaria,	134.	40	606
Cannabis, Hemp,	104.	10	892
Caper-Spurge, sp. of Euphorbia,			
Capsella, Shepherd's Purse,	12.	17	114
Cardamine, Bitter-Cress,	12.	5	121
Cardinal-Flower, sp. of Lobelia,			
Carduus, Plumeless Thistle,	59.	69	942
Carex, Sedge,	133.	16	821
Carnation, Dianthus,	21.	1	194

Carolina Allspice, Calycanthus,	40.	1	57
Carphephorus,	59.	4	934
Carpinus, Hornbeam, Iron-wood,	107.	5	792
Carrion-Flower, sp. of Smilax,			
Carrot, Daucus,	52.	5	294
Carya, Hickory,	106.	2	783
Cassandra, Leather-Leaf,	62.	8	431
Cassia, Senna,	38.	32	231
Cassiope,	62.	9	428
Castanea, Chestnut,	107.	2	788
Castileja, Painted Cup,	74.	21	465
Catalpa, Indian Bean,	72.	3	328
Catbrier, Smilax,	125.	1	866
Catchfly, Silene,	21.	4	197
Catgut, sp. of Tephrosia,			
Cat-Mint, sp. of Nepeta,			
Catnip, sp. of Nepeta,			
Cat-tail, Typha,	113.	1	814
Cat-tail Flag, Typha,	113.	1	814
Cat's-tail Grass, Phleum,	134.	4	579
Caulophyllum, Blue Cohosh,	5.	2	141
Ceanothus, New-Jersey Tea,	34.	3	288
Cedronella,	77.	22	485
Celandine, Chelidonium,	10.	4	15
Celandine Poppy, Stylophorum,	10.	3	14
Celastrus, Staff-Tree,	35.	1	162
Celtis, Nettle-Tree,	104.	3	663
Cenchrus, Bur-Grass,	134.	61	605
Centaurea, Star-Thistle,	59.	66	938
Centaury, Erythræa,	83.	2	349
Centrosema, Spurred Butterfly-Pea,	38.	28	253
Centunculus, Chaffweed,	70.	9	348
Chamælirium,	127.	10	889
Cinna, Wood Reed-Grass,	134.	9	585
Circæa, Enchanter's Nightshade,	43.	6	273
Cirsium, Plumed Thistle,	59.	68	942
Cissus, subgenus of Vitis,	33.	1	158
Cladium, Twig-Rush,	133.	14	561
Cladrastis, Yellow-wood,	38.	30	233
Claytonia, Spring Beauty,	22.	4	152
Clearweed, Pilea,	104.	7	825
Cleavers, Galium,	56.	1	529
Clematis, Virgin's-Bower,	1.	2	33
Clethra, White Alder, Sweet Pepperbush,	62.	12	173
Climbing Fumitory, Adlumia,	11.	1	222
Climbing Hempweed, Mikania,	59.	8	916
Clinopodium, Basil,	77.	12	496
Clintonia,	126.	6	722

ADDENDA.

Cephalanthus, Button-Bush,	56.	4	527
Cerastium, Mouse-ear Chickweed,	21.	12	186
Cerasus, Cherry,	39.	1	53
Ceratophyllum, Hornwort,	99.	1	838
Ceratoschœnus, Horned-Rush,	133.	12	563
Cercis, Judas-Tree,	38.	31	233
Chærophyllum, Chervil,	52.	23	313
Chaff-seed, Schwalbea,	74.	22	464
Chaffweed, Centunculus,	70.	9	348
Cheat, sp. of Bromus,			
Checkerberry, Gaultheria,	62.	6	422
Chelidonium, Celandine,	10.	4	15
Chelone, Turtle-head, Snake-head,	74.	6	476
Chenopodina, Sea-Goosefoot,	90.	8	683
Chenopodium, Goosefoot,	90.	2	683
Cherry, Cerasus, Prunus,	39.	1	53
Chervil, Chærophyllum,	52.	23	313
Chestnut, Castanea,	107.	2	788
Chickweed, Stellaria,	21.	10	200
Chickweed-Wintergreen, Trientalis,	70.	4	137
Chimaphila, Pipsissewa,	62.	24	173
Chinquapin, sp. of Castanea,			
Chiogenes, Creeping Snowberry,	62.	3	512
Chionanthus, Fringe-Tree,	86.	3	100
Chives, sp. of Allium,			
Chokeberry, sp. of Pyrus,			
Chrysogonum,	59.	27	983
Chrysopsis, Golden Aster,	59.	22	957
Chrysosplenium, Golden Saxifrage,	50.	8	689
Cichorium, Succory,	59.	73	990
Cicuta, Water-Hemlock,	52.	20	307
Cimicifuga, Bugbane,	1.	21	18
Cinquefoil, Potentilla,	39.	11	64

INDEX. 351

Clitoria, Butterfly-Pea,	38. 27	254
Clover, Trifolium,	38. 4	256
Clotbur, Xanthium,	59. 32	931
Cloud-berry, sp. of Rubus,		
Club-Rush, Scirpus,	133. 6	558
Cnicus, Blessed Thistle,	59. 67	941
Cnidoscolus, Spurge-Nettle,[1]	102. 2	839
Cocculus,	4. 1	858
Cocklebur, Xanthium,	59. 32	931
Cock's-foot Grass, Dactylis,	134. 27	635
Cockspur-Thorn, sp. of Cratægus,		
Cohosh, Actæa,	1. 20	18
Colic-root, Aletris,	121. 3	749
Collinsia, Innocence,	74. 5	478
Collinsonia, Horsebalm, Stone-wort,	77. 15	444
Columbine, Aquilegia,	1. 15	42
Comandra, Bastard Toad-Flax,	96. 1	699
Comfrey, Symphytum,	78. 3	381
Commelyna, Day-flower,	130. 1	227
Common Darnel, sp. of Lolium,		
Compass-Plant, sp. of Silphium,		
Comptonia, Sweet Fern,	108. 2	793
Cone-flower, Rudbeckia,	59. 38	979
Conioselinum, Hemlock-Parsley,	52. 13	502
Conium, Poison Hemlock,	52. 25	314
Conobea,	74. 9	479
Conoclinium, Mist-flower,	59. 9	917
Conopholis, Squaw-root,	73. 2	455
Convallaria, Lily of the Valley,	126. 5	725
Convolvulus, Bindweed,	81. 3	404
Coptis, Goldthread,	1. 13	46
Coral-Berry, sp. of Symphoricarpus,		
Corallorrhiza, Coral-root,	119. 15	766
Coral-root, Corallorrhiza,	119. 15	766
Cord-Grass, Spartina,	134. 16	578
Corema, Broom-Crowberry,	103. 2	851
Coreopsis, Tickseed,	59. 42	937
Corky White Elm, sp. of Ulmus,		
Corn-Cockle, Agrostema,	21. 5	207
Cornel, Cornus,	54. 1	277
Corn-Salad, Fedia,	57. 2	535
Cornus, Cornel,	54. 1	277
Corpse-Plant, Monotropa,	62. 27	163
Corydalis,	11. 3	223
Corylus, Hazelnut,	107. 4	789
Cosmanthus, Subgenus of Phacelia,		
Cotton-Grass, Eriophorum,	133. 7	558
Cotton-Rose, Filago,	59. 61	929

Cottonwood, sp. of Populus,
Couch-Grass, sp. of Triticum,
Cowbane, Cicuta and Archemora,
Cowberry, sp. of Vaccinium,
Cowherb, Vaccaria, ... 21. 3 195
Cow-Parsnip, Heracleum, ... 52. 7 290
Cowslip, Primula, ... 70. 1 362
Cow-Wheat, Melampyrum, ... 74.-26 471
Crab-Apple, sp. of Pyrus,
Crab-Grass, Eleusine, ... 134. 22 604
Craneberry, Oxycoccus, ... 62. 2 514
Craneberry-Tree, sp. of Viburnum,
Crane-Fly Orchis, Tipularia, ... 119. 11 761
Cranesbill, Geranium, ... 28. 1 168
Crantzia, ... 52. 2 292
Cratægus, Hawthorn, ... 39. 16 56
Creeping Snowberry, Chiogenes, . . . 62. 3 512
Creeping Wintergreen, Gaultheria, . . . 62. 6 422
Crotalaria, Rattle-box, ... 38. 2 237
Croton, ... 102. 6 808
Crotonopsis, ... 102. 7 808
Crowberry, Empetrum, ... 103. 1 851
Crowbeard, Verbesina, ... 59. 44 966
Crowfoot, Ranunculus, ... 1. 8 81
Cryptotæuia, Honewort, ... 52. 22 307
Ctenium, Toothache-Grass, ... 134. 17 607
Cuckoo-flower, sp. of Cardamine,
Cudweed, Gnaphalium, ... 59. 59 980
Culver's Physic, sp. of Veronica,
Culver's-Root, sp. of Veronica,
Cunila, Dittany, ... 77. 6 445
Cuphea, ... 42. 4 138
Cup-Plant, sp. of Silphium,
Cupressus, White Cedar, ... 111. 5 781
Cupseed, Calycocarpum, ... 4. 3 867
Currant, some sp. of Ribes,
Cuscuta, Dodder, ... 81. 7 344
Cut-Grass, sp. of Leersia,
Cycloloma, Winged Pigweed, ... 90. 1 681
Cynodon, Bermuda-Grass, ... 134. 20 573
Cynoglossum, Hound's-Tongue, . . . 78. 9 379
Cynthia, ... 59. 75 991
Cyperus, Galingale, ... 133. 1 554
Cypress, Cupressus, ... 111. 5 781
Cypress-Vine, Quamoclit, ... 81. 1 403
Cypripedium, Lady's-Slipper, ... 119. 17 758

D

Dactylis, Cock's-foot,		134. 27	635
Dactyloctenium, Egyptian Grass,		134. 21	604
Daisy, Bellis,		59. 18	963
Dalea,		38. 8	240
Dalibarda,		39. 13	63
Dandelion, Taraxacum,		59. 80	999
Danthonia, Wild Oat-grass,		134. 48	628
Dangleberry, sp. of Gaylussacia,			
Darnel, Lolium,		134. 42	611
Dasystoma, False Foxglove,		74. 20	336
Date-Plum, Diospyros,		66. 1	421
Datura, Thorn-Apple,		82. 5	399
Daucus, Carrot,		52. 5	294
Day-flower, Commelyna,		130. 1	227
Day-Lily, Hemerocallis,		126. 7	725
Dead-Nettle, Lamium,		77. 31	506
Deerberry, sp. of Vaccinium,			
Deer-Grass, Rhexia,		41. 1	280
Delphinium, Larkspur,		1. 16	42
Dentaria, Pepper-root,		12. 4	120
Deschampsia, subgenus of Aira,		134. 47	620
Desmanthus,		38. 35	151
Desmodium, Tick-Trefoil,		38. 17	246
Dewberry, sp. of Rubus,			
Devil's-Bit, Chamælirium,		127. 10	889
Devil-Wood, Olea,		86. 2	329
Dianthera, Water-Willow,		75. 1	449
Dianthus, Pink,		21. 1	194
Diapensia,		80. 3	407
Diarrhena,		134. 26	633
Dicentra, Dutchman's Breeches,		11. 2	222
Dichondra,		81. 6	410
Dichromena,		133. 11	560
Diervilla, Bush-Honeysuckle,		55. 4	518
Diodia, Buttonweed,		56. 3	532
Dioscorea, Yam,		124. 1	867
Diospyros, Date-Plum,		66. 1	421
Diphyllcia, Umbrella-Leaf,		5. 3	142
Diplopappus, Double-bristled Aster,		59. 16	954
Dipsacus, Teasel,		58. 1	527
Dipteracanthus,		75. 2	338
Dirca, Leatherwood,		94. 1	660
Discopleura, Mock Bishop-weed,		52. 19	308
Ditch-Grass, Ruppia,		115. 4	543
Ditch-Stonecrop, Penthorum,		49. 1	271

Dittany, Cunila,		77. 6	445
Dock, Rumex,		92. 5	714
Dodecatheon, American Cowslip,		70. 3	361
Dodder, Cuscuta,		81. 7	344
Dogbane, Apocynum,		84. 3	413
Dog's-tail, Eleusine,		134. 22	604
Dog's-tooth Violet, Erythronium,		126. 12	739
Dogwood, Cornus,		54. 1	277
Door-weed, sp. of Polygonum,			
Draba, Whitlow-Grass,		12. 12	110
Dracocephalum, Dragon-head,		77. 21	486
Dragon-Arum, Arisæma,		112. 1	812
Dragon-head, Dracocephalum,		77. 21	486
Dragon-root, sp. of Arisæma,			
Drop-seed Grass, Sporolobus and Muhlenbergia,			
Drosera, Sundew,		17. 1	188
Dryas,		39. 8	65
Duck's-meat, Lemna,		114. 1	541
Duckweed, Lemna,		114. 1	541
Dulichium,		133. 3	553
Dutchman's Breeches, Dicentra,		11. 2	222
Dutchman's Pipe, sp. of Aristolochia,			
Dwarf Cornel, sp. of Cornus,			
Dwarf Dandelion, Krigia,		59. 74	991
Dyer's Weed or Rocket, Reseda,		14. 1	7
Dyer's Greenweed, Genista,		38. 3	237
Dysodia, Fetid Marigold,		59. 45	947

E

Eatonia,		134. 29	635
Echinacea, Purple Cone-flower,		59. 37	977
Echinodorus,		116. 4	267
Echinospermum, Stickseed,		78. 8	379
Echinocystis, Wild Balsam Apple,		48. 2	806
Echium, Viper's Bugloss, Blue Weed,		78. 1	377
Eclipta,		59. 34	972
Eel-Grass, Zostera and Vallisneria,			
Egyptian Grass, Dactyloctenium,		134. 21	604
Elatine, Waterwort,		20. 1	85
Elder, Sambucus,		55. 6	520
Elecampane, Inula,		59. 28	960
Eleocharis, Spike-Rush,		133. 5	557
Elephantopus, Elephant's-foot,		59. 2	914
Elephant's-foot, Elephantopus,		59. 2	914
Eleusine, Crab-Grass,		134. 22	604
Ellisia,		79. 3	368
Elm, Ulmus,		104. 1	758

INDEX. 355

Elodea, Marsh St. John's-wort,		19. 3	193
Elymus, Lyme-Grass,		134. 45	609
Empetrum, Crowberry,		103. 1	851
Enchanter's Nightshade, Circæa,		43. 6	273
Enslenia,		85. 3	416
Epigæa, Trailing Arbutus,		62. 5	425
Epilobium, Willow-herb,		43. 1	282
Epiphegus, Beech-drops,		73. 1	800
Eragrostis,		134. 34	638
Erechthites, Fireweed,		59. 62	928
Erianthus, Woolly Beard-Grass,		134. 63	600
Erigenia, Harbinger-of-Spring,		52. 27	311
Erigeron, Fleabane,		59. 15	920
Eriocaulon, Pipewort,		132. 1	803
Eriophorum, Cotton-Grass,		133. 7	558
Erodium, Storksbill,		28. 2	147
Eryngium, Button Snake-root,		52. 4	293
Erysimum, Treacle Mustard,		12. 9	125
Erythræa, Centaury,		83. 2	349
Erythronium, Dog's-tooth Violet,		126. 12	748
Eulophus,		56. 26	314
Euonymus, Spindle-Tree,		35. 2	103
Eupatorium, Thoroughwort,		59. 7	917
Euphorbia, Spurge,		102. 1	538
Euphrasia, Eye-bright,		74. 23	471
Euxolus, False Amaranth,		91. 2	649
Evening Primrose, Œnothera,		43. 2	283
Everlasting, Antennaria,		59. 60	930
Everlasting Pea, Lathyrus,		38. 21	242
Eye-bright, Euphrasia,		74. 23	471

F

Fagopyrum, Buckwheat,		92. 3	687
Fagus, Beech,		107. 3	788
False Asphodel, Tofielda,		127. 11	263
False Bugbane, Trautvetteria,		1. 7	37
False Gromwell, Onosmodium,		78. 4	283
False Flax, Camelina,		12. 15	113
False Dandelion, Pyrrhopappus,		59. 81	994
False Dragon-head, Physostegia,		77. 24	505
False Foxglove, Dasystoma,		74. 20	336
False Goat's-beard, Astilbe,		50. 1	190
False Hellebore, Veratrum,		127. 6	734
False Honeysuckle, Azalea,		62. 16	395
False Indigo, Amorpha and Baptisia,			
False Jessamine, Gelsemium,		74. 27	409
False Lettuce, Mulgedium,		59. 84	993

False Loosestrife, Ludwigia,	43.	5	277
False Mermaid, Flœrkea,	30.	1	94
False Mistletoe, Phoradendron,	97.	1	841
False Orchis, Platanthera,	119.	3	763
False Pennyroyal, Isanthus,	77.	3	340
False Pimpernel, Ilysanthus,	74.	12	451
False Redtrop, sp. of Poa,			
False Rice, Leersia,	134.	1	567
False Rocket, Iodanthus,	12.	2	121
False Rue-Anemone, Isopyrum,	1.	10	40
False Solomon's Seal, Smilacina,	126.	3	746
False Spikenard, sp. of Smilacina,			
False Water-Dropwort, Tiedemannia,	52.	10	300
False Wintergreen, Pyrola,	62.	22	174
Featherfoil, Hottonia,	70.	11	358
Feather Geranium, sp. of Chenopodium,			
Feather-Grass, Stipa,	134.	14	583
Fedia, Corn-Salad,	57.	2	535
Fescue-Grass, Festuca,	134.	36	633
Fetid Horehound, Ballota,	77.	32	504
Festuca, Fescue-Grass,	134.	36	633
Fever-Bush, Benzoin,	93.	3	703
Feverfew, Matricaria,	59.	56	962
Feverwort, Triosteum,	55.	5	522
Field-Sorrel, sp. of Rumex,			
Figwort, Scrophularia,	74.	4	475
Filago, Cotton-Rose,	59.	61	929
Filbert, Corylus,	107.	4	789
Fimbristylis,	133.	8	559
Finger-Grass, sp. of Panicum,			
Fir, Abies,	111.	2	779
Fireweed, Erechthites,	59.	62	928
Five finger, Potentilla,	39.	11	64
Flax, Linum,	26.	1	185
Fleabane, Erigeron,	59.	15	920
Floating Heart, Limnanthemum,	83.	9	358
Flœrkea, False Mermaid,	30.	1	94
Flower-de-Luce, Iris,	123.	1	705
Fly-catch Grass, sp. of Leersia,			
Fly-Poison, Amianthium,	127.	7	734
Fog-Fruit, Lippia,	76.	2	457
Fool's Parsley, Æthusa,	42.	14	309
Forked Chickweed, Anychia,	21.	17	691
Forget-me-not, Myosotis,	78.	7	384
Forestiera,	86.	5	854
Forsteronia,	84.	2	418
Fothergilla,	51.	2	50
Foxtail-Grass, Alopecurus,	134.	3	569

INDEX. 35

Fragaria, Strawberry,	39. 12	6?
Frangula, subgenus of Rhamnus		
Frasera, American Columbo,	83. 3	35?
Fraxinus, Ash,	86. 4	84?
French Mulberry, Callicarpa,	76. 3	33?
Fringe-Tree, Chionanthus,	86. 3	10?
Frœlichia,	91. 6	67?
Frogs-bit, Limnobium,	117. 1	88?
Frostweed, Helianthemum,	16. 1	2?
Fuirena, Umbrella-Grass,	133. 9	64?
Fumaria, Fumitory,	11. 4	22?
Fumitory, Fumaria,	11. 4	22?

G

Galactia, Milk-Pea,	38. 25	254
Galax,	63. 1	147
Galeopsis, Hemp-Nettle,	77. 28	502
Galingale, Cyperus,	133. 1	554
Galinsoga,	59. 57	968
Galium, Bedstraw, Cleavers,	56. 1	529
Gama-Grass, Tripsacum,	134. 62	565
Garget, Phytolacca,	89. 1	652
Garlic, Allium,	126. 10	722
Gaultheria, Creeping Wintergreen,	62. 6	422
Gaura,	43. 3	283
Gaylussacia, Huckleberry,	62. 1	513
Gelsemium, Yellow False Jessamine,	74. 27	409
Genista, Wood-Waxen,	38. 8	237
Gentian, Gentiana,	83. 5	376
Gentiana, Gentian,	83. 5	376
Geranium, Cranesbill,	28. 1	168
Gerardia,	74. 19	338
Germander, Teucrium,	77. 1	493
Geum, Avens,	39. 9	65
Giant Hyssop, Lophanthus,	77. 19	484
Gill, Glechoma,	77. 20	487
Gillenia, Indian Physic,	39. 3	317
Ginseng, Panax,	53. 2	287
Glasswort, Salicornia,	90. 7	544
Glaucium, Horn-Poppy,	10. 5	15
Glaux, Sea-Milkwort,	70. 7	677
Glechoma, Gill, Ground-Ivy,	77. 20	487
Gleditschia, Honey-Locust,	38. 34	92
Globe-flower, Trollius,	1. 12	46
Glyceria, Manna-Grass,	134. 81	632
Gnaphalium, Cudweed,	59. 59	930
Goat's-Beard, sp. of Spiræa,		

Goat's-Rue, sp. of Tephrosia,
Golden Aster, Chrysopsis, ... 59. 22 957
Golden Club, Orontium, . . . 112. 5 673
Golden Rod, Solidago, ... 59. 20 959
Golden Saxifrage, Chrysosplenium, . 50. 8 689
Goldthread, Coptis, ... 1. 13   46
Gonolobus, ... 85. 4 415
Good-King-Henry, sp. of Blitum,
Goodyera, Rattlesnake-Plantain, . . 119. 4 770
Gooseberry, some sp. of Ribes,
Goosefoot, Chenopodium, . . . 90. 2 683
Goose-Grass, sp. of Galium,
Gordonia, Loblolly Bay, . . . 25. 2   76
Grape, Vitis, ... 33. 1 158
Grass of Parnassus, Parnassia, . . 18. 1 183
Grass-wrack, Zostera, . . . 115. 3 809
Gratiola, Hedge-Hyssop, . . . 74. 11 451
Great Burnet, Sanguisorba, . . . 39. 5 700
Great Laurel, sp. of Rhododendron,
Greek Valerian, Polemonium, . . 80. 1 396
Greenbrier, Smilax, ... 125. 1 866
Green Milkweed, Acerates, . . . 85. 2 417
Green Violet, Solea, ... 15. 1 225
Gromwell, Lithospermum, . . . 78. 5 384
Grossularia, subgenus of Ribes, . . 46. 1 286
Ground Cherry, Physalis, . . . 82. 2 391
Ground Hemlock, Taxus, . . . 111. 8 850
Ground Ivy, Glechoma, . . . 77. 20 487
Ground Laurel, Epigæa, . . . 62. 5 425
Groundnut, Apios, ... 38. 23 249
Groundsel, Senecio, ... 59. 64 921
Groundsel-Tree, Baccharis, . . . 59. 25 927
Grove Sandwort, Alsine, . . . 21. 7 205
Gymnadenia, Naked Gland Orchis, . 119. 2 763
Gymnocladus, Kentucky Coffee-Tree, . 38. 33 847
Gymnopogon, Naked Beard-Grass, . 134. 19 596
Gymnostichum, Bottle-Brush-Grass, . 134. 46 612

H

Hackberry, Celtis, ... 104. 3 663
Hair-Grass, Aira, ... 134. 47 621
Hairy Pipewort, Lachnocaulon, . . 132. 3 820
Halenia, Spurred Gentian, . . . 83. 4 351
Halesia, Silver-bell Tree, . . . 65. 2 511
Hamamelis, Witch-Hazel, . . . 51. 1 129
Harbinger-of-Spring, Erigenia, . . 52. 27 311
Hardhack, sp. of Spiræa,

Harebell, sp. of Campanula,		
Hawkbit, Leontodon,	59. 76	98
Hawkweed, Hieracium,	59. 77	99
Hawthorn, Cratægus,	39. 16	5
Hazelnut, Corylus,	107. 4	78
Heal-all, Brunella,	77. 25	49
Heart's-ease, Viola,	15. 2	22
Hedeoma, Mock Pennyroyal,	77. 14	44
Hedgehog-Grass, Cenchrus,	134. 61	60
Hedge-Hyssop, Gratiola,	74. 11	45
Hedge-Mustard, Sisymbrium,	12. 10	12
Hedge-Nettle, Stachys,	77. 29	50
Hedysarum,	38. 16	24
Helenium, False Sunflower,	59. 47	94
Helianthemum, Rock-Rose,	16. 1	2
Helianthus, Sunflower,	59. 40	98
Heliophytum, Indian Heliotrope,	78. 11	38
Heliopsis, Ox-eye,	59. 36	97
Heliotrope, Heliotropium,	78. 10	38
Heliotropium, Heliotrope,	78. 10	38
Hellebore, Helleborus,	1. 14	4
Helleborus, Hellebore,	1. 14	4
Helonias,	127. 9	73
Hemerocallis, Day-Lily,	126. 7	72
Hemianthus, Micranthemum,		
Hemicarpha,	133. 4	55
Hemlock, Conium,	52. 25	31
Hemlock-Parsley, Conioselinum,	52. 13	30
Hemlock-Spruce, sp. of Abies,		
Hemp, Cannabis,	104. 10	89
Hemp-Nettle, Galeopsis,	77. 28	50
Hempweed, Mikania,	59. 8	91
Henbane, Hyoscyamus,	82. 4	39
Hepatica, Liver-Leaf,	1. 5	3
Heracleum, Cow-Parsnip,	52. 7	29
Hercules' Club, sp. of Aralia,		
Herds-Grass, Phleum,	134. 4	57
Herpestis,	74. 10	46
Heteranthera, Mud-Plantain,	129. 2	71
Heuchera, Alum-root,	50. 5	28
Hibiscus, Rose-Mallow,	23. 9	7
Hickory, Carya,	106. 2	78
Hieracium, Hawkweed,	59. 77	99
Hierochloa, Holy-Grass,	134. 53	61
Highwater Shrub, Iva,	59. 30	93
Hippuris, Mare's-Tail,	43. 9	54
Hoary Pea, Tephrosia,	38. 13	26
Hobble-Bush, sp. of Viburnum,		

Hog-Peanut, Amphicarpæa, ... 38. 26 252
Hogweed, sp. of Ambrosia, ...
Holcus, Meadow Soft-Grass, ... 134. 52 617
Holly, Ilex, ... 64. 1 130
Holosteum, Jagged Chickweed, ... 21. 11 180
Holy-Grass, Hierochloa, ... 134. 53 616
Honey-Locust, Gleditschia, ... 38. 34   92
Honewort, Cryptotænia, ... 52. 22 307
Honkenya, Sea-Sandwort, ... 21. 6 205
Hop, Humulus, ... 104. 11 895
Hop-Hornbeam, Ostrya, ... 107. 6 792
Hop-Tree, Ptelea, ... 31. 2   92
Hordeum, Barley, ... 134. 44 577
Horehound, Marrubium, ... 77. 27 500
Hornbeam, Carpinus, ... 107. 5 792
Horned Pondweed, Zanichellia, . . . 115. 2 818
Horned Rush, Ceratoschœnus, ... 133. 12 563
Horn-Poppy, Glaucium, ... 10. 5   15
Hornwort, Ceratophyllum, ... 99. 1 838
Horse-Balm, Collinsonia, ... 77. 15 444
Horse-Chestnut, Æsculus, ... 36. 2 230
Horse-Gentian, Triosteum, ... 55. 5 522
Horse-Mint, Monarda, ... 77. 17 445
Horse-Nettle, sp. of Solanum, ...
Horseradish, Armoracia, ... 12. 13 111
Horse-Sugar, Symplocos, ... 65. 3   76
Horseweed, sp. of Erigeron, ...
Hottonia, Featherfoil, ... 70. 11 358
Hound's-Tongue, Cynoglossum, . . . 78. 9 379
Houstonia, Bluets, ... 56. 7 553
Hudsonia, ... 16. 2   22
Humulus, Hop, ... 104. 11 895
Huckleberry, Gaylussacia, ... 62. 1 513
Hydrastis, Orange-root, ... 1. 19   39
Hydrocotyle, Water-Pennywort, . . . 52. 1 292
Hydrophyllum, Water-Leaf, ... 79. 1 367
Hymenopappus, ... 59. 46 914
Hyoscyamus, Henbane, ... 82. 4 898
Hypericum, St. John's-wort, ... 19. 2   74
                                              & 199
Hypoxis, Star-Grass, ... 120. 4 751
Hyssop, Hyssopus, ... 77. 7 489
Hyssopus, Hyssop, ... 77. 7 489

I

Ilex, Holly, ... 64. 1 130
Ilysanthes, False Pimpernel, ... 74. 12 451

INDEX.   361

Impatiens, Balsam,		29. 1	224
Indian Bean, Catalpa,		72. 3	328
Indian Chickweed, Mollugo,		21. 20	692
Indian Cucumber-root, Medeola,		123. 3	721
Indian Currant, sp. of Symphoricarpus,			
Indian Fig, Opuntia,		45. 1	51
Indian Grass, Sorghum,		134. 65	602
Indian Heliotrope, Heliophytum,		78. 11	386
Indian Hemp, Apocynum,		84. 3	413
Indian Physic, Gillenia,		39. 3	317
Indian Pipe, Monotropa,		62. 27	163
Indian Plantain, Cacalia,		59. 63	921
Indian Poke, sp. of Veratrum,			
Indian Rice, Zizania,		134. 2	567
Indian Turnip, sp. of Arisæma,			
Inkberry, sp. of Prinos,			
Inula, Elecampane,		59. 23	960
Iodanthus, False Rocket,		12. 2	121
Ipomœa, Morning Glory,		81. 2	404
Iresine,		91. 5	667
Iris, Flower-de-Luce,		123. 1	705
Iron-Weed, Vernonia,		59. 1	924
Iron-Wood, Ostrya,		107. 6	792
Isanthus, False Pennyroyal,		77. 3	340
Isopyrum, False Rue-Anemone,		1. 10	40
Itea,		50. 9	162
Iva, Marsh-Elder,		59. 30	935

J

Jack-in-the-pulpit, Arisæma,		112. 1	550
Jacob's Ladder, Polemonium,		80. 1	396
Jagged Chickweed, Holosteum,		21. 11	180
Jamestown-Weed, Datura,		82. 5	399
Jeffersonia, Twin-Leaf,		5. 4	137
Jerusalem Oak, sp. of Chenopodium,			
Jerusalem Sage, Phlomis,		77. 33	506
Jewel-Weed, Impatiens,		29. 1	224
Joe-Pye-Weed, sp. of Eupatorium,			
Joint-Grass, sp. of Paspalum,			
Joint-Weed, sp. of Polygonum,			
Judas'-Tree, Cercis,		38. 31	233
Juglans, Walnut,		106. 1	783
Juncus, Bog-Rush,		128. 3	741
June-berry, Amelanchier,		39. 18	55
Juniper, Juniperus,		111. 7	850
Juniperus, Juniper,		111. 7	850
Jussiæa,		43. 4	282

K

Kœleria,	134. 28	637
Kalmia,	62. 14	424
Kentucky Coffee-Tree, Gymnocladus,	38. 33	847
Kidney-Bean, Phaseolus,	38. 22	249
Kinnikinnik, sp. of Cornus,		
Knawel, Scleranthus,	21. 19	691
Knapweed, sp. of Centaurca,		
Knot-Grass, sp. of Polygonum,		
Knotweed, Polygonum,	92. 1	687
Kosteletzkya,	23. 8	73
Krigia, Dwarf Dandelion,	59. 74	991
Kuhnia,	59. 6	925
Kyllingia,	133. 2	554

L

Labrador Tea, Ledum,	62. 19	153
Lachnanthes, Red-Root,	121. 1	709
Lachnocaulon, Hairy Pipewort,	132. 3	820
Lactuca, Lettuce,	59. 83	999
Ladies' Tresses, Spiranthes,	119. 5	770
Lady's Mantle, Alchemilla,	39. 6	754
Lady's Slipper, Cypripedium,	119. 17	758
Lady's Thumb, sp. of Polygonum,		
Lambkill, sp. of Kalmia,		
Lamb-Lettuce, Fedia,	57. 2	535
Lamb's Quarters, sp. of Chenopodium,		
Lamium, Dead Nettle,	77. 31	506
Lampsana, Nipple-wort,	59. 72	987
Laportea, Wood-Nettle,	104. 6	831
Lappa, Burdock,	59. 71	940
Larch, Larix,	111. 3	778
Larix, Larch,	111. 3	778
Larkspur, Delphinium,	1. 16	42
Lathyrus, Vetchling,	38. 21	242
Laurentinus, Viburnum,	55. 7	520
Lead-Plant, sp. of Amorpha,		
Leaf-Cup, Polymnia,	59. 26	982
Leather-Leaf, Cassandra,	62. 8	431
Leather-Flower, sp. of Clematis,		
Leather-wood, Dirca,	94. 1	660
Leavenworthia,	12. 3	123
Lechea, Pinweed,	16. 8	97
Ledum, Labrador Tea,	62. 19	153
Leersia, False Rice,	134. 1	567

INDEX. 363

Leiophyllum, Sand-Myrtle,		62. 21	171
Lemna, Duckmeat,		114. 1	541
Leontodon, Hawkbit,		59. 76	988
Leonurus, Motherwort,		77. 30	503
Lepachys,		59. 39	979
Lepidium, Pepperwort,		12. 16	115
Leptochloa,		134. 23	624
Leptopoda,		59. 48	943
Lepturus,		134. 41	570
Lespedeza, Bush-Clover,		38. 18	246
Lettuce, Lactuca,		59. 83	999
Leucanthemum, Ox-Eye Daisy,		59. 55	963
Leucothoë,		62. 7	431
Lever-Wood, Ostrya,		107. 6	792
Liatris, Button Sneak-root,		59. 5	924
Ligusticum, Lovage,		52. 15	310
Ligustrum, Privet,		86. 1	329
Lilium, Lily,		126. 11	744
Lily, Lilium,		126. 11	744
Lily of the Valley, Convallaria,		126. 5	725
Limnanthemum, Floating Heart,		83. 9	358
Limnobium, American Frog's-Bit,		117. 1	885
Limosella, Mudwort,		74. 14	334
Linaria, Toad-Flax,		74. 2	462
Linden, Tilia,		24. 1	75
Linnæa, Twin-Flower,		55. 1	525
Linum, Flax,		26. 1	185
Lion's-foot, sp. of Nabalus,			
Liparis, Twayblade,		119. 14	773
Lippia, Fog-Fruit,		76. 2	457
Liquidambar, Sweet Gum-Tree,		51. 3	796
Liriodendron, Tulip-Tree,		2. 2	47
Listera, Twayblade,		119. 6	769
Lithospermum, Gromwell,		78. 5	384
Lizard's-Tail, Saururus,		98. 1	548
Live-forever, sp. of Sedum,			
Liverleaf, Hepatica,		1. 5	35
Lobelia,		60. 1	521
Locust-Tree, Robinia,		38. 11	260
Loblolly Bay, Gordonia,		25. 2	76
Loiseleuria, Alpine Azalea,		62. 20	409
Lolium, Darnel,		134. 42	611
Long Moss, Tillandsia,		122. 1	89
Lonicera, Honeysuckle,		55. 3	519
Loosestrife, Lythrum, Lysimachia, and Naumburgia,			
Lophanthus, Giant Hyssop,		77. 19	484
Lophiola, Crest-Flower,		121. 2	749
Lopseed, Phryma,		76. 4	457

Louscwort, Pedicularis, ... 74. 25 466
Lovage, Ligusticum, ... 52. 15 810
Lucerne, sp. of Medicago, ...
Ludwigia, False Loosestrife, ... 43. 5 277
Lungwort, Mertensia, ... 78. 6 382
Lupine, Lupinus, ... 38. 1 238
Lupinus, Wild Lupine, ... 38. 1 238
Luzula, Wood-Rush, ... 128. 2 741
Lycopsis, Bugloss, ... 78. 2 381
Lycopus, Water-Horehound, ... 77. 5 325
Lygodesmia, ... 59. 82 996
Lyme-Grass, Elymus, ... 134. 45 609
Lysimachia, Yellow Loosestrife, . . . 70. 5 373
Lythrum, Purple Loosestrife, ... 42. 2 143

M

Magnolia, ... 2. 1 47
Majanthemum, ... 126. 4 674
Mallow, Malva, ... 23. 2 71
Malus, subgenus of Pyrus, ... 89. 17 56
Malva, Mallow, ... 23. 2 71
Mandrake, Podophyllum, ... 5. 5 217
Manna-Grass, Glyceria, ... 134. 31 632
Man of the Earth, sp. of Ipomœa, . . .
Maple, Acer, ... 36. 3 167
Mare's-tail, Hippuris, ... 43. 9 545
Marrubium, Horehound, ... 77. 27 500
Marshallia, ... 59. 50 933
Marsh-Elder, Iva, ... 59. 24 928
Marsh-Grass, Spartina, ... 134. 16 578
Marsh-Mallow, Althæa, ... 23. 1 70
Marsh-Marigold, Caltha, ... 1. 11 40
Marsh-Rosemary, Statice, ... 69. 1 188
Marsh, St. John's-wort, Elodea, . . . 19. 3 198
Martynia, Unicorn-Plant, ... 72. 4 450
Maruta, Mayweed, ... 59. 52 977
Matricaria, Wild Chamomile, . . . 59. 56 962
Mayaca, ... 131. 1 89
May-Apple, Podophyllum, ... 5. 5 217
May-Flower, Epigæa, ... 62. 5 425
May-weed, Maruta, ... 59. 52 977
Meadow-Beauty, Rhexia, ... 41. 1 280
Meadow-Grass, Poa, ... 134. 33 638
Meadow-Parsnip, Thaspium, ... 52. 16 310
Meadow-Rue, Thalictrum, ... 1. 6 36
Meadow Soft-Grass, Holcus, ... 134. 52 617
Meadow-Sweet, Spiræa, ... 89. 2 58

INDEX. 365

Meconopsis, Yellow Poppy,	10.	3	14
Medeola, Indian Cucumber-Root,	125.	3	721
Medicago, Medick,	38.	6	258
Medick, Medicago,	38.	6	258
Melampyrum, Cow-Wheat,	74.	26	471
Melanthium, Bunch-Flower,	127.	4	733
Melica, Melic-Grass,	134.	30	627
Melic-Grass, Melica,	134.	30	627
Melilot, Melilotus,	38.	5	259
Melilotus, Melilot, Sweet Clover,	38.	5	259
Melissa, Balm,	77.	13	495
Melothria,	48.	3	507
Menispermum, Moonseed,	4.	2	858
Mentha, Mint,	77.	4	342
Mentzelia,	44.	1	52
Menyanthes, Buckbean,	83.	8	360
Menziesia,	62.	15	432
Mermaid-weed, Proserpinaca,	43.	7	645
Mertensia, Smooth Lungwort,	78.	6	382
Mexican Tea, sp. of Chenopodium,			
Micranthemum,	74.	13	448
Microstylis, Adder's-Mouth,	119.	13	772
Mignoncttc, Reseda,	14.	1	7
Mikania, Climbing Hempweed,	59.	8	916
Milfoil, sp. of Achillea,			
Milium, Millet-Grass,	134.	56	590
Milk-Pea, Galactia,	38.	25	254
Milkweed, Asclepias,	85.	1	417
Milkwort, Polygala,	37.	1.	228
Milk-Vetch, Astragalus,	38.	14	261
Millet, sp. of Setaria,			
Millet-Grass, Milium,	134.	56	590
Mimulus, Monkey-flower,	74.	8	477
Mint, Mentha,	77.	4	342
Mist-flower, Conoclinium,	59.	9	917
Mitchella, Partridge-berry,	56.	5	530
Mitella, Bishop's-Cap,	50.	6	192
Mitreola, Mitrewort,	56.	8	406
Mitrewort, Mitreola,	56.	8	406
Mocasson-flower, Cypripedium,	119.	17	758
Mocker-nut, sp. of Carya,			
Mock-Orange, Philadelphus,	50.	11	54
Mock-Pennyroyal, Hedeoma,	77.	14	442
Modiola,	23.	7	214
Mœhringia,	21.	9	203
Mœnchia,	21.	13	134
Mollugo, Indian Chickweed,	21.	20	692
Monarda, Horse-Mint,	77.	17	445

366   PRACTICAL BOTANY.

Monesis, One-flowered Pyrola,	62. 23	174
Monkey-flower, Mimulus,	74. 8	477
Monkshood, Aconite,	1. 17	43
Monotropa, Indian Pipe,	62. 27	163
Montelia,	91. 3	901
Moonseed, Menispermum,	4. 2	858
Moose-wood, Dirca,	94. 1	660
Morning-Glory, Pharbitis and Ipomœa,	81. 2	404
Morus, Mulberry,	104. 4	795
Motherwort, Leonurus,	77. 30	503
Mountain-Ash, sp. of Pyrus,		
Mountain-Holly, Nemopanthes,	64. 3	130
Mountain-Mint, Pycnanthemum,	77. 8	492
Mountain-Rice, Oryzopsis,	134. 13	586
Mountain-Sorrel, Oxyria,	92. 4	85
Mouse-ear, Myosotis,	78. 7	384
Mouse-ear Chickweed, Cerastium,	21. 12	186
Mouse-tail, Myosurus,	1. 9	266
Mud-Plantain, Heteranthera,	129. 2	711
Mudwort, Limosella,	74. 14	334
Mugwort, Common Artemisia,		
Muhlenbergia, Drop-seed Grass,	134. 10	587
Mulberry, Morus,	104. 4	795
Mulgedium, Blue Lettuce,	59. 84	993
Mullein, Verbascum,	74. 1	397
Mullein-Foxglove, Seymeria,	74. 18	336
Muskit-Grass, Bouteloua,	134. 18	596
Musquash-root, sp. of Cicuta,		
Mustard, Sinapis,	12. 11	122
Myosotis, Forget-me-not,	78. 7	384
Myosurus, Mouse-tail,	1. 9	266
Myrica, Wax-Myrtle,	108. 1	874
Myriophyllum, Water-Milfoil,	43. 8	828

N

Nabalus, Rattlesnake-root,	59. 78	996
Najad, Najas,	115. 1	818
Najas, Najad,	115. 1	818
Naked Beard-Grass, Gymnopogon,	134. 19	596
Naked Broom-Rape, Aphyllon,	73. 3	454
Naked-Gland-Orchis, Gymnadenia,	119. 2	763
Napæa, Glade-Mallow,	23. 4	214
Nardosmia, Sweet Colt's-foot,	59. 10	951
Narthecium, Bog-Asphodel,	128. 1	742
Nasturtium, Watercress,	12. 1	112
Naumburgia, Tufted Loosestrife,	70. 6	143
Neckweed, sp. of Veronica,		

Negundo, Box-Elder,		36. 4	846
Nelumbium, Sacred Bean,		6. 1	26
Nemopanthes, Mountain Holly,		64. 3	130
Nemophila,		79. 2	367
Nepeta, Cat-Mint,		77. 20	487
Nesæa, Swamp Loosestrife,		42. 3	171
Nettle, Urtica,		104. 5	832
Nettle-Tree, Celtis,		104. 3	663
New-Jersey Tea, Ceanothus,		34. 3	288
Nicandra, Apple of Peru,		82. 3	391
Nicotiana, Tobacco,		82. 6	400
Nimble Will, sp. of Muhlenbergia,			
Nightshade, Solanum,		82. 1	390
Nine-Bark, sp. of Spiræa,			
Nondo, sp. of Ligusticum,			
Nonesuch, sp. of Medicago,			
Nuphar, Spatter-Dock,		8. 2	24
Nut-Grass, sp. of Cyperus,			
Nut-Rush, Scleria,		133. 15	821
Nymphæa, Water-Lily,		8. 1	24
Nyssa, Sour Gum-Tree, Pepperidge,		54. 2	859

O

Oak, Quercus,		107. 1	789
Oat, Avena,		134. 50	625
Oat-Grass, Arrhenatherum,		134. 51	618
Obione, Sand-Orache,		90. 6	836
Obolaria,		83. 7	352
Œnothera, Evening Primrose,		43. 2	283
Oil-nut, Pyrularia,		96. 2	869
Oldenlandia, Bluets,		56. 6	515
Olea, Olive,		86. 2	329
Olive, Olea,		86. 2	329
Onion, Allium,		126. 10	722
Onopordon, Cotton-Thistle,		59. 70	939
Onosmodium, False Gromwell,		78. 4	383
Opuntia, Indian Fig,		45. 1	51
Orache, Atriplex,		90. 5	826
Orange-root, Hydrastis,		1. 19	39
Orange-Grass, sp. of Hypericum,			
Orchard-Grass, Dactylis,		134. 27	635
Orchis, Showy Orchis,		119. 1	762
Origanum, Wild Marjoram,		77. 9	491
Ornithogalum, Star of Bethlehem,		126. 8	745
Orontium, Golden Club,		112. 5	673
Orpine, Sedum,		49. 2	272
Oryzopsis, Mountain Rice,		134. 13	586

Osier, Salix,	110.	1	874
Osmorrhiza, Sweet Cicely,	52.	24	312
Ostrya, Hop-Hornbeam,	107.	6	792
Oswego-Tea, sp. of Monarda,			
Oxalis, Wood-Sorrel,	27.	1	208
Ox-eye, Heliopsis,	59.	36	971
Ox-eye Daisy, Leucanthemum,	59.	55	963
Oxybaphus,	88.	1	538
Oxycoccus, Cranberry,	62.	2	514
Oxydendrum, Sorrel-Tree,	62.	11	429
Oxyria, Mountain-Sorrel,	92.	4	85

P

Pachysandra,	102.	9	828
Pæpalanthus,	132.	2	803
Painted-Cup, Castileja,	74.	21	465
Panax, Ginseng,	53.	2	287
Pancratium,	120.	2	750
Panicum, Panic-Grass,	134.	59	601
Panic-Grass, Panicum,	134.	59	601
Papaver, Poppy,	10.	1	12
Pappoose-Root, Caullophylum,	5.	2	141
Parietaria, Pellitory,	104.	9	667
Parnassia, Grass-of-Parnassus,	18.	1	183
Paronychia, Whitlow-wort,	21.	18	691
Parsley-Piert, Alchemilla,	39.	6	754
Parsnip, Pastinaca,	52.	8	298
Parthenium,	59.	29	975
Partridge-berry, Mitchella,	56.	5	530
Partridge-Pea, sp. of Cassia,			
Paspalum,	134.	58	573
Pasque-flower, Pulsatilla,	1.	3	32
Passiflora, Passion-flower,	47.	1	148
Pastinaca, Parsnip,	52.	8	298
Pear, Pyrus,	39.	17	56
Pearl-wort, Sagina,	21.	14	134
Pecan-nut, sp. of Hickory,			
Pedicularis, Louse-wort,	74.	25	466
Pellitory, Parietaria,	104.	9	667
Peltandra, Arrow-Arum,	112.	2	813
Pencil-flower, Stylosanthes,	38.	19	238
Pennywort, Hydrocotyle,	52.	1	292
Penthorum, Dutch Stone-crop,	49.	1	271
Pentstemon, Beard-Tongue,	74.	7	476
Pepperbush, Clethra,	62.	12	173
Pepper-Grass, Lepidium,	12.	16	115
Pepperidge, Nyssa,	54.	2	359

INDEX. 369

Peppermint, sp. of Mentha,			
Pepper-root, Dentaria,		12. 4	120
Pepper-wort, Lepidium,		12. 16	115
Periploca,		85. 5	414
Persea, Alligator Pear,		93. 1	704
Persimmon, Diospyros,		66. 1	421
Petalostemon, Prairie Clover,		38. 9	240
Phacelia,		79. 4	369
Phalaris, Canary Grass,		134. 55	594
Pharbitis, Morning-Glory,		81. 2	404
Phaseolus, Kidney-bean,		38. 22	249
Phelipæa, Broomrape,		73. 4	455
Philadelphus, Mock-Orange,		50. 11	54
Phleum, Cat's-tail-Grass,		134. 4	579
Phlomis, Jerusalem Sage,		77. 33	506
Phlox,		80. 2	408
Phoradendron, False Mistletoe,		97. 1	841
Phragmites, Reed,		134. 39	615
Phryma, Lopseed,		76. 4	457
Phyllanthus,		102. 8	823
Phyllodoce,		62. 13	432
Physalis, Ground Cherry,		82. 2	391
Physostegia, False Dragon-Head,		77. 24	505
Phytolacca, Pokeweed,		89. 1	652
Pickerel-Weed, Pontederia,		129. 1	756
Pigeon-berry, Phytolacca,		89. 1	052
Pig-nut, sp. of Carya,			
Pigweed, Chenopodium and some sp. of Amaranth,			
Pilea, Richweed, Clearweed,		104. 7	825
Pimpernel, Anagallis,		70. 8	371
Pine, Pinus,		111. 1	779
Pine-drops, Pterospora,		62. 25	433
Pine-sap, Monotropa,		62. 27	163
Pine-weed, sp. of Hypericum,			
Pinguicula, Butterwort,		71. 2	439
Pink, Dianthus,		21. 1	194
Pink-root, Spigelia,		56. 9	406
Pinus, Pine,		111. 1	779
Pinweed, Lechea,		16. 3	97
Pinxter-flower, sp. of Azalea,			
Pipe-Vine, sp. of Aristolochia,			
Pipewort, Eriocaulon,		132. 1	803
Pipsissewa, Chimaphila,		62. 24	173
Pitcher-Plant, Sarracenia,		9. 1	19
Planera, Planer-Tree,		104. 2	663
Planer-Tree, Planera,		104. 2	663
Plane-Tree, Platanus,		105. 1	796
Plantago, Plantain,		68. 1	354

Plantain, Plantago,	68.	1	354
Platanthera, False Orchis,	119.	3	763
Platanus, Plane-Tree,	105.	1	796
Pleurisy-root, sp. of Asclepias,			
Pluchea, Marsh-Fleabane,	59.	24	928
Plum, Prunus in part,	39.	1	53
Poa, Meadow or Spear-Grass,	134.	33	638
Podophyllum, May-Apple,	5.	5	217
Podostemon, Riverweed,	101.	1	549
Pogonia,	119.	8	773
Poison Hemlock, Conium,	39.	1	314
Poison Ivy, sp. of Rhus,			
Poison-Oak, Poison Ivy,			
Poison Sumach, sp. of Rhus,			
Poke, Phytolacca,	89.	1	652
Polanisia,	13.	1	127
Polemonium, Jacob's Ladder,	80.	1	396
Polygala, Milkwort,	37.	1	228
Polygonatum, Solomon's Seal,	126.	2	719
Polygonella,	92.	2	686
Polygonum, Knotweed,	92.	1	687
Polymnia, Leaf-cup,	59.	26	982
Polypogon, Beard-Grass,	134.	8	579
Polypremum,	56.	10	354
Polytænia,	52.	6	297
Pond-Spice, Tetranthera,	98.	4	703
Pondweed, Potamogeton,	115.	5	653
Pontederia, Pickerelweed,	129.	1	756
Poor Man's Weather-Glass, Anagallis,	70.	8	371
Poplar, Populus,	110.	2	873
Poppy, Papaver,	10.	1	12
Populus, Poplar, Aspen.	110.	2	873
Porcupine-Grass, sp. of Stipa,			
Portulaca, Purslane,	22.	2	206
Potamogeton, Pondweed,	115.	5	653
Potentilla, Cinquefoil,	39.	11	64
Poverty-Grass, sp. of Aristida,			
Prairie Clover, Petalostemon,	38.	9	240
Prairie Dock, sp. of Silphium,			
Prickly Ash, Xanthoxylum,	39.	1	844
Prickly Pear, Opuntia,	45.	1	51
Prickly Poppy, Argemone,	10.	2	13
Prim, Ligustrum,	86.	1	329
Primrose, Primula,	70.	1	362
Primula, Primrose,	70.	1	362
Prince's Feather, sp. of Amarantus,			
Prince's Pine, sp. of Chimaphila,			
Prinos, Winterberry and Inkberry,	64.	2	862

INDEX.

Privet, Ligustrum,		86. 1	329
Prosartes,		127. 2	740
Proserpinaca, Mermaid-weed,		43. 7	645
Prunella, see Brunella,			
Prunus, Plum and Cherry,		39. 1	53
Psilocarya, Bald-Rush,		133. 10	562
Psoralea,		38. 7	256
Ptelea, Shrubby Trefoil,		31. 2	92
Pterospora, Pine-drops,		62. 25	433
Puccoon, Lithospermum,		78. 5	384
Pulsatilla, Pasque-flower,		1. 3	32
Purslane, Portulaca,		22. 2	206
Pully-root, Aplectrum,		119. 16	766
Pycnanthemum, Mountain-Mint,		77. 8	492
Pyrola, False Wintergreen,		62. 22	174
Pyrrhopappus, False Dandelion,		59. 81	994
Pyrularia, Oil-nut,		96. 2	869
Pyrus, Pear and Apple,		39. 17	56
Pyxidanthera,		80. 4	395

Q

Quaking-Grass, Briza,		134. 35	629
Quamash, Scilla,		126. 9	747
Quamoclit,		81. 1	403
Queen of the Prairie, sp. of Spiræa,			
Quercus Oak,		107. 1	789
Quick-Grass, sp. of Triticum,			
Quitch-Grass, Quick-Grass,			

R

Radish, Raphanus,		12. 21	106
Ragweed, Ambrosia,		59. 31	965
Ragwort, some sp. of Senecio,			
Ram's-Head, sp. of Cypripedium,			
Ramsted, Common Toad-Flax,			
Ranunculus, Crowfoot,		1. 8	31
Raphanus, Radish,		12. 21	106
Raspberry, some sp. of Rubus,			
Rattle-box, Crotalaria,		38. 2	237
Rattlesnake-Grass, sp. of Glyceria,			
Rattlesnake-Master, sp. of Eryngium,			
Rattlesnake-Plantain, Goodyera,		119. 4	770
Rattlesnake-Root, Nabalus,		59. 78	996
Rattlesnake-Weed, sp. of Hieracium,			
Ray-Grass, sp. of Lolium,			
Red-Bay, Persea,		93. 1	704

PRACTICAL BOTANY.

Red-Bud, Cercis,	38.	31	233
Red-Root, Ceanothus and Lachnanthes,			
Red-Top, sp. of Agrostis,			
Reed, Phragmites,	134.	39	615
Reed Bent-Grass, Calamagrostis,	134.	12	575
Reed-Mace, sp. of Typha,			
Reed Meadow-Grass, sp. of Glyceria,			
Reseda, Dyer's Rocket,	14.	1	7
Rhamnus, Buckthorn,	34.	2	103
Rhexia, Deer-Grass,	41.	1	280
Rhinanthus, Yellow Rattle,	74.	24	468
Rhododendron, Rose-Bay,	62.	17	426
Rhodora,	62.	18	434
Rhus, Sumach,	32.	1˙	178
Rhynchosia,	38.	24	251
Rhynchospora, Beak-Rush,	133.	13	563
Rhytiglossa, Water-Willow,	75.˙	1	449
Ribes, Currant and Gooseberry,	46.	1	286
Rib-Grass, Plantago,	68.	1	354
Richweed, Piléa,	104.	7	825
Ripple-Grass, sp. of Plantago,			
Riverweed, Podostemon,	101.	1	549
Robinia, Locust-Tree,	38.	11	260
Robin's Plantain, sp. of Erigeron,			
Rock-Cress, Arabis,	2.	6	119
Rock-Rose, Helianthemum,	16	1	22
Roman Wormwood, sp. of Ambrosia,			
Rosa, Rose,	39.	15	59
Rose, Rosa,	39.	15	59
Rose-Bay, Rhododendron,	62.	17	426
Rose-Mallow, Hibiscus,	23.	9	73
Rosin-Plant, Silphium,	59.	28	975
Rosin-weed, sp. of Silphium,			
Roubieva,	90.	3	681
Rubus, Bramble,	39.	14	62
Rudbeckia, Cone-flower,	59.	38	979
Rue-Anemone, sp. of Thalictrum,			
Rumex, Dock-Sorrel,	92.	5	714
Ruppia, Ditch-Grass,	115.	4	543
Rush-Grass, Vilfa,	134.	5	593
Rush-Salt Grass, sp. of Spartina,			
Rye-Grass, sp. of Lolium,			

S

Sabbatia, American Centaury,	83.	1	375
Sacred Bean, Nelumbium,	6.	1	26
Sage, Salvia,	77.	16	444

Sagina, Pearlwort,	21. 14	134
Sagittaria, Arrow-head,	116. 5	802
St. Andrew's Cross, sp. of Ascyrum,		
St. John's-wort, Hypericum,	19. 2	'74 & 199
St. Peter's-wort, sp. of Ascyrum,		
Salicornia, Glasswort,	90. 7	544
Salix, Willow,	110. 1	874
Salsola, Saltwort,	90. 9	676
Salt Marsh-Grass, sp. of Spartina,		
Salt Reed-Grass, sp. of Spartina,		
Saltwort, Salsola,	90. 9	676
Salvia, Sage,	77. 16	444
Sambucus, Elder,	55. 6	520
Samolus, Water-Pimpernel,	70. 10	365
Samphire, Salicornia,	90. 7	544
Sand-Grass, Uralepis,	134. 25	631
Sand-Myrtle, Leiophyllum,	62. 21	171
Sand-Orache, Obione,	90. 6	836
Sandwort, Arenaria,	21. 8	203
Sanguinaria, Blood-root,	10. 6	11
Sanguisorba, Great Burnet,	39. 5	700
Sanicle, Sanicula,	52. 3	293
Sanicula, Sanicle,	52. 3	293
Saponaria, Soapwort,	21. 2	195
Sarracenia, Pitcher-Plant,	9. 1	19
Sarsaparilla, Wild Aralia,	53. 1	287
Sassafras,	93. 2	702
Satureja, Savory,	77. 11	492
Saururus, Lizard's-tail,	98. 1	548
Savin, sp. of Juniperus,		
Savory, Satureja,	77. 11	492
Saxifraga, Saxifrage,	50. 2	192
Saxifrage, Saxifraga,	50. 2	192
Scheuchzeria,	116. 2	728
Schollera, Water-Star-Grass,	129. 3	711
Schrankia, Sensitive Brier,	38. 26	418
Schwalbea, Chaff-seed,	74. 22	464
Schweinitzia, Sweet Pine-sap,	62. 26	433
Scilla, Squill,	126. 9	747
Scirpus, Bulrush,	133. 6	558
Scleranthus, Knawel,	21. 19	691
Scleria, Nut-Rush,	133. 15	821
Sclerolepis,	59. 3	913
Scoke, Phytolacca,	89. 1	652
Scorpion-Grass, Myosotis,	78. 7	384
Scrophularia, Figwort,	74. 7	475
Scutch-Grass, Cynodon,	134. 20	573

Scutellaria, Skull-cap,		77. 26	498
Sea-Goosefoot, Chenopodina,		90. 8	683
Sea-Lavender, Statice,		69. 1	188
Sea-Milkwort, Glaux,		70. 7	677
Sea-Ox-eye, Borrichia,		59. 35	972
Sea-Purslane, Sesuvium,		22. 1	50
Sea-Rocket, Cakile,		12. 20	106
Sea-Sand-Reed, sp. of Calamagrostis,			
Sea-Sandwort, Honkenya,		21. 6	205
Sea-Spear-Grass, sp. of Glyceria,			
Sedge, Carex,		138. 16	821
Sedum, Stone-crop,		49. 2	272
Seed-Box, sp. of Ludwigia,			
Self-Heal, Brunella,		77. 25	498
Senebiera, Watercress,		12. 19	115
Seneca-Grass, sp. of Hierochloa,			
Seneca Snakeroot, sp. of Polygala,			
Senecio, Groundsel,		59. 64	921
Senna, Cassia,		38. 32	281
Sensitive Brier, Schrankia,		38. 36	418
Sensitive Joint-Vetch, Æschyomene,		38. 15	245
Sericocarpus, White-topped Aster,		59. 13	955
Service-berry, sp. of Amelanchier,			
Sesame-Grass, Tripsacum,		134. 62	565
Sesuvium, Sea-Purslane,		22. 1	50
Setaria, Bristly Foxtail-Grass,		134. 60	601
Seymeria,		74. 18	336
Shad-Bush, sp. of Amelanchier,			
Shag-Bark Hickory, sp. of Carya,			
Sheep-Sorrel, sp. of Rumex,			
Sheep-berry, sp. of Viburnum,			
Shell-Bark, sp. of Carya,			
Shell-Flower, Chelone,		74. 6	476
Shepherdia,		95. 1	854
Shepherd's Purse, Capsella,		12. 17	114
Shin-Leaf, sp. of Pyrola,			
Shrubby Bitter-Sweet, Celastrus,		35. 1	162
Shrubby Trefoil, Ptelea,		31. 2	92
Shrub Yellow-root, Xanthorrhiza,		1. 18	270
Sibbaldia,		39. 7	269
Sickle-Pod, sp. of Arabis,			
Sicyos, One-seeded Star-Cucumber,		48. 1	807
Sida,		23. 5	72
Side-Saddle-Flower, Sarracenia,		9. 1	19
Silene, Catchfly, Campion,		21. 4	197
Silkweed, Asclepias,		85. 1	417
Silphium, Rosin-Plant,		59. 28	975
Silver-Bell-Tree, Halesia,		65. 2	511

INDEX. 375

Silver-Weed, sp. of Potentilla,		
Sinapis, Mustard,	12. 11	122
Sisymbrium, Hedge-Mustard,	12. 10	125
Sisyrinchium, Blue-eyed Grass,	123. 2	709
Sium, Water-Parsnip,	52. 21	308
Skull-Cap, Scutellaria,	77. 26	498
Skunk-Cabbage, Symplocarpus,	55. 2	673
Sloe, sp. of Prunus,		
Smartweed, sp. of Polygonum,		
Smilacina, False Solomon's Seal,	126. 3	746
Smilax, Greenbrier,	125. 1	866
Sneak-head, Chelone,	74. 6	476
Snapdragon, Antirrhinum,	74. 3	462
Sneezeweed, Helenium,	59. 47	948
Sneezewort, sp. of Achillea,		
Snow-Ball-Tree, sp. of Viburnum,		
Snowberry, Symphoricarpus,	55. 2	529
Snowdrop, Halesia,	65. 2	511
Soapwort, Saponaria,	21. 2	195
Solanum, Nightshade,	82. 1	390
Solea, Green Violet,	15. 1	225
Solidago, Golden-Rod,	59. 20	959
Solomon's Seal, Polygonatum,	126. 2	719
Sonchus, Sow-Thistle,	59. 85	1000
Sorghum, Broomcorn,	134. 65	602
Sorrel, Rumex,	92. 5	714
Sorrel-Tree, Oxydendrum,	62. 11	429
Sour Gum-Tree, Nyssa,	54. 2	859
Sour-Wood, Oxydendrum,	62. 11	429
Southern Buckthorn, Bumelia,	67. 1	389
Sow-Thistle, Sonchus,	59. 85	1000
Spanish Bajonet, Yucca,	126. 13	727
Spanish Needles, sp. of Bidens,		
Sparganium, Bur-Reed,	113. 2	814
Spartina, Cord-Grass,	134. 16	578
Spatter-Dock, Nuphar,	8. 2	24
Spear-Grass, Poa,	134. 33	688
Spearmint, sp. of Mentha,		
Spearwort, sp. of Ranunculus,		
Specularia, Venus' Looking-Glass,	61. 2	524
Speedwell, Veronica,	74. 16	326
Spergula, Spurrey,	21. 16	187
Spergularia, Spurrey-Sandwort,	21. 15	201
Spermacoce, Button-Weed,	56. 2	522
Spice-Bush, sp. of Benzoin,		
Spider-wort, Tradescantia,	130. 2	88
Spigelia, Pink-root,	56. 9	406
Spike-Grass, Brizopyrum and Uniola,		

Spikenard, sp. of Aralia,			
Spike-Rush, Eleocharis,	133.	5	557
Spindle-Tree, Euonymus,	35.	2	103
Spiræa, Meadow-Sweet,	89.	2	58
Spiranthes, Ladies' Tresses,	119.	5	770
Spoonwood, sp. of Kalmia,			
Sporolobus, Dropseed-Grass,	134.	6	593
Spotted Cow-bane, Cicuta,	52.	20	307
Spotted Wintergreen, sp. of Chimaphila,			
Spring-Beauty, Claytonia,	22.	4	152
Spruce, Abies,	111.	2	779
Spurge, Euphorbia,	102.	1	538
Spurge-Nettle, Cnidoscolus,	102.	2	839
Spurred Gentian, Halenia,	83.	4	351
Spurrey, Spergula,	21.	16	187
Spurrey-Sandwort, Spergularia,	21.	15	201
Squaw-Root, Conopholis,	73.	2	455
Squaw-weed, sp. of Senecio,			
Squill, Scilla,	126.	9	747
Squirrel-Corn, sp. of Dicentra,			
Squirrel-tail-Grass, sp. of Hordeum,			
Stachys, Hedge-Nettle,	77.	29	503
Staff-Tree, Celastrus,	35.	1	162
Stagger-Bush, sp. of Andromeda,			
Staphylea, Bladdernut,	36.	1	178
Star-Cucumber, Sicyos,	48.	1	807
Star-Flower, Trientalis,	70.	4	137
Star-Grass, Aletris and Hypoxis,			
Star-of-Bethlehem, Ornithogalum,	126.	8	745
Star-Thistle, Centaurea,	59.	66	938
Star-wort, Stellaria,	21.	10	200
Statice, Marsh-Rosemary,	69.	1	188
Steeple-Bush, sp. of Spiræa,			
Stellaria, Star-wort,	21.	10	200
Stenanthium, subgenus of Veratrum,			
Stickseed, Echinospermum,	78.	8	379
Stillingia,	102.	5	824
Stipa, Feather-Grass,	134.	14	583
Stonecrop, Sedum,	49.	2	272
Stone-root, Collinsonia,	77.	15	444
Storax, Styrax,	65.	1	422
Storksbill, Erodium,	28.	2	147
Strawberry, Fragaria,	30.	12	64
Strawberry-Bush, sp. of Euonymus,			
Streptopus, Twisted Stalk,	127.	3	719
Stuartia,	25.	1	67
Stylisma,	81.	5	401
Stylophorum, Celandine Poppy,	10.	3	14

INDEX. 377

Stylosanthus, Pencil-flower,	38. 19	288
Styrax, Storax,	65. 1	422
Subularia, Awlwort,	12. 18	111
Succory, Cichorium,	59. 73	990
Sugar-berry, Celtis,	104. 3	663
Sullivantia,	50. 4	181
Sumach, Rhus,	32. 1	178
Summer-Haw, sp. of Cratægus,		
Summer-Savory, Satureja,	77. 11	492
Sundew, Drosera,	17. 1	188
Sun-drops, sp. of Œnothera,		
Sunflower, Helianthus,	59. 40	980
Supple-Jack, Berchemia,	34. 1	160
Swamp-Honeysuckle, sp. of Azalen,		
Swamp-Loosestrife, Nesæa,	42. 3	171
Sweetbrier, sp. of Rosa,		
Sweet Cicely, Osmorrhiza,	52. 24	312
Sweet Colt's-foot, Nardosmia,	59. 10	951
Sweet Fern, Comptonia,	108. 2	793
Sweet Flag, Acorus,	112. 6	715
Sweet Gale, sp. of Myrica,		
Sweet Gum-Tree, Liquidambar,	51. 3	796
Sweet Leaf, Symplocos,	65. 3	76
Sweet Pepperbush, sp. of Clethra,		
Sweet Scabious, sp. of Erigeron,		
Sweet-scented Shrub, Calycanthus,	40. 1	57
Sweet-scented Vernal Grass, Anthoxanthum,	134. 54	598
Swine-Cress, Senebiera,	12. 19	115
Sweet Pine-sap, Schweinitzia,	62. 26	483
Sycamore, Platanus,	105. 1	796
Symphoricarpus, Snowberry,	55. 2	529
Symphytum, Comfrey,	78. 3	381
Symplocarpus, Skunk-Cabbage,	112. 4	673
Symplocos, Sweet Leaf,	65. 3	76
Synandra,	77. 23	499
Synthyris,	74. 15	826

T

Talinum,	22. 3	215
Tall Red-Top, Tricuspis,	134. 24	631
Tamarack, Larix,	111. 3	778
Tanacetum, Tansy,	59. 57	911
Tansy, Tanacetum,	59. 57	911
Tansy-Mustard, sp. of Sisymbrium,		
Tape-Grass, Vallisneria,	117. 3	886
Taraxacum, Dandelion,	59. 80	999
Tare, Vicia,	38. 20	242

Taxodium, Bald Cypress,		111. 6	780
Taxus, Yew,		111. 8	850
Tea-berry, Gaultheria,		62. 6	422
Tear-Thumb, sp. of Polygonum,			
Teasel, Dipsacus,		58. 1	527
Tecoma, Trumpet-flower,		72. 2	333
Tephrosia, Hoary Pea,		38. 13	261
Tetragonotheca,		59. 33	971
Tetranthera, Pond-Spice,		98. 4	703
Teucrium, Germander,		77. 1	498
Thalictrum, Meadow-Rue,		1. 6	87
Thaspium, Meadow-Parsnip,		52. 16	310
Thimbleberry, sp. of Rubus,			
Thin-Grass, sp. of Agrostis,			
Thorn-Apple, Datura,		82. 5	399
Three-leaved Nightshade, Trillium,		125. 2	90
Three-seeded Mercury, Acalypha,		102. 3	839
Three-thorned Acacia, sp. of Gleditschia,			
Thorough-wax, Bupleurum,		52. 18	305
Thoroughwort, Eupatorium,		59. 7	917
Thuja, Arbor-Vitæ,		111. 4	781
Thyme, Thymus,		77. 10	491
Thymus, Thyme,		77. 10	491
Tiarella, False Mitrewort,		50. 7	193
Tickseed, Coreopsis,		59. 42	937
Tickseed-Sunflower, sp. of Coreopsis,			
Tick-Trefoil, Desmodium,		38. 17	246
Tiedemannia, False Water-Dropwort,		52. 10	300
Tilia, Linden,		24. 1	75
Tillæa,		49. 1	272
Tillandsia, Long-Moss,		122. 1	86
Timothy, sp. of Phleum,			
Tipularia, Crane-fly Orchis,		119. 11	761
Toad-Flax, Linaria,		74. 2	462
Tobacco, Nicotiana,		82. 6	400
Tofielda, False Asphodel,		127. 11	735
Toothache-Grass, Ctenium,		134. 17	607
Tooth-wort, Dentaria,		12. 4	120
Tower-Mustard, Turritis,		12. 7	117
Tradescantia, Spider-wort,		130. 2	88
Tragia,		102. 4	824
Trailing Arbutus, Epigæa,		62. 5	425
Trautvetteria, False Bugbane,		1. 7	87
Treacle-Mustard, Erysimum,		12. 9	125
Tread-Softly, Cnidoscolus,		102. 2	839
Trefoil, Trifolium,		38. 4	256
Trichelostylis, subgenus of Fimbristylis,			
Trichostema, Blue Curls,		77. 2	340

INDEX. 379

Tricuspis, Tall Red-top, ... 134. 24 631
Trientalis, Chickweed-Wintergreen, . . . 70. 4 137
Trifolium, Trefoil, ... 33. 4 256
Triglochin, Arrow-Grass, ... 116. 1 728
Trillium, Three-leaved Nightshade, . . . 125. 2   90
Triosteum, Horse-Gentian, ... 55. 5 522
Triple-awned Grass, Aristida, ... 134. 15 582
Tripsacum, Gama-Grass, ... 134. 62 565
Trisetum, ... 134. 49 622
Triticum, Wheat, ... 134. 43 611
Trollius, Globe-flower, ...   1. 12   46
Troximon, ... 59. 79 1000
Trumpet-flower, Tecoma, ... 72. 2 333
Trumpets, sp. of Sarracenia, ...
Trumpet-Weed, sp. of Eupatorium, . . .
Tufted Loosestrife, Naumburgia, . . . 70. 6 148
Tulip-Tree, Liriodendron, ...   2. 2   47
Tupelo, Nyssa, ... 54. 2 859
Turritis, Tower-Mustard, ... 12. 7 117
Tussilago, Colt's-foot, ... 59. 11 951
Turtle-head, Chelone, ... 74. 6 476
Tway blade, Listera and Liparis, ...
Twig-Rush, Cladium, ... 133. 14 561
Twin-flower, Linnæa, ... 55. 1 525
Twin-leaf, Jeffersonia, ...   5. 4 137
Twisted-Stalk, Streptopus, ... 127. 3 719
Typha, Cat's-tail, ... 113. 1 814

U

Ulmus, Elm, ... 104. 1 753
Umbrella-Grass, Fuirena, ... 133. 9 646
Umbrella-Leaf, Diphylleia, ...   5. 3 142
Umbrella-Tree, sp. of Magnolia, ...
Unicorn-Plant, Martynia, ... 72. 4 450
Uniola, Spike-Grass, ... 134. 38 636
Uralepis, Sand-Grass, ... 134. 25 631
Urtica, Nettle, ... 104. 5 832
Utricularia, Bladderwort, ... 71. 1 439
Uvularia, Bellwort, ... 127. 1 740

V

Vaccaria, Cow-herb, ... 21. 3 195
Vaccinium, Bil-, Blue- and Crane-berry, . . 62. 2 514
Valerian, Valeriana, ... 57. 1 535
Valeriana, Valerian, ... 57. 1 535
Valerianella, Fedia, Corn-Salad, Lamb-Lettuce, . . 57. 2 535

Vallisneria, Tape-Grass, ... 117. 3 886
Vanilla-Grass, sp. of Hierochloa, . . .
Vanilla-Plant, sp. of Liatris, ...
Velvet-Grass, Holcus, ... 134. 52 617
Velvet-Leaf, Abutilon, ... 23. 6 72
Venus' Looking-Glass, Specularia, . . . 61. 2 524
Veratrum, False Hellebore, . . . 127. 6 734.
Verbascum, Mullein, ... 74. 1 397
Verbena, Vervain, ... 76. 1 339
Verbesina, Crownbeard, : 59. 44 966
Vernal-Grass, Anthoxanthum, . . . 134. 54 598
Vernonia, Iron-Weed, ... 59. 1 924
Veronica, Speedwell, ... 74. 16 326
Vervain, Verbena, ... 76. 1 339
Vesicaria, Bladder-Pod, ... 12. 14 113
Vetch, Vicia, ... 38. 20 242
Vetchling, Lathyrus, ... 38. 21 242
Viburnum, Arrow-wood, ... 55. 7 520
Vicia, Vetch, Tare, ... 38. 20 242
Vilfa, Rush-Grass, ... 134. 5 593
Viola, Violet, ... 15. 2 225
Violet, Viola, ... 15. 2 225
Viper's Bugloss, Echium, ... 78. 1 377
Virgaurea, some sp. of Solidago, . . .
Virginian Cowslip, sp. of Mertensia, . . .
Virginian Creeper, Ampelopsis, . . 33. 2 157
Virginia Snake-root, sp. of Aristolochia, . .
Virgin's-Bower, Clematis, . . . 1. 2 83
Vitis, Grape, ... 33. 1 158
Vitis-Idæa Cowberry, sp. of Vaccinium, . .

W

Waahoo, sp. of Euonymus, ...
Wake-Robin, sp. of Trillium, . . .
Waldsteinia, Barren Strawberry, . . . 39. 10 63
Walnut, Juglans, ... 106. 1 783
Wart-Cress, Senebiera, ... 12. 19 115
Washington Thorn, sp. of Cratægus, . .
Water-Arum, Calla, ... 112. 3 813
Water-Beech, Carpinus, . . . 107. 5 792
Water-Chinquapin, Nelumbium, . . 6. 1 26
Watercress, Senebiera, ... 12. 19 115
Water-Dropwort, Tiedemannia, . . . 52. 10 300
Water-Hemlock, Cicuta, . . . 52. 20 307
Water-Hemp, Acnida, ... 91. 4 901
Water-Horehound, Lycopus, . . . 77. 5 325
Water-Leaf, Hydrophyllum, . . . 79. 1 367

INDEX. 381

ater-Lily, Nymphæa,	8.	1	24
ater-Locust, sp. of Gleditschia,			
ater-Marigold, sp. of Bidens,			
ater-Milfoil, Myriophyllum,	43.	8	828
ater-Nymph, Nymphæa,	8.	1	24
ater-Parsnip, Sium,	52.	21	308
ater-Pennywort, Hydrocotyle,	52.	1	292
ater-Pepper, sp. of Polygonum,			
ater-Pimpernel, Samolus,	70.	10	365
ater-Plantain, Alisma,	116.	3	267
ater-Purslane, sp. of Ludwigia,			
ater-Rice, Zizania,	134.	2	567
ater-Shield, Brassenia,	7.	1	263
ater-Star-Grass, Schollera,	129.	3	711
ater-Star-wort, Callitriche,	100.	1	817
ater-Violet, Hottonia,	70.	11	358
ater-Willow, Dianthera,	75.	1	449
ater-weed, Anacharis,	117.	2	886
ater-wort, Elatine,	20.	1	85
ax-Myrtle, Myrica,	108.	1	874
ax-work, Celastrus,	35.	1	162
ayfaring-Tree, sp. of Viburnum,			
estern Wall-flower, sp. of Eryngium,			
haboo, sp. of Ulmus,			
heat, Triticum,	134.	43	611
heat-Grass, sp. of Triticum,			
hin, Genista,	38.	3	237
hite Alder, Clethra,	62.	12	173
hite Cedar, Cupressus,	111.	5	781
hite Daisy, Leucanthemum,	59.	55	963
hite Grass, Leersia,	134.	1	567
hite Weed, Leucanthemum,	59.	55	963
hite Lettuce, sp. of Nabalus,			
hite Snake-root, sp. of Eupatorium,			
hite Swamp-Honeysuckle, sp. of Azalea,			
hite Thorn, Cratægus,	39.	16	56
hitlow-Grass, Draba,	12.	12	110
hitlow-wor , Paronychia,	21.	18	691
ild Allspic , sp. of Benzoin,			
ild Balsam-Apple, Echinocystis,	48.	2	806
ild Bean, Apios,	38.	23	249
ild Chamomile, Matricaria,	59.	56	962
ild Comfrey, sp. of Cynoglossum,			
ild Elder, sp. of Aralia,			
ild False Indigo, Baptisia,	38.	29	232
ild Ginger, Asarum,	87.	1	644
ild Hyacinth, Scilla,	126.	9	747
ild Ipecac, sp. of Euphorbia,			

Wild Leek, Allium,	126. 10	722
Wild Liquorice, sp. of Galium,		
Wild Marjoram, Origanum,	77. 9	491
Wild Oat-Grass, Danthonia,	134. 48	628
Wild Potato-vine, sp. of Ipomœa,		
Wild Rod, sp. of Viburnum,		
Wild Rosemary, sp. of Andromeda,		
Wild Rye, Elymus,	134. 45	609
Wild Sarsaparilla, sp. of Aralia,	58. 1	287
Wild Sensitive Plant, sp. of Cassia,		
Wild Snake-root, sp. of Eupatorium,		
Wild Sweet William, sp. of Phlox,		
Willow, Salix,	110. 1	874
Willow-herb, Epilobium,	43. 1	282
Windflower, Anemone,	1. 4	36
Winged Elm, sp. of Ulmus,		
Winged Pigweed, Cycloloma,	90. 1	681
Winterberry, sp. of Prinos,		
Winter-Cress, Barbarea,	12. 8	124
Wintergreen, Gaultheria and Pyrola,		
Wire-Grass, Eleusine,	134. 22	604
Wistaria,	88. 12	250
Witch-Hazel, Hamamelis,	51. 1	129
Woad-Waxen, Genista,	38. 3	237
Wolfberry, sp. of Symphoricarpus,		
Wolfsbane, Aconitum,	1. 17	43
Wood-Anemone, sp. of Anemone,		
Woodbine, Lonicera,	55. 3	519
Wood-Grass, Sorghum,	134. 65	602
Wood-Nettle, Laportea,	104. 6	831
Wood-Reed-Grass, Cinna,	134. 9	585
Wood-Rush, Luzula,	128. 2	741
Wood-Sage, Teucrium,	77. 1	493
Wood-Sorrel, Oxalis,	27. 1	203
Wood-Waxen, Genista,	38. 3	237
Wool-Grass, sp. of Scirpus,		
Woolly Beard-Grass, Erianthus,	134. 63	600
Worm-Grass, Spigelia,	56. 9	406
Wormseed, sp. of Chenopodium,		
Wormseed-Mustard, sp. of Erysimum,		
Wormwood, Artemisia,	59. 57	910
Wound-wort, sp. of Stachys,		

X

Xanthium, Clotbur,	59. 82	932
Xanthoxylum, Prickly Ash,	81. 1	844
Xanthorrhiza, Shrub-Yellow-root,	1. 18	270

Xerophyllum,	127. 8	737
Xyris, Yellow-eyed Grass,	131. 2	89

Y

Yam, Dioscorea,	124. 1	867
Yard-Grass, Eleusine,	134. 22	604
Yarrow, Achillea,	59. 54	967
Yaupon, sp. of Ilex,		
Yellow-eyed Grass, Xyris,	131. 2	89
Yellow False Jessamine, Gelsemium,	74. 27	409
Yellow Pond-Lily, Nuphar,	8. 2	24
Yellow Puccoon, Hydrastis,	1. 19	39
Yellow Rattle, Rhinanthus,	74. 24	468
Yellow-wood, Cladrastis,	38. 30	233
Yew, Taxus,	111. 8	850
Yucca, Bear-Grass,	126. 13	727

Z

Zanichellia, Horned Pondweed,	115. 2	818
Zanthorrhiza, see Xanthorrhiza,		
Zanthoxylum, Xanthoxylum,		
Zephyranthus, Atamasco-Lily,	120. 1	752
Zizania, Water-Rice,	134. 2	567
Zizia,	52. 17	305
Zostera, Grass-Wrack,	115. 3	809
Zygadenus,	127. 5	733

PUBLISHED BY HENRY HOLT & CO.

AUSTIN'S (JOHN) JURISPRUDENCE. Lectures on Jurisprudence, the Philosophy of Positive Law. By the late John Austin, of the Inner Temple, Barrister-at-Law. Abridged from the larger work for the use of students. By ROBERT CAMPBELL, 1 vol. 8vo. $3.00.

MAINE'S (SIR HENRY SUMNER) WORKS. Ancient Law: Its Connection with the Early History of Society, and its Relation to Modern Ideas. By HENRY SUMNER MAINE. 8vo. $3.50.

"History read from the point of Law, and Law studied by the light of History. It is consequently a book that addresses itself as much to the general student as to the lawyer."—*Westminster Review.*

"Mr. Maine's profound work on Ancient Law in its relation to modern ideas."—*John Stuart Mill.*

"Sir Henry Maine's great work on Ancient Law."—*Nation.*

"A text-book for all English students of jurisprudence."—*Saturday Review.*

Lectures on the Early History of Institutions. 8vo. $3.50.

"In the power of tracing analogies between different institutions, in the capacity for seeing the bearing of obscure and neglected facts, he surpasses any living writer."—*Nation.*

Village Communities and Miscellanies. 8vo. $3.50.

MILL'S (JOHN STUART) MISCELLANEOUS WORKS. Uniform Library edition. 8vo. Tinted and laid paper, $2.50 per volume (except volume on Comte). The 12 volumes in a box, $29.00; half calf or half morocco, $59.00.

Three Essays on Religion. 1 vol.
The Autobiography. 1 vol.
Dissertations and Discussions. 5 vols.
Considerations on Representative Government. 1 vol.
Examination of Sir William Hamilton's Philosophy. 2 vols.
On Liberty: The Subjection of Women. Both in 1 vol.
Comte's Positive Philosophy. 1 vol. $1.50.

CHEAP EDITIONS.

The Subjection of Women. 12mo. $1.25.
Principles of Political Economy. 12mo. $2.50.
Memorial Volume. John Stuart Mill: His Life and Works. Twelve Sketches, as follows: (His Life, by J. R. Fox Bourne; His Career in the India House, by W. T. Thornton; His Moral Character, by Herbert Spencer; His Botanical Studies, by Henry Turner; His Place as a Critic, by W. Minto; His Work in Philosophy, by J. H. Levy; His Studies in Morals and Jurisprudence, by W. A. Hunter; His Work in Political Economy, by J. E. Cairnes; His Influence at the Universities, by Henry Fawcett; His Influence as a Practical Politician, by Mrs. Fawcett; His Relation to Positivism, by Frederick Harrison; His Position as a Philosopher, by W. A. Hunter. 16mo. $1.00.

STEPHEN'S (J. F.) LIBERTY, EQUALITY, FRATERNITY. Large 12mo. $2.00.

A criticism of "Mill on Liberty."

"One of the most thorough overhaulings of the moral, religious, and political bases of society which they have recently received. . . . . . Everybody who wants to see all the recent attempts to set things right analyzed by a master-hand, and in English which stirs the blood, will have a great treat in reading him."—*Nation.*

"One of the most valuable contributions to political philosophy which have been published in recent times."—*London Saturday Review.*

PUBLISHED BY HENRY HOLT & CO.

BRINTON'S (D. G.) WORKS. The Myths of the New World. A Treatise on the Symbolism and Mythology of the Red Race of America. Second edition, large 12mo, $2.50. Large-paper (first) edition, $6.00.

"The philosophical spirit in which it is written is deserving of unstinted praise, and justifies the belief that in whatever Dr. Brinton may in future contribute to the literature of Comparative Mythology, he will continue to reflect credit upon himself and his country."—*N. A. Review.*

The Religious Sentiment, its Source and Aim. A contribution to the science and philosophy of religion. Large 12mo. $2.50. (*Just ready.*)

CONWAY'S (M. D.) Sacred Anthology. 8vo. $4.00.

"He deserves our hearty thanks for the trouble he has taken in collecting these gems, and stringing them together for the use of those who have no access to the originals, and we trust that his book will arouse a more general interest in a long-neglected and even despised branch of literature, the Sacred Books of the East."—*Prof.* MAX MULLER.

DEUTSCH'S (E.) LITERARY REMAINS. With a Brief Memoir. 8vo. $4.00.

CONTENTS:—The Talmud—Islam—Egypt. Ancient and Modern—Hermes Trismegistus—Judeo-Arabic Metaphysics—Semitic Palæography—Renan's "Les Apôtres"—Worship of Baalim in Israel—The Œcumenical Council—Apostolicæ Sedes ?—Roman Passion Drama—Semitic Languages—Samaritan Pentateuch—The Targums—Book of Jashar—Arabic Poetry.

"A noble monument of study and erudition."—*N. Y. Tribune.*

GOULD'S (REV. S. B.) LEGENDS of the PATRIARCHS AND PROPHETS. Crown 8vo. $2.00.

"There are few Bible readers who have not at some time wished for just such a volume."—*Congregationalist.*

HAWEIS' (REV. H. R.) THOUGHTS FOR THE TIMES. Sermons by the author of "Music and Morals." 12mo. $1.50.

"They have a special interest as exhibiting the treatment which old-fashioned orthodoxy is just now undergoing at the hands of the liberal clergy."—*Pall Mall Gazette.*

MARTINEAU'S (JAMES) WORKS.

Essays, Theological and Philosophical. 2 vols. 8vo. $5.00.

Mr. Martineau's contributions to the *Prospective, Westminster, National,* and other Reviews attracted the attention of the best minds in both England and America, and produced a marked and favorable impression upon men of all denominations.

The New Affinities of Faith. 12mo, paper. 25 cents.

STRAUSS' (D. F.) THE OLD FAITH AND THE NEW. 12mo. $2.00.

"Will make its mark upon the time, not so much as an attack upon what we venerate as an apology for those who honestly differ from the majority of their brothers."—*Atlantic Monthly.*

TYLOR'S (E. B.) PRIMITIVE CULTURE: Researches into the Development of Mythology, Philosophy, Religion, Art, and Custom. 2 vols., 8vo. $5.00.

"One of the most remarkable and interesting books of the present day. . . . It takes up man . . . at the remotest periods of which we have any knowledge, and traces his intellectual growth . . . from that time forth. . . Admirably written, often with great humor, and at times with eloquence, and never with a dull line. . . . The many students of Darwin and Spencer in this country cannot do better than to supplement the books of those writers by . . . these really delightful volumes."—*Atlantic Monthly.*

"One of the few erudite treatises which are at once truly great and thoroughly entertaining."—*North American Review.*

ADAMS' (PROF. C. K.) DEMOCRACY AND MONARCHY IN FRANCE, from the Inception of the Great Revolution to the Overthrow of the Second Empire. By Prof. C. K. Adams, of the University of Michigan. Large 12mo. Cloth. $2.50.

"A valuable example of the scientific mode of dealing with political problems."—*Nation.*

"Full of shrewd and suggestive criticism, and few readers will peruse it without gaining new ideas on the subject."—*London Saturday Review.*

"Remarkably lucid writing."—*London Academy.*

CHESNEY'S (C. C.) MILITARY BIOGRAPHY. Essays in Military Biography. By CHARLES CORNWALLIS CHESNEY, Colonel in the British Army, Lieutenant-Colonel in the Royal Engineers. Large 12mo. $2.50.

"Very able."—*Nation.*

"Uncommonly entertaining and instructive."—*N. Y. Evening Post.*

"Full of interest, not only to the professional soldier, but to the general reader."—*Boston Globe.*

FAMILY RECORD ALBUM. In Blanks classified on a New System. Large quarto, gilt edges, 328 pages. Cloth, $5. Half Morocco, $8. Full Morocco, $15. Levant or Russia, $25.

THIS BOOK for keeping family records has been made because the editor felt the need of it and supposed that many others felt the same need.

The pages are of eight kinds, called respectively Family, Genealogical, Tabular, Biographical Heirloom, Domestic Economy, Travel, and Miscellaneous.

FREEMAN'S (EDWARD A.) HISTORICAL COURSE. A series of historical works on a plan entirely different from that of any before published for the general reader or educational purposes. It embodies the results of the latest scholarship in comparative philology, mythology, and the philosophy of history. Uniform volumes. 16mo.

   1. **General Sketch of History.** By EDWARD A. FREEMAN, D.C.L. 16mo. $1.25.
   2. **History of England.** By Miss EDITH THOMPSON. $1.00.
   3. **History of Scotland.** By MARGARET MACARTHUR. $1.00.
   4. **History of Italy.** By Rev. W. HUNT, M.A. $1.00.
   5. **History of Germany.** By JAMES SIME. $1.00.
   6. **History of the United States.** By J. A. DOYLE. With Maps by Francis A. Walker, Prof. in Yale College. $1.40.
   7. **History of France.** By Rev. J. R. GREEN. (*In preparation.*)
   8. **History of Greece.** By J. ANNAN BRYCE. (*In preparation.*)

"Useful not only for school study, but for the library."—*Boston Advertiser.*

SUMNER'S (PROF. W. G.) HISTORY OF AMERICAN CURRENCY. With Chapters on the English Bank Restriction and Austrian Paper Money. To which is appended "The Bullion Report." Large 12mo. With diagrams. $3.00.

"The historical information which it contains has never been brought together before within the compass of a single work."—*N. Y. Tribune.*

"A most valuable collection of facts, thoroughly digested and properly arranged. . . . Has a freshness and vivacity rarely found in works of the kind."—*Atlantic Monthly.*

PUBLISHED BY HENRY HOLT & CO.

TAINE'S (HIPPOLYTE ADOLPHE) WORKS. UNIFORM LIBRARY EDITION. 13 vols., large 12mo. $2.50 per vol. *Each vol. sold separately.*

In box, per set, cloth, $32.50; half calf, $65.00; tree calf, $78.00.

"The paper, print, and binding of this series leave nothing to be desired in point of taste and attractiveness."—*Nation.*

The Ancient Régime. Being the first of a series of works on "*Les Origines de la France Contemporaine.*" Translated by John Durand. Large 12mo. $2.50.

History of English Literature. Translated by H. VAN LAUN. From new stereotype plates. 3 vols., $7.50; half calf, $15.00.

On Intelligence. Translated by T. D. HAYE. 2 vols. $5.00.

Italy. (Rome and Naples.) By JOHN DURAND. $2.50.

Italy. (Florence and Venice.) By JOHN DURAND. $2.50.

Notes on Paris. THE LIFE AND OPINIONS OF M. FREDERICK THOMAS GRAINDORGE. Tr. by JOHN AUSTIN STEVENS. $2.50.

Notes on England. Translated, with an introductory chapter, by W. F. RAE. With a portrait of the author. Large 12mo. $2.50.

A Tour through the Pyrenees. Tr. by J. Safford Fiske. $2.50.

Lectures on Art. Translated by JOHN DURAND. First Series. (Containing the Philosophy of Art; The Ideal in Art.) $2.50.

Lectures on Art. Translated by JOHN DURAND. Second Series. (Containing the Philosophy of Art in Italy; The Philosophy of Art in the Netherlands; The Philosophy of Art in Greece.) $2.50.

SEPARATE EDITIONS.

A Tour through the Pyrenees. Translated by J. Safford Fiske. Illustrated by GUSTAVE DORÉ. Square 8vo, gilt, $10.00. Full levant morocco, $20.00.

A superb presentation volume, with nearly 250 illustrations in Doré's early, careful manner. The illustrations are not confined to the scenery, but also refer to many of the adventures, tragic and grotesque, that beset the traveller, and, what is of more interest and importance, many of the legends of that historic and romantic country from Froissart and the other old Chroniclers.

"It is rarely that a book is printed in so perfect a combination of the book-making trinity, author, artist, and publisher, as is this superb work."—*N. Y. Evening Mail.*

"A marvel of beauty."—*Boston Transcript.*

The Philosophy of Art in Italy. Translated by JOHN DURAND. 16mo, $1.25.

Art in the Netherlands. Translated by JOHN DURAND. 16mo. $1.25.

The Class-room Taine. History of English Literature, by H. A. Taine. Abridged from the translation of H. VAN LAUN, and edited, with chronological table, notes, and index, by JOHN FISKE, Lecturer and Assistant Librarian in Harvard University. Large 12mo. $2.50.

PUBLISHED BY HENRY HOLT & CO.

BOSWELL'S (J.) LIFE OF JOHNSON, including a Tour to the Hebrides. Condensed by C. H. Jones. (*In press.*)

COLERIDGE'S BIOGRAPHIA LITERARIA. 2 vols. large 12mo. $5.00.

CRITICAL AND SOCIAL ESSAYS. Reprinted from the *New York Nation.* 16mo. $1.50.

"The best thinking and temper of the time."—*N. A. Review.*

HADLEY'S (PROF. JAS.) ESSAYS, Philological and Critical. With an introduction by W. D. WHITNEY. 8vo. $5.00.

"Certainly no teacher or student of English can afford to be without these essays. . . . Few books have as much in them to praise, and as little to find fault with."—*Nation.*

"Rarely have we read a book which gives us so high a conception of the writer's whole nature."—*London Athenæum.*

HOUGHTON'S (LORD) MONOGRAPHS, Personal and Social. By LORD HOUGHTON (Richard Monckton Milnes). 12mo. With Portraits of WALTER SAVAGE LANDOR, CHARLES BULLER, HARRIET LADY ASHBURTON, and SULEIMAN PASHA. 12mo. $2.00.

"An extremely agreeable volume. . . . . . He writes so as to adorn everything which he touches."—*London Athenæum.*

Monographs Political and Literary. (*In preparation.*)

LIBRARY OF FOREIGN POETRY. 16mo. Gilt top and side stamp, bevelled edges.

    I. **Herz's King René's Daughter.** $1.25.
    II. **Tegnér's Frithiof's Saga.** $1.50.
    III. **Lessing's Nathan the Wise.** $1.50.
    IV. **Selections from the Kalevala.** $1.50.
    V. **Heine's Book of Songs.** $1.50.
    VI. **Goethe's Poems and Ballads.** $1.50.

MARTINEAU'S (H.) BIOGRAPHICAL SKETCHES. Biographical Sketches. By HARRIET MARTINEAU. 8vo. $1.50.

There are over fifty eminent persons "sketched" in this volume.

"Miss Martineau's large literary power, and her fine intellectual training, makes these litttle sketches more instructive, and constitute them more generally works of art, than many more ambitious and diffuse biographies."—*Fortnightly Review.*

MOSCHELES' (IGNATZ) RECENT MUSIC AND MUSICIANS. 12mo. (Amateur Series.) $2.00.

"Full of pleasant gossip. The diary and letters between them contain notices and criticisms on almost every musical celebrity of the last half century."—*Pall Mall Gazette.*

SAINTE-BEUVE'S (C. A.) ENGLISH PORTRAITS. Selected and translated from the "Causeries du Lundi." With an Introductory Chapter on Sainte-Beuve's Life and Writings. 12mo. $2.00.

CONTENTS:—Sainte-Beuve's Life—His Writings—General Comments—Mary Queen of Scots—Lord Chesterfield—Benjamin Franklin—Edward Gibbon—William Cowper—English Literature by H. Taine—Pope as a Poet.

"A charming volume, and one that may be made a companion, in the confident assurance that the better we know it the better we shall enjoy it."—*Boston Advertiser*

WAGNER'S (R.) ART AND LIFE THEORIES. Selected from his Writings, and translated by EDWARD L. BURLINGAME. With a preface, a catalogue of Wagner's published works, and drawings of the Bayreuth Opera House. 12mo. (Amateur Series.) $2.00.

"A more beautiful and every way delightful gift-book for Christmas or any other season would be difficult to conceive."—*Boston Advertiser.*

PUBLISHED BY HENRY HOLT & CO.

GAUTIER'S (THEOPHILE) WORKS. A Winter in Russia. Translated from the French by M. M. RIPLEY. 12mo. $2.00.

"The book is a charming one, and nothing approaching it in merit has been written on the outward face of things in Russia."—*Nation.*

"We do not remember when we have taken up a more fascinating book."—*Boston Gazette.*

Constantinople. Translated from the French by Robert Howe Gould, M.A. 12mo. $2.00.

"It is never too late in the day to reproduce the sparkling descriptions and acute reflections of so brilliant a master of style as the present author."—*N. Y. Tribune.*

JONES' (C. H.) AFRICA: the History of Exploration and Adventure as given in the leading authorities from Herodotus to Livingstone. By C. H. Jones. With Map and Illustrations. 8vo. $5.00.

"A cyclopædia of African exploration, and a useful substitute in the library for the whole list of costly original works on that subject."—*Boston Advertiser.*

"This volume contains the quintessence of a whole library. . . . What makes it peculiarly valuable is its combination of so much material which is inaccessible to the general reader. The excellent map, showing the routes of the leading explorers, and the numerous illustrations increase the value and interest of the book."—*Boston Globe.*

MORELET'S (ARTHUR) TRAVELS IN CENTRAL AMERICA. Including Accounts of some Regions Unexplored since the Conquest. Introduction and Notes by E. GEO. SQUIER. Post 8vo. Illus. $2.00.

"One of the most interesting books of travel we have read for a long time. . . . His descriptions are evidently truthful, as he seems penetrated with true scientific spirit."—*Nation.*

PUMPELLY'S (R.) AMERICA AND ASIA. Notes of a Five Years' Journey Around the World, and of Residence in Arizona, Japan and China. By RAPHAEL PUMPELLY, Professor in Harvard University, and some time Mining Engineer in the employ of the Chinese and Japanese Governments. With maps, woodcuts, and lithographic facsimiles of Japanese color-printing. Fine edition, royal 8vo, tinted paper, gilt side, $5.00. Cheap edition, post 8vo, plain, $2.50.

"One of the most interesting books of travel we have ever read . . . We have great admiration of the book, and feel great respect for the author for his intelligence, humanity, manliness, and philosophic spirit, which are conspicuous throughout his writings."—*Nation.*

"Crowded with entertainment and instruction. A careful reading of it will give more real acquaintance with both the physical geography and the ethnology of the northern temperate regions of both hemispheres than perhaps any other book in existence."—*N. Y. Evening Post.*

STILLMAN'S (W. J.) CRETAN INSURRECTION OF 1866-7-8. By W. J. STILLMAN, late U. S. Consul in Crete. 12mo. $1.50.

WHIST (SHORT WHIST). Edited by J. L. Baldwin. The Standard adopted by the London Clubs. And a Treatise on the Game, by J. C. 18mo, appropriately decorated, $1.00.

"Having been for thirty-six years a player and lover of the game, we commend the book to a beginner desirous of playing well."—*Boston Commonwealth.*

www.ingramcontent.com/pod-product-compliance
Lightning Source LLC
Chambersburg PA
CBHW051242300426
44114CB00011B/857